Studies in Computational Intelligence   **1089**

Series Editor

Janusz Kacprzyk, *Polish Academy of Sciences, Warsaw, Poland*

The series "Studies in Computational Intelligence" (SCI) publishes new developments and advances in the various areas of computational intelligence—quickly and with a high quality. The intent is to cover the theory, applications, and design methods of computational intelligence, as embedded in the fields of engineering, computer science, physics and life sciences, as well as the methodologies behind them. The series contains monographs, lecture notes and edited volumes in computational intelligence spanning the areas of neural networks, connectionist systems, genetic algorithms, evolutionary computation, artificial intelligence, cellular automata, self-organizing systems, soft computing, fuzzy systems, and hybrid intelligent systems. Of particular value to both the contributors and the readership are the short publication timeframe and the world-wide distribution, which enable both wide and rapid dissemination of research output.

Indexed by SCOPUS, DBLP, WTI Frankfurt eG, zbMATH, SCImago.

All books published in the series are submitted for consideration in Web of Science.

Lars Braubach · Kai Jander · Costin Bădică
Editors

Intelligent Distributed Computing XV

 Springer

Editors
Lars Braubach
Bremen University of Applied Sciences
Bremen, Germany

Kai Jander
Brandenburg University of Applied Sciences
Brandenburg, Germany

Costin Bădică 🆔
University of Craiova
Craiova, Romania

ISSN 1860-949X ISSN 1860-9503 (electronic)
Studies in Computational Intelligence
ISBN 978-3-031-32016-3 ISBN 978-3-031-29104-3 (eBook)
https://doi.org/10.1007/978-3-031-29104-3

This Springer imprint is published by the registered company Springer Nature Switzerland AG
The registered company address is: Gewerbestrasse 11, 6330 Cham, Switzerland

Preface

Distributed intelligent computing has emerged from the combination of the fields of distributed computing and artificial intelligence. This combination is especially interesting and challenging because many real-world scenarios and applications require skills in both dimensions.

The book includes the peer-reviewed proceedings of the 15th International Symposium on Intelligent Distributed Computing (IDC 2022). The conference was organized in a cooperation between City University of Applied Sciences Bremen and Brandenburg University of Applied Sciences and scheduled to be held in Bremen, Germany, from September 14 to 15, 2022, but was rescheduled to be held online due to complications in the aftermath of the COVSARS2 pandemic that had already impacted the previous symposium in 2021, as well as outbreak of war in Ukraine beginning in February 2022.

This 2022 edition of the IDC continues the legacy of the previous IDC Symposium Series as an initiative of two research groups: (i) Systems Research Institute, Polish Academy of Sciences, Warsaw, Poland, and (ii) Department of Software Engineering, University of Craiova, Craiova, Romania.

The IDC 2022 event comprised the following seven sessions: (1) Machine Learning 1, (2) Distributed Systems, Agents & Microservices, (3) Text & Research 1, (4) Text & Research 2, (5) Machine Learning 2, (6) Social Systems, and (7) Smart Cities. The proceedings book contains contributions of 24 regular and 9 short papers selected from 52 received submissions from 14 countries.

IDC 2022 enjoyed outstanding keynote speeches by distinguished invited speakers: Prof. Sahin Albayrak, Head of the Distributed Artificial Intelligence Laboratory (DAI-Labor) at TU Berlin, Germany, and Prof. Felix Sasaki, Chief Expert for Knowledge Graph and Semantic Technology at SAP SE. Prof. Albayrak reported on novel approaches on autonomous vehicles researched and field tested in Berlin while Prof. Sasaki gave an introduction to knowledge graphs and their application in the context of data reasoning.

December 2022

Lars Braubach
Kai Jander
Costin Bădică
General Chairs, IDC 2022

Organization

IDC 2022 was organized by the Department of Computer Science, City University of Applied Sciences Bremen, Germany, together with the Department of Business, Brandenburg University of Applied Sciences, Germany.

General Chairs

Lars Braubach	City University of Applied Sciences Bremen, Germany
Kai Jander	Brandenburg University of Applied Sciences, Germany
Costin Bădică	University of Craiova, Romania

Steering Committee

Costin Bădică	University of Craiova, Romania
Janusz Kacprzyk	Polish Academy of Sciences, Poland
Javier Del Ser	University of the Basque Country, Spain
David Camacho	Universidad Autonoma de Madrid, Spain
Paulo Novais	University of Minho, Portugal
Filip Zavoral Charles	University Prague, Czech Republic
Frances Brazier	Delft University of Technology, The Netherlands
George A. Papadopoulos	University of Cyprus, Cyprus
Giancarlo Fortino	University of Calabria, Italy
Kees Nieuwenhuis	Thales Research & Technology, The Netherlands
Marcin Paprzycki	Polish Academy of Sciences, Poland
Michele Malgeri	University of Catania, Italy
Mohammad Essaaidi Abdelmalek Essaadi	University in Tetouan, Morocco
Mirjana Ivanovic	University of Novi Sad, Serbia
Amal El Fallah Seghrouchni	University Pierre and Marie Curie, France
Igor Kotenko	ITMO University, Russia
Domenico Rosaci	University Mediterranea of Reggio Calabria, Italy
Giuseppe M. L. Sarne	University Mediterranea of Reggio Calabria, Italy
Mario Versaci	University Mediterranea of Reggio Calabria, Italy

Program Committee

S. Abreu
A. Alonso-Betanzos
C. Bădică
A. Bădică
B. Bagula
F. Baiardi
N. Bassiliades
D. Bednárek
D. Berbecaru
S. Bobek
D. Brdjanin
M. Brezovan
G. Cabri
V. Carchiolo
D. Carneiro
A. Chechulin
C. Chesñevar
Y. Chevalier
F. Cicirelli
D. Cojocaru
M. Colhon
R. Collier
G. Coviello
P. Davidsson
M. De Vos
P. Demestichas
V. Desnitsky
G. DiFatta
I. Dionysiou
E. Doynikova
J. Fähndrich
S. Fidanova
I. Fister
G. Fortino
L. Fotia
D. Gajić
M. Ganzha
N. Garanina
J. Garcia Blas
S. Gesing
D. Gifu
L. Gladkov
A. Gonzalez-Pardo
V. Gorodetsky

R. Gravina
A. Groza
C. Herpson
K. Hindriks
M. Hölbl
G. Ilieva
G. Jezic
J. Jung
V. Jusas
H. Karatza
D. Katsaros
P. Kefalas
A. Kiss
P. Koprinkova-Hristova
I. Kotenko
G. Kougka
K. Kravari
V. Kurbalija
T. Küster
W. Lamersdorf
F. Leon
A. Liotta
I. Lirkov
D. Logofătu
M. Lujak
D. Macedo
J. Machado
M. Malgeri
G. Mangioni
S. Mariani
M. Marinescu
G. Marreiros
E. Martinez
C. Mastroianni
F. Messina
M. Montangero
R. Montella
G. J. Nalepa
M. Nistor
P. Novais
V. Oleshchuk
P. Oliveira
A. Omicini
E. Onaindia

M. Oprea
E. Osaba
F. Otero
G. Papadopoulos
M. Paprzycki
A. Pereira
D. Petcu
A. Poggi
F. Pop
C. B. Pop
E. Popescu
M. Postorino
R. Precup
M. Radovanović
W. Renz
A. Ricci
S. Ristov
D. Rosaci
I. Saenko
I. Sakellariou
A. Sapienza
I. Satoh
M. Savic
I. Savvas
R. Schumann
D. Selisteanu
R. Shaikh
Y. Shichkina
I. Spence
G. Spezzano
M. Strecker
J. Sudeikat
P. Thulasiraman
I. J. Timm
V. Toporkov
M. Torquati
R. Unland
S. Venticinque
J. Villadsen
G. Weiss
J. Yaghob
C. Zamfirescu
F. Zavoral

Contents

Machine Learning

Distributed Systems, Agents and Microservices

Text and Research

Social Systems

Smart Cities

Machine Learning

Using Artificial Neural Networks to Model Initial Recruitment of Mediterranean Pine Forests

Lidia Fotia[1]($\boxtimes$), Manuel Esteban Lucas-Borja[2], Domenico Rosaci[3], Giuseppe M. L. Sarné[4], and Demetrio Antonio Zema[5]

[1] Department DIEM, University of Salerno, Via Giovanni Paolo II, 132, 84084 Fisciano, SA, Italy
lfotia@unisa.it

[2] Departamento de Ciencia y Tecnología Agroforestal y Genética, Universidad de Castilla La Mancha,
Campus Universitario s/n, 02071 Albacete, Spain
manuelesteban.lucas@uclm.es

[3] Department DIIES, University Mediterranea of Reggio Calabria, via Graziella snc, loc. Feo di Vito, 98123 Reggio Calabria, Italy
domenico.rosaci@unirc.it

[4] Department of Psychology, University of Milan Bicocca, P.za della Ateneo Nuovo, 1, 20126 Milan, Italy
giuseppe.sarne@unimib.it

[5] Department od Agricolture, University "Mediterranea" of Reggio Calabria, Località Feo di Vito, 89122 Reggio Calabria, Italy
dzema@unirc.it

Abstract. Artificial Neural Networks (NNs) have been recognized as a powerful tool for automatically learning complex relationships in data. In this paper, we propose to apply such a tool for modeling forest regeneration, a possibility not yet investigated in the literature. In order to evaluate the capability of NNs to simulate initial recruitment of pine species in Mediterranean forests, a feed-forward multi-layer neural network has been applied to seed germination and seedling survival of four pine species under three soil conditions, with or without seed protection, in Castilla La Mancha (Central-Eastern Spain). The experimental campaign has shown good performance in predicting the two pine initial recruitment stages. The proposed approach may help to predict the success of natural regeneration in Mediterranean pine forests under different basal areas and management strategies.

1 Introduction

Forests provide multiple ecosystem with important functions and services to humanity including, among the others, production, climate regulation via carbon interception and water regulation by preventing soil erosion and flooding [2,13, 18,21]. On this context, a goal in the forest management strategies should rely on warrantying in the ecosystem a persistence and an adequate level of stability [16].

Natural regeneration is usually recognized as a slow and unpredictable process, because of the complex interactions occurring between the success of tree seedling establishment and the site factors. This is peculiar of Mediterranean forests, where fire events have endowed many species with an adaptive mechanisms that allow them to persist and regenerate forests after recurrent fires, particularly frequent in the later decades [3,15]. Moreover, in these areas drought and soil desiccation are primary constraints to natural regeneration, as a consequence of the Mediterranean climate which is characterized by long and dry summer periods and, therefore, represents an hostile factor [7,14,22].

Prediction models are used worldwide for several applications in the environmental field. The performance of such models is tightly dependent on the availability of adequate input datasets, and their results are validated under environmental scenarios as similar as possible to those under which the models must operate. On this context, and in order to support successful forest restoration projects, it is important to careful predict the initial recruitment of Pinus trees with respect to the both tree and site management characteristics mentioned above. This requires the availability of reliable models that must be able to simulate the regeneration process under variable environmental conditions.

To this aim, many different model approaches have been proposed in the literature to predict tree yield or forest natural regeneration [12,17,19]. Particularly, in the last decades data-driven models, such as the Artificial Neural Networks (NNs), have been object of an increasing popularity for a large variety of applications [1,10]. Compared to traditional prediction models, NNs require to be provided with a certain amount of input information about the processes to be modeled, which are simulated by using a mathematical form of learning invariant with respect to the application [20,23]. Moreover, NNs are able to provide useful results also when the input dateset is incomplete or noisy, and are characterized by the ability to work also in highly dynamic scenarios. A fundamental characteristic of a supervisioned NN is of learning the relationship between input and output data from examples provided with a dataset in a training phase and, after that the trained NN will be validated in a specific phase, it does not require further reprogramming when data in input vary [4,8,9].

In the scenario described above, this study proposes an NN modeling seed germination and plant survival of four pine species (i.e. Pinus nigra, Pinus pinaster, Pinus halepensis and Pinus sylvestris) under three soil conditions (i.e. scalped, post-fire and without action) and under with or without seed protection [5,11]. After an analysis on the effects of these soil conditions and management techniques on regeneration, the NN has been trained, optimized and tested using 10-year observations (time window: 2006–2016) of the modeled variables in forest stands of Castilla La Mancha (Central-Eastern Spain). The proposed approach may be of help in predicting the success of natural regeneration in Mediterranean pine forests with respect to different basal areas and management strategies.

2 The Artificial Neural Networks Stage

2.1 An Overview

In this work, a feedforward multi-layer neural network has been exploited to simulate the pine initial recruitment.

Multi-layer neural networks (thereafter only NNs for simplicity's sake) are composed by a set N of n nodes and a set A of m arcs. In turn, the nodes are partitioned into L groups, called layers, with $L > 2$. The first one is the *input layer* I formed by n_I nodes. Then, each of the $L - 2$ *hidden layer* Ht, with t $= 1, \cdots, L - 2$, consists of a set of n_{Ht} nodes and, finally, a set of O of n_O nodes forms the output layer. More in detail, each node belonging to the *input layer* I is connected by an edge directed to each node of the first *hidden layer*; the input layer nodes perform the only task of passively distributing the input data at all the nodes of the underlying layer. All the nodes belonging to a *hidden layer* are fully connected by edges with all the nodes belonging to the nearest layer and each node of the last $H(L - 2)$ *hidden layer* is connected with each other node belonging to the *output layer* (denoted by O). Furthermore, for each edge of the network, we denote by i (resp., j) the source (resp., destination) node and we associate a real value W_{ij}, named *weight*, with the edge.

Shortly, the NN is used to reproduce a real function. In particular, nodes of the *input layer* receives a set of input values, while the nodes belonging to the output layer returns the output (calculated by the NN), associated with that set of input values. *Hidden layer* nodes are associated with intermediate results of the computation. More specifically, a NN computes its output values as follows. Both of each *hidden* and *output layers* nodes are provided by a function f (that for sake of simplicity we assume to be the same for all the nodes belonging to the *hidden* and *output layers*), which is called *activation function*, and with a parameter Θ named *bias*. The node j of the first hidden layer $H1$ computes its associated output value (i.e. y_j^{H1}) as the sum of the values in input (i.e. x_i^I) coming from the *input layer* nodes, suitably weighted by the respective W_{ij}, that are the weights associated with the connections between the nodes i and j as in Eq. 1.

$$y_j^{H1} = f(\sum_{i=1}^{I} W_{ij} \cdot x_i^I - \Theta) \tag{1}$$

Then each node j of each *hidden layer* computes its output value y by exploiting the inputs coming from the overlying input layer (i.e. x), see Eq. 5. Analogously, each node of the *output layer* computes its associated output value y by exploiting the inputs coming from the last hidden layer, see Eq. 3.

$$y_j^{Hi} = f(\sum_{i=1}^{H} W_{ij} \cdot x_i^{H(i-1)} - \Theta) \tag{2}$$

$$y_j^{O} = f(\sum_{i=1}^{H} W_{ij} \cdot x_i^{H(L-2)} - \Theta) \tag{3}$$

The weight W_{ij} and the activation function parameters are suitably set by a *training algorithm* that tries to learn how approximating the correct output with a desired precision (i.e. error). Training algorithms can be unsupervised or supervised. In the first case, the NN autonomously learns the functional dependence between an input and its correct output. Differently, a supervised training algorithm takes advantage from the availability of a training dataset where for each input its correct output is provided. By measuring the error between the correct and the computed NN outputs then the NN tunes its parameters in order to minimize this error. When the NN training reaches the desired precision in reproducing the outputs of the training dataset, it has to be validated by means of a test session where are used data unknown to the NN. If also in this case the percentage of error between correct and NN output is satisfactory, then the NN can be considered ready to work. Many types of activation function (usually non-linear) can be used with the above NN model. The most known are the well-known sigmoid function and the hyperbolic tangent.

2.2   The Dataset

The dataset consists of 3614 tuples formed from six attributes, namely: i) forest species, ii) basal area, iii) seed protection, iv) soil condition, v) seed germination rate and vi) plant survival rate.

The first four attributes are used as NN inputs, while the last two attributes are NN outputs.

Data have been pre-processed to obtain a suitable dataset for the NN training. In particular, both the values of forest species and soil condition have been transformed into a real value belonging to the domain $[0; 1]$. Respectively, for the first attribute the following values have been adopted, i.e. 0.25 for Pinus nigra, 0.5 for Pinus pinaster, 0.75 for Pinus sylvestris and 1 for Pinus halepensis. For the second attribute, it has been assumed 0 for Scalped soil, $0, 5$ for No action and 1 for Post-fire. Finally, the seed protection attributed assumed values 0 for NO and 1 for YES. Then, the data set has been normalized in the range from 0 to 1 by means of:

$$X_n = \frac{X - X_{min}}{X_{max} - X_{min}} \tag{4}$$

where X is the value to normalize, X_n is the normalized value, X_{min} is the minimum value of X and X_{max} is the maximum value of X.

2.3   The Neural Network Setting

In this paper, to map input data into the appropriate output, a feedforward multilayer perceptron NN has been adopted. It consists of multiple layers of nodes structured as a directed graph, with each layer fully connected to the underlying one. The NN has been trained with a back-propagation with momentum algorithm by using the dataset previously described. Where the momentum is a real value added to speed up the process of learning and to improve the performance and efficiency.

For determining the best NN topology, the mostly performing activation function and the parameter setting, some preliminary tests have been carried out (see the next Section for more detail). However, the maximum admitted NN error, the learning rate and the momentum have been set to 0.0001, 0.2 and 0.7, respectively.

To measure the NN error, the Mean Percent Absolute Error (MAPE) has been computed as:

$$MAPE = \frac{100}{n} \cdot \sum_{i=1}^{n} \frac{|A_t - F_t|}{A_t}\%$$

(5)

where A_t are the actual output and F_t the corresponding predictions.

3   Results and Discussion

For the training, validation and experimental phases, the tool Deeplearning4j [6], a JAVA framework supporting deep learning algorithms and Multi Layer Perceptron architecture, has been exploited.

An exhaustive preliminary test session was carried out in order to identify the best NN topology and setting. Initially, we tried to train a unique NN for both the two outputs by varying topology and the activation functions (e.g. hyperbolic tangent and sigmoid functions) of the neurons but without obtaining satisfactory results.

Next, we simplified the problem by using two separate NNs, one for each desired output. First, we trained the NN for the germination rate output. From preliminary tests, the best NN topology has been identified in a four layer NN, i.e. input (i.e. 4), 2 hidden layer (each one with 50 neurons) and output (i.e. 1). With respect to the activation functions, in these attempts, the best results were obtained by using as activation functions hyperbolic tangents for all the neurons in the hidden layer/layers and sigmoids for that belonging to the output layer, while as learning rule a conventional *back-propagation with momentum* was exploited. More in detail, by adopting a single hidden layer then the error, in terms of MAPE, varied from 8% with 20 neurons to 7.5% with 1000 neurons, while in the adopted NN configuration with two hidden layer above described, the error stopped at 7.33% with only 50 neurons for hidden layer. Note as we tested also other NNs with more hidden layers without that the error decreased but only increasing the NN computational complexity.

For the first NN, dealing with the germination rate, some of the results in terms of MAPE obtained in our attempts are presented in Table 1 for different hidden layers configuration and by adopting hyperbolic tangents as neuron activation functions. In Fig. 1, for the best NN configuration are depicted the results in terms of MAPE obtained by using hyperbolic tangents (blue) or sigmoid (red) as activation functions of the hidden layer neurons.

Table 1. Germination rate. MAPE for different configuration of the first NN using hyperbolic tangent as activation function.

Neurons for Hidden Layer	Number of Hidden Layers			
	1	2	3	4
20	0,0808	0,0743	0,0734	0,1267
50	0,0777	**0,0733**	0,0733	0,0738
100	0,0759	0,0742	0,0740	0,0752

Then, we realized a further preliminary stage in order to identify the best configuration and setting also for the second NN tackling the survival rate.

We verified as the minimum MAPE (i.e. 1.24%) has been obtained by using a NN having 3 hidden layers, with 20 neurons for each of them. Note that also increasing the number of hidden layers or the number of neurons for hidden layer results do not improved at all. Also for the second NN, hyperbolic tangents have been adopted as activation functions for all the neurons in the hidden layers and a sigmoid function for that belonging to the output layer. Similarly to the first NN, as learning rule the *back-propagation with momentum* was exploited.

Also for the second NN, aimed to deal with the survival rate, some of the results in terms of MAPE obtained in our attempts are presented in Table 2 for different hidden layers configuration and hyperbolic tangents adopted for activation function of their neurons. In Fig. 2, for the best configuration of this NN the results, in terms of MAPE, obtained by using hyperbolic tangents (blue) or sigmoid (red) as activation function of the hidden layer neurons are depicted.

Table 2. Survival rate. MAPE for different configuration of the first NN using hyperbolic tangent as activation function.

Neurons for Hidden Layer	Number of Hidden Layers			
	1	2	3	4
20	0,0153	0,0149	**0,0124**	0,0124
50	0,0142	0,0144	0,0153	0,0154
100	0,0139	0,0156	0,0157	0,0147

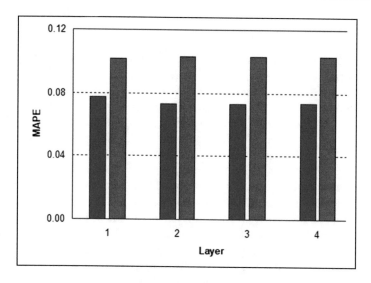

Fig. 1. Germination rate. MAPE for the best NN configuration equipping hidden layer neuron with hyperbolic tangent (blue) or sigmoid (red). (Color figure online)

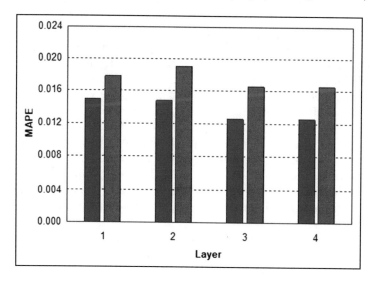

Fig. 2. Survival rate. MAPE for the best NN configuration equipping hidden layer neuron with hyperbolic tangent (blue) or sigmoid (red). (Color figure online)

Finally, by comparing the NNs' outputs with the field observations, we observed very low differences in the mean values of seed germination plant survival were achieved between field observations and the corresponding NNs outputs. From the results of our tests these differences were lower than 3.2% and show as this approach is able to predict both the stages of pine initial recruitment.

4 Conclusions

This study has investigated on the possibility to apply Artificial Neural Networks to model initial recruitment of four pine species in Castilla La Mancha (Central-Eastern Spain). A precious data set of 3614 tuples consisting of six attributes, namely *i*) forest species, *ii*) basal area, *iii*) seed protection, *iv*) soil condition, *v*) seed germination rate and *vi*) plant survival rate has been used and where the last two parameters are the desired outputs of the NN. The best performance was obtained by exploiting two different NNs. The first NN for dealing with the seed germination rate and its topology, resulted formed from four inputs, two hidden layers with 50 neurons for each layer, and one output, obtaining a MAPE of 7.33%. The second NN for dealing with the seed survival rate and its topology, resulted formed from four inputs, three hidden layers with 30 neurons for each layer, and one output, obtaining a MAPE of 1.24%. Summarizing, the main differences in seed emergence and plant survival were between field observations and the corresponding simulations were lower than 3.2%. Based on these preliminary results, we can assert that NNs show a good ability to predict both phases of early pine recruitment.

Note as the proposed NNs have been validated in environmental conditions that are similar at those in which the models must be used. We must highlight that several different prediction models are already used in the literature and their simulations of modeled processes generally produce results quantitatively comparable with those obtained by our NN-based approach, provided that suitable datasets of input parameters are available. However, the main limitations of these models are that *i*) a priori knowledge of the underlying process or assumptions of the target function structure is required, *ii*) the accuracy of NNs are higher that other statistical techniques, particularly when the problem is poorly defined and *iii*) the model procedure is lower than NN when the problem is extremely complex (i.e. natural regeneration processes). Conversely, the main advantages of using the NN are that *i*) the training phase is more efficient than the calibration of statistical models and that *ii*) once trained the network could easily improve in time their results by continuing the training on further available data.

We conclude that this modeling approach using NNs should be further promoted in order to better predict pine species initial seedling recruitment under different climate change scenarios.

References

1. Abiodun, O.I., Jantan, A., Omolara, A.E., Dada, K.V., Mohamed, N.A., Arshad, H.: State-of-the-art in artificial neural network applications: a survey. Heliyon 4(11), e00938 (2018)
2. Albuquerque, R.W., et al.: Mapping key indicators of forest restoration in the amazon using a low-cost drone and artificial intelligence. Remote Sens. 14(4), 830 (2022)

3. Alcañiz, M., Úbeda, X., Cerdà, A.: A 13-year approach to understand the effect of prescribed fires and livestock grazing on soil chemical properties in Tivissa, NE Iberian Peninsula. Forests **11**(9), 1013 (2020)

4. Bayat, M., Ghorbanpour, M., Zare, R., Jaafari, A., Pham, B.T.: Application of artificial neural networks for predicting tree survival and mortality in the Hyrcanian forest of Iran. Comput. Electron. Agric. **164**, 104929 (2019)

5. Borlea, I.-D., Precup, R.-E., Borlea, A.-B.: Improvement of k-means cluster quality by post processing resulted clusters. Procedia Comput. Sci. **199**, 63–70 (2022)

6. Deeplearning4j (2022). https://deeplearning4j.konduit.ai

7. Fotia, L., Lucas-Borja, M.E., Rosaci, D., Sarné, G.M.L., Zema, D.A.: An artificial neural network to simulate surface runoff and soil erosion in burned forests. In: Camacho, D., Rosaci, D., Sarné, G.M.L., Versaci, M. (eds.) IDC 2021. SCI, pp. 113–122. Springer, Cham (2022). https://doi.org/10.1007/978-3-030-96627-0_11

8. Gholami, V., Booij, M.J., Nikzad Tehrani, E., Hadian, M.A.: Spatial soil erosion estimation using an artificial neural network (ANN) and field plot data. Catena **163**, 210–218 (2018)

9. Haykin, S.: A comprehensive foundation. Neural Netw. **2**(2004), 41 (2004)

10. Islam, M., Chen, G., Jin, S.: An overview of neural network. Am. J. Neural Netw. Appl. **5**(1), 7–11 (2019)

11. Jander, K., Braubach, L., Lamersdorf, W.: Distributed monitoring and workflow management for goal-oriented workflows. Concurr. Comput.: Pract. Experience **28**(4), 1324–1335 (2016)

12. Kim, M., et al.: Seed dispersal models for natural regeneration: a review and prospects. Forests **13**(5), 659 (2022)

13. Liu, Z., Peng, C., Work, T., Candau, J.-N., DesRochers, A., Kneeshaw, D.: Application of machine-learning methods in forest ecology: recent progress and future challenges. Environ. Rev. **26**(4), 339–350 (2018)

14. Lucas-Borja, M.E., Heydari, M., Miralles, I., Zema, D.A., Manso, R.: Effects of skidding operations after tree harvesting and soil scarification by felled trees on initial seedling emergence of Spanish black pine (Pinus nigra Arn. ssp. salzmannii). Forests **11**(7), 767 (2020)

15. Lucas-Borja, M.E., et al.: Post-fire restoration with contour-felled log debris increases early recruitment of Spanish black pine (Pinus nigra Arn. ssp. salzmannii) in Mediterranean forests. Restoration Ecol. **29**(4), e13338 (2021)

16. Lucas-Borja, M.E., Hedo, J., de Santiago, Yu., Yang, Y.S., Candel-Pérez, D.: Nutrient, metal contents and microbiological properties of litter and soil along a tree age gradient in Mediterranean forest ecosystems. Sci. Total Environ. **650**, 749–758 (2019)

17. Manso, R., Pukkala, T., Pardos, M., Miina, J., Calama, R.: Modelling Pinus pinea forest management to attain natural regeneration under present and future climatic scenarios. Can. J. For. Res. **44**(3), 250–262 (2014)

18. Perry, D.A., Oren, R., Hart, S.C.: Forest Ecosystems. JHU Press (2008)

19. Rogers, R., Johnson, P.S.: Approaches to modeling natural regeneration in oak-dominated forests. Forest Ecol. Manag. **106**(1), 45–54 (1998)

20. Sazeides, C.I., Christopoulou, A., Fyllas, N.M.: Coupling photosynthetic measurements with biometric data to estimate gross primary productivity (GPP) in Mediterranean pine forests of different post-fire age. Forests **12**(9), 1256 (2021)

21. Shivaprakash, K.N., et al.: Potential for artificial intelligence (AI) and machine learning (ML) applications in biodiversity conservation, managing forests, and related services in India. Sustainability **14**(12), 7154 (2022)

22. Zema, D.A., Lucas-Borja, M.E., Fotia, L., Rosaci, D., Sarnè, G.M.L., Zimbone, S.M.: Predicting the hydrological response of a forest after wildfire and soil treatments using an artificial neural network. Comput. Electron. Agric. **170**, 105280 (2020)
23. Zhou, R., Dasheng, W., Zhou, R., Fang, L., Zheng, X., Lou, X.: Estimation of DBH at forest stand level based on multi-parameters and generalized regression neural network. Forests **10**(9), 778 (2019)

Applying Machine Learning Methods to Detect Abnormal User Behavior in a University Data Center

Igor Kotenko[✉] and Igor Saenko

SPC RAS, 39, 14th Liniya, St. Petersburg, Russia
{ivkote,ibsaen}@comsec.spb.ru

Abstract. Anomaly detection in the work of data center users is an important step in ensuring data center security. Such anomalies can be caused by both SQL injection attacks and user attempts to violate access control rules. One of the most effective approaches to detect abnormal user behavior in data centers is the use of machine learning methods. The paper explores the possibilities of using various machine learning models (classifiers) to detect such anomalies. A feature of the problem being solved is its focus on the university data center, whose databases have a non-normalized structure. In this case, the problem of reducing the dimension of the feature space for machine learning arises. The paper proposes an algorithm for generating a dataset based on typing the data table names. The issues of software implementation of the proposed approach are considered. The experimental results obtained on seven classifiers confirmed the high efficiency of the proposed approach. They showed that the decision tree, the k-nearest neighbors' method and the multilayer neural network have the highest efficiency in the problem being solved.

1 Introduction

Nowadays the popularity of using data centers (DCs) in control systems has significantly increased [1]. Data centers provide their users with the possibility of joint sustainable and timely use of information resources in the interests of solving various problems [2, 3]. For this reason, DCs are the primary targets for internal and external security violators to obtain information or disrupt centers [4, 5].

Different methods of anomaly detection can be used in the creation of DC information security systems. As a rule, anomalies are detected in the network traffic with the help of various network security tools (e.g., intrusion detection systems, firewalls, antiviruses). However, traffic anomalies do not fully reflect abnormal (anomalous) DC user behavior. It manifests itself in the form of user requests to databases with incorrect, anomalous requests that allow to make malicious changes in the content of databases or to obtain unauthorized information from databases. Such queries are a special kind of computer attacks - SQL injection [6]. In addition, anomalous queries may look normal and not contain SQL injection, but access to forbidden areas of the database. Protection against such accesses is usually assigned to the database access control system. However, for complex databases with a large number of data tables, creating an access control system that completely bans abnormal queries is a very difficult task.

The proposed approach, based on machine learning, is designed to help to fulfill this task. At the same time, it is proposed to use database logs as the initial data, on the basis of which the datasets used in machine learning methods are formed. These logs record the texts of queries the users accessed the databases. If the databases are based on the relational model, then queries are written in SQL. However, the proposed approach is not strictly tied to SQL and can be applied to any types of databases, such as NoSQL. The paper discusses the possibility of using machine learning methods to detect abnormal user behavior in DCs stored information and solved problems related to the university educational process. The choice of this type of DCs is explained by two reasons. First, the educational process databases contain a very large number of data tables. As a result, if you use table names to form a feature space, then the number of features will be very large. This makes the application of machine learning impossible or extremely difficult. Secondly, there are a large number of insider security threats in the university DC. Students should be considered as potential insiders. Therefore, the development of an additional security frontier for the university DC that allows detecting anomalous requests to DC databases is an urgent task. At the same time, it should be noted that a few known works are currently devoted to the topic of detecting or analyzing the possible malicious behavior of DC users.

The main contribution of the paper is as follows: (1) an original formal statement of the problem of detecting anomalous actions of DC users is proposed, oriented to the use of machine learning methods; (2) proposed to form a feature space, which determines the structure of training and testing datasets from the three categories of features - the keywords of the SQL language, signatures specific to SQL injection, and table names of data; (3) a heuristic approach is proposed to reduce the dimension of the original feature space and an algorithm that implements it; (4) the implementation of the proposed approach was carried out using a variety of the most well-known machine learning models; (5) an experimental evaluation of the proposed approach was carried out, which confirmed its effectiveness and high efficiency. The simplest approach to evaluate models (split into a fixed training and test dataset) was used. More complex approaches, such as those based on cross-validation, belong to the directions of further research.

The paper is structured as follows. Section 2 provides an overview of related works. Section 3 discusses the theoretical foundations of the proposed approach. Section 4 describes the details of the software implementation of the proposed approach and the generation of a dataset for its experimental evaluation. Experimental results and their discussion are considered in Sect. 5. Section 6 contains conclusions and directions for further research.

2   Related Work

Works related to the topic of detecting anomalies in the operation of DCs can be divided into two groups: (1) detecting anomalies in the operation of DCs and (2) detecting computer attacks such as SQL injection. Among the works of the first group, the works [7–12] should be marked. [7] notes that log entries in DC are stochastic and non-stationary in nature. Therefore, this work proposes an approach in which features are extracted from time windows and used to develop and update an evolving Gaussian Fuzzy Classifier on the fly. [8] proposes to use the word2vec algorithm to extract features from logs, and to use an LSTM neural network to detect anomalies. In [9], it is proposed to use unsupervised machine learning methods to determine the normal and abnormal behavior of DC cooling systems. The issues of preventing hostile influence on the detection of anomalies in DC are considered in [10]. This work proposes a linear regression-based optimization framework with the ability to poison data in the training phase. [11] suggests to detect anomalies by judging the deviation of predicted data and true data in DC operating using various machine learning methods. [12] uses linear regression and random forest to not only detect but also classify attacks in DC network traffic.

Despite the good results obtained in these anomaly detection works, it should be noted that these works did not consider anomalies in SQL queries and SQL injection detection. The works of the second group are devoted to this, for example [13–17]. [13] emphasizes that SQL injections became possible due to lack of validation of input queries. This paper presents an approach which detects a query token with reserved words-based lexicon to detect SQL injections. [14] proposes a robust semantic query ensemble learning model for SQL injection prediction. The proposed learning model used a set of nine basic classifiers designed to provide maximum prediction based on the voting ensemble. In [15], it is proposed to use multivariate statistical tests to detect anomalies in the behavior of DC users. In [16], a method for detecting SQL injections in web applications based on the Elastic-Pooling convolutional neural network is presented. [17] proposes an approach to detecting SQL injections based on the analysis of the response and state of a web application during various attacks. [18] considers a machine learning-based approach to prevent SQL attacks, in which over 20 different classifiers are tested and the top five are selected. The results of this work were used in our study in selecting classifiers.

3   Theoretical Background

By a university DC we mean a DC with several databases, which are used in the interests of organizing and conducting the educational process. We will assume that the behavior of DC users consists in accessing the databases available in the DC using queries written in some language (e.g., SQL). Databases can be stored on nodes of a distributed file system (e.g., HDFS). Then parallel anomaly detection will be possible. In the paper, however, we do not focus on this. Database queries are recorded in the database management system (DBMS) logs (e.g., PostgreSQL, MySQL and others). The log consists of separate records. Each record reflects the fact of some user accessing the database and contains the following fields: date, time, user identifier and SQL query text. Therefore, the task of detecting abnormal behavior of university DC users comes down to detecting anomalous SQL queries to DC databases, which leads to searching for anomalous records in the logs.

If we imagine the DBMS log as a dataset consisting of records, then a possible technique for analyzing such a dataset for anomaly detection involves the following possible steps: 1) the formation of a set of features that characterize SQL queries; 2) transformation of the log dataset into a dataset, the records of which contain the values of the generated features; 3) formation of a training sample on which the machine learning process will be carried out; 4) the use of trained tools to directly identify anomalous requests.

The initial data of the task are:

1) a set of DC users $\mathbf{U} = \{U_1, U_2, ..., U_N\}$ and logs $\mathbf{L} = \{L_1, L_2, ..., L_M\}$;
2) each log is represented as a set of records $L_m = \{l_{mi}\}$;
3) each log record is represented as a tuple $l_{mi} = <Date_{mi}, TimeStamp_{mi}, User_{mi}, Op_{mi}>$, where $Date_{mi}$ is the date of the i-th query in the m-th log; $TimeStamp_{mi}$ is the request timestamp; $User_{mi} \in \mathbf{U}$ is the query user; Op_{mi} is the SQL statement;
4) each SQL statement can be represented as $Op_{mi} = <Operator, \{Tables\}, \{Fields\}, \{Values\}>$, where $Operator$ is the SQL operator; $\{Tables\}$ is a set of table names that are present in the SQL statement; $\{Fields\}$ is a set of field names; $\{Values\}$ is a set of field values;
5) machine learning models (binary classifiers), which are most often used to detect anomalies in data sets [19, 20];
6) requirements for detecting attacks such as SQL injection: probability of correct attack detection: $P_{det} \geq 0.95$; probability of missing an attack: $P_{mis} \leq 0.1$.

To calculate these probabilities, it is proposed to use the following formulas: $P_{det} \approx TP/(TP + FP + FN)$, $P_{mis} \approx FN/(FN + TP)$, where TP is the number of correctly detected anomalies in the dataset (True Positive); FP is the number of False Positive; FN is the number of False Negative.

As a result of solving the task, it is required to develop a feature space model that characterizes the normal and anomalous activities of database users in the absence and presence of attacks, and a method for detecting anomalous SQL queries based on the use of binary classification models.

In the proposed feature space model, the features are divided into three categories. The features of the first category determine the number of occurrences of a keyword in the SQL query. A total of 30 keywords were selected, such as SELECT, INSERT, CREATE, etc. The second category of features was the number of occurrences of certain signatures specific to SQL injection. The third category was formed by the number of occurrences of data table names. The DC University database used to experimentally evaluate the proposed approach contained over 4,000 data tables. This is due to the non-normalized nature of its structure. Approximately 2,000 tables contained faculty data, one table per faculty member. Each course, discipline, and study group also had its own data table.

Due to such a large number of tables, we made the following two decisions. First, we decided not to use field names and values when forming the feature space. Second, we decided to reduce the number of table names in the feature space by replacing them with a generic type name. So, all table names for teachers were replaced with the type name "Teacher_Table", table names for study groups were replaced with "Group_Table", etc. Thus, it was possible to reduce the number of table names taken into account to 141.

The composition of the generated feature space is presented in Table 1. It shows that the total number of features has become 181. Of these, 30 feature belong to the first category, 10 - to the second category and 141 - to the third category.

Table 1. The composition of the feature space.

#	Category	Name	Description
1	1	SELECT_COUNT	Number of occurrences of SELECT
2		INSERT_COUNT	Number of occurrences of INSERT
...		...	...
30		HAVING_COUNT	Number of occurrences of HAVING
31	2	Execute_COUNT	Number of occurrences of "Execute"
32		"1=1"_COUNT	Number of occurrences of "1=1"
...		...	...
40		txtUserId_COUNT	Number of occurrences of "txtUserId"
41	3	Table_1_COUNT	Number of occurrences of data table name 1
42		Table_2_COUNT	Number of occurrences of data table name 2
...		...	...
181		Table_141_COUNT	Number of occurrences of data table name 141

4   Implementation

To implement the approach, the Python v.3.8.8 was used with the following libraries: sklearn, numpy, pandas, matplotlib, Scipy, Re, Pylab, Math. The computing environment was organized on a Jupyter notebook. The following models were subject to study as classifiers: Support Vector Machine (SVM), Decision Tree (DT), Regression Logistic Classifier (RLC), Random Forest (RF), Gaussian Naïve Bayes (GNB); k-Nearest Neighbor (KNN), and Neural Network (NN). The hyperparameters of the training models were selected. In particular, for kNN it was found that the best value is $k = 5$. The university DC used DBMS PostgreSQL v.13.4 running under Ubuntu v.13.4 to create the database.

To form a dataset used to train binary classifiers, a fragment of the log of the university DC database was selected, showing the work of users with the database for 15 min. In total, this fragment initially contained 82192 instructions. Figure 1 shows a view of several of the instructions included in this log.

```
2022-01-18 12:44:09.749 UTC [1531176] LOG:  database system is ready to accept connections
2022-01-18 12:44:10.986 UTC [1531188] postgres@template1 LOG:  statement:
2022-01-18 12:44:37.957 UTC [1531211] postgres@2122 LOG:  statement: SELECT DISTINCT "groups" FROM
"p_uplan_pmk" ORDER BY "groups"
...
2022-01-18 12:45:01.514 UTC [1531230] postgres@2122 LOG:  statement: SELECT "potok_num" FROM
"p_group" WHERE groups='1123'
2022-01-18 12:45:01.514 UTC [1531230] postgres@2122 LOG:  statement: SELECT "groups" FROM "p_group"
WHERE potok_num='112'
2022-01-18 12:45:01.567 UTC [1531234] postgres@2122 LOG:  statement: SELECT "23" FROM "1123_2021"
ORDER BY count;
...
2022-01-18 12:46:41.250 UTC [1531276] postgres@2122 LOG:  statement: select * from pg_tables where
tablename='IvanovDA';
2022-01-18 12:46:41.254 UTC [1531276] postgres@2122 LOG:  statement: select * from "IvanovDA" where
"count"='2'
...
2022-01-18 12:47:20.926 UTC [1531276] postgres@2122 LOG:  statement: select "n_aud" from "D-0406" where
"prep"='5' and "groups"='5811'
...
2022-01-18 13:00:33.497 UTC [1531656] postgres@2122 LOG:  statement: SELECT "groups" FROM "p_group"
WHERE potok_num='534'
2022-01-18 13:00:33.498 UTC [1531587] postgres@2122 LOG:  statement: select "n_aud" from "D-2006" where
"prep"='38' and "groups"='3882'
```

Fig. 1. Log's fragment of the university DC

Figure 1 shows that the recording of this fragment was made on March 18, 2022. It started at 12:44:09 and ended at 13:00:33. Several users worked with the database and had IDs 1174, 1187, 1211, 1230, 1234, 1276, 1565, and 1587. The queries accessed various data tables. So, the query with time 12:44:37 turned to the system table "p_learn_plan" (it contained planning information on the educational process). The name of this table appears in the statement after the word FROM. Other queries addressed the following tables: "p_group", "1123", "pg_tables", "IvanovDA", "D-0406", "D-2006". The p_group and pg_tables tables were system ones. They were created by the system when the database itself was created. Other tables are non-system created by users using the CREATE command while working with the database. The table "IvanovDA" contains data about the teacher D.A. Ivanov. Table "1123" contains data on study group # 1123. Tables "D-0406" and "D-2006" contain data on academic disciplines, which have identifiers D-0406 and D-2006.

Since the database contained a very large number of tables with data on teachers, study groups and academic disciplines, it was decided to type such tables, that is, replace them with names of types.

This procedure has become one of the initial steps of the developed dataset formation algorithm. In total, such an algorithm contains the following 5 steps.

Step 1. Extracting all table names from the log fragment and forming a set of table names used in the fragment. In total, 310 table names were extracted.
Step 2. Generation of a set of new, typical table names. For the names of tables with data about teachers, the name "Teacher_Table" was used, for the names of tables with data about groups - "Group_Table", etc. In total, this set included 141 names, including the names of system tables.

Step 3. Replacing the original table names in the log fragment with generic names. At the same time, the number of instructions in the fragment was still 82192.

Step 4. Formation of the initial dataset in CSV format. For each instruction, a CSV record was created from the fragment. The fields of this record were the features that were included in the feature space model and shown in Table 1. In addition, the `Result` field was included in the initial version of the dataset, which value played the role of a label for a normal or abnormal record. It was assumed that if `Result = 0`, then the record is normal, and if `Result = 1` - abnormal.

Step 5. Removing duplicate records from the dataset and introducing anomalies into it. Since the date and time of the query were excluded from the feature space at this stage of the study (this was done deliberately to test the effectiveness of machine learning on SQL query structures), a large number of duplicate records appeared in the dataset. At this stage such records were deleted. After their removal, only 1,026 records remained in the dataset. In addition, a few randomly selected records were changed to match various possible anomalies (SQL injection and access violation attempts). A total of 50 such records were modified and marked as anomalous in the `Result` field. An example of this modification is writing the WHERE field in the SQL query as follows: WHERE "UserId = 105" or "1=1". This entry allows an attacker to get the entire data table.

The dataset formed using the algorithm described above was further analyzed using the selected binary classifiers.

5   Experimental Results

Experimental studies were divided into two stages: the training stage and the testing stage. The first stage was devoted to training classifiers and checking the accuracy. At the second stage, the trained classifiers were used to directly detect anomalies. At the first stage the dataset was divided into two parts. The first part, intended for training, included 80% of the entries. The second part, intended for control testing, included the remaining 20% of the entries.

The results of control testing are presented in Fig. 2. We can see that the DT classifier showed the highest accuracy (Fig. 2-b). It detected both normal and anomalous recordings without errors. Quite good results were also demonstrated by KNN (Fig. 2-c), NN (Fig. 2-e) and RF (Fig. 2-f). They detected all normal records without errors and had from 10 to 25 percent errors in detecting anomalies. RLC (Fig. 2-a) and SVM (Fig. 2-g) detected all normal records without errors, but were unable to correctly detect any of the anomalies. GNB (Fig. 2-d) correctly detected all anomalies, but had a very large number of errors in detecting normal records (29%).

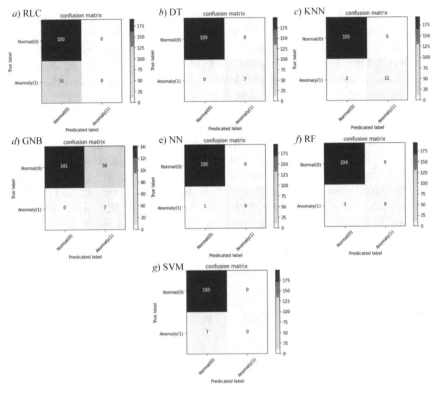

Fig. 2. Test results on various classifiers

At the second stage, the trained classifiers were used to test a new dataset generated from the log. A new log fragment contained 78880 entries written during 40 min. After algorithm considered above processed it, 1852 records remained in the new dataset. 50 anomalies were added to this dataset by modifying the entries.

The testing results are presented in Table 2.

Table 2. Results of testing a new dataset

Classifier	TN	TP	FN	FP	P_{del}	P_{mis}
RLC	1807	5	45	0	0.10	0.90
DT	**1807**	**50**	**0**	**0**	**1.00**	**0.00**
KNN	**1807**	**48**	**2**	**0**	**0.96**	**0.04**
GNB	1256	50	0	551	0.08	0.00
NN	**1807**	**48**	**2**	**0**	**0.96**	**0.04**
RF	1807	38	12	0	0.76	0.24
SVM	1807	1	49	0	0.02	0.98

As can be seen from Table 2, only 3 from 7 considered classifiers (DT, KNN and NN) satisfy requirements for P_{del} and P_{mis}. The DT classifier is the best. The GBN classifier fulfills the requirements for P_{mis}, but has very poor values for P_{del}. The remaining classifiers (RLC, RF, and SVM) do not meet the requirements for P_{del} and P_{mis}. Perhaps this is due to the fact that a very small training sample was used. However, in order to obtain a training sample of several tens or even hundreds of thousands of records, it is necessary to use a log fragment corresponding to one day of work. We consider these experiments as a further research.

For the same reason, studies of the temporal characteristics of the learning and testing turned out to be ineffective. On the computer used in the experiments, these times were within 1–3 s, which corresponds to computational noise. We expect that in further studies it will be possible to build time dependencies for all used classifiers.

6 Conclusion

The paper presents the task statement, algorithms, issues of implementation and the results of an experimental evaluation of a new approach to detecting anomalous queries to SQL databases, based on a heuristic algorithm for reducing the dimension of a feature space and using binary classification methods. The initial data of the task are logs, database users, selected binary classifiers, and requirements for the accuracy of detecting anomalous SQL queries. The result of the approach implementation is a feature space model, presented as a set of normal and abnormal data records containing the values of the generated features, and a technique for searching for abnormal queries. An experimental evaluation of the proposed approach was carried out on real data sets generated during the work of university DC users with the database of the educational process, using a set of binary classifiers. The evaluation results confirmed the effectiveness of the proposed approach and its high efficiency.

Further research is aimed at improving the accuracy of detecting anomalous SQL queries by improving the parameters of classifiers and combining them.

Acknowledgements. This research is being supported by the grant of RSF #21-71-20078 in SPC RAS.

References

1. Alqahtani, J., Alanazi, S., Hamdaoui, B.: Traffic behavior in cloud data centers: a survey. In: 2020 International Wireless Communications and Mobile Computing (IWCMC), pp. 2106–2111 (2020)
2. Welsh, T., Benkhelifa, E.: On resilience in cloud computing: a survey of techniques across the cloud domain. ACM Comput. Surv. **53**(3), 59 (2021)
3. Mujib, M., Sari, R.F.: Performance evaluation of data center network with network micro-segmentation. In: 2020 12th International Conference ICITEE, pp. 27–32 (2020)
4. Klymash, M., Shpur, O., Lavriv, O., Peleh, N.: Information security in virtualized data center network. In: 2019 3rd International Conference on Advanced Information and Communications Technologies (AICT), pp. 419–422 (2019)

5. Paiusescu, L., Barbulescu, M., Vraciu, V., Carabas, M., Cuza, A.I.: Efficient datacenters management for network and security operations. In: 2018 17th RoEduNet Conference: Networking in Education and Research (RoEduNet), pp. 1–5 (2018)

6. Marashdeh, Z., Suwais, K., Alia, M.: A survey on SQL injection attack: detection and challenges. In: 2021 International Conference ICIT, pp. 957–962 (2021)

7. Decker, L., Leite, D., Giommi, L., Bonacorsi, D.: Real-time anomaly detection in data centers for log-based predictive maintenance using an evolving fuzzy-rule-based approach. In: 2020 IEEE International Conference on Fuzzy Systems (FUZZ-IEEE), pp. 1–8 (2020)

8. Shahid, N., Ali Shah, M.: Anomaly detection in system logs in the sphere of digital economy. In: Competitive Advantage in the Digital Economy, pp. 185–190 (2021)

9. Nanekaran, N.P., Esmalifalak, M., Narimani, M.: Fast anomaly detection in micro data centers using machine learning techniques. In: 2020 IEEE 18th International Conference on Industrial Informatics (INDIN), pp. 86–93 (2020)

10. Deka, P.K., Bhuyan, M.H., Kadobayashi, Y., Elmroth, E.: Adversarial impact on anomaly detection in cloud datacenters. In: 2019 IEEE 24th Pacific Rim International Symposium on Dependable Computing (PRDC), pp. 188–18809 (2019)

11. Chen, J., Wang, L., Hu, Q.: Machine learning-based anomaly detection of ganglia monitoring data in HEP data center. In: EPJ Web Conference, vol. 245, p. 07061 (2020)

12. Salman, T., Bhamare, D., Erbad, A., Jain, R., Samaka, M.: Machine learning for anomaly detection and categorization in multi-cloud environments. In: 2017 IEEE 4th International Conference on Cyber Security and Cloud Computing (CSCloud), pp. 97–103 (2017)

13. Hlaing, Z.C.S.S., Khaing, M.: A detection and prevention technique on SQL injection attacks. In: 2020 IEEE Conference on Computer Applications, pp. 1–6 (2020)

14. Gowtham, M., Pramod, H.B.: Semantic query-featured ensemble learning model for SQL-injection attack detection in IoT-ecosystems. IEEE Trans. Reliab. 71, 1057–1074 (2022)

15. Prarthana, T.S., Gangadharₒ, N.D.: User behaviour anomaly detection in multidimensional data. In: 2017 IEEE International Conference on Cloud Computing in Emerging Markets (CCEM), pp. 3–10 (2017)

16. Xie, X., Ren, C., Fu, Y., Xu, J., Guo, J.: SQL injection detection for web applications based on elastic-pooling CNN. IEEE Access 7, 151475–151481 (2019)

17. Xiao, Z., Zhou, Z., Yang, W., Deng, C.: An approach for SQL injection detection based on behavior and response analysis. In: 2017 IEEE 9th International Conference on Communication Software and Networks (ICCSN), pp. 1437–1442 (2017)

18. Hasan, M., Balbahaith, Z., Tarique, M.: Detection of SQL injection attacks: a machine learning approach. In: 2019 International Conference on Electrical and Computing Technologies and Applications (ICECTA), pp. 1–6 (2019)

19. Branitskiy, A.A., Kotenko, I.V.: Analysis and classification of methods for network attack detection. SPIIRAS Proc. 2(45), 207–244 (2016)

20. Kotenko, I., Saenko, I., Branitskiy, A.: Detection of distributed cyber attacks based on weighed ensemble of classifiers and big data processing architecture. In: IEEE Conference on Computer Communications Workshops, IEEE INFOCOM 2019, pp. 1–6 (2019)

Dynamic Management of Distributed Machine Learning Projects

Filipe Oliveira[1], André Alves[1], Hugo Moço[1], José Monteiro[1], Óscar Oliveira[1], Davide Carneiro[1(✉)], and Paulo Novais[2]

[1] CIICESI, ESTG, Politécnico do Porto, Felgueiras, Portugal
{fvol,afba,hmsm,jmgm,oao,dcarneiro}@estg.ipp.pt
[2] Algoritmi Center/Department of Informatics, University of Minho, Braga, Portugal
pjon@di.uminho.pt

Abstract. Given the new requirements of Machine Learning problems in the last years, especially in what concerns the volume, diversity and speed of data, new approaches are needed to deal with the associated challenges. In this paper we describe CEDEs - a distributed learning system that runs on top of an Hadoop cluster and takes advantage of blocks, replication and balancing. CEDEs trains models in a distributed manner following the principle of data locality, and is able to change parts of the model through an optimization module, thus allowing a model to evolve over time as the data changes. This paper describes its generic architecture, details the implementation of the first modules, and provides a first validation.

1 Introduction

Despite all the recent advances in the field of Machine Learning (ML) and related supporting areas, many new challenges keep emerging [1]. These stem mostly from the volume and diversity of data in current ML problems, as well as from the need to deliver services in real-time, which led to the emergence of streaming data, streaming analytics [2] and streaming ML [3].

ML applications nowadays require such a volume that datasets must be distributed across clusters. Consequently, ML algorithms need to learn in a distributed manner, from multiple sources of data. Moreover, these data change over time, leading to the need for models to be updated or fully retrained, which has an increasingly significant cost on the organizations' infrastructure.

In this paper we describe the initial work being developed in the context of the CEDEs project - Continuously Evolving Distributed Ensembles. CEDEs aims to build an environment for the distributed training of ML models, in which the models can evolve over time as data change, in a cost-effective way. Therefore, not only does it address the issue of learning from large datasets, but also that of learning continuously from streaming data.

CEDEs takes advantage of existing block-based distributed file systems, such as the Hadoop Distributed File System (HDFS) [4], to parallelize and distribute

© The Author(s), under exclusive license to Springer Nature Switzerland AG 2023
L. Braubach et al. (Eds.): IDC 2022, SCI 1089, pp. 23–32, 2023.
https://doi.org/10.1007/978-3-031-29104-3_3

learning tasks, following the principle of data locality. That is, the computation is moved to where the data reside, rather than the other way around [5]. Moreover, instead of more traditional models, CEDEs uses Ensembles [6]: a base model is trained for a block of a dataset, where the data resides. Then, Ensembles are built in real time by combining available models according to criteria such as their performance or the state of the nodes where the base models reside. This is done by the optimization module that adapts the Ensemble as data changes.

The system also makes use of two other mechanisms: replication and balancing. Replication allows to automatically create replicas of blocks so that the same block is available in multiple locations simultaneously. This increases the use of storage space, but makes it easier for the optimization module to find suitable nodes for carrying out tasks. Balancing, on the other hand, distributes the blocks evenly across the cluster, so that the load when executing tasks is appropriately distributed.

Finally, CEDEs also stores the base models themselves in the HDFS, which are then replicated. This means that, not only, the Ensemble itself is distributed, which speeds up predictions, but also that, thank to replication, multiple nodes will have the same base models available for making predictions. Once again, it is up to the optimization module to decide which base models to use and in which nodes to make predictions. This paper describes the implementation, operation and testing/validation of the core modules of this system.

2   Conceptualization

This section details the proposed system in terms of its main components and functionalities. It first describes the underlying data model (Sect. 2.1), and then the proposed architecture (Sect. 2.2).

2.1   Data Model

The data model that supports an instance of the proposed system is composed of seven main entities, as depicted in Fig. 1, which are described next.

Each instance of the system is expected to address a specific Machine Learning problem or domain (e.g. Healthcare, Fraud Detection, Finance). While the system allows multiple organizations to be part of an instance, they are expected to be in some way related to the domain of that instance. Moreover, organizations can perform different roles, depending on how they use the available services. For example, an organization may primarily be a contributor of data, when it contributes with data and, consequently, models, but makes little to no use of predictive services. Alternatively, an organization may essentially be a consumer, when it contributes with little to no data, but makes a heavy use of predictive services. Any intermediary range is also possible. Given that this characterization is volatile through time (i.e. the stance of each organization may change), it is not part of the data model. Instead, this is derived from the blockchain, as detailed further below.

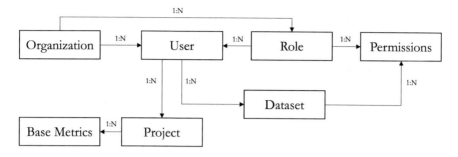

Fig. 1. Simplified view of the Data Model that supports the proposed system.

Each organization may have multiple users, who are the central entity in the data model. A user may create and edit ML projects and import or access datasets. The access of each user to each dataset is governed by a system of roles and permissions. Each organization may define multiple roles, according to its internal organization (e.g. Data Scientist, ML Engineer, Data Engineer). Each user may then have one or more roles inside her/his organization. Permissions are then attributed on the basis of user roles. Moreover, when a user uploads a new dataset, she/he becomes the owner of the dataset. This is important to determine the provenance of the data.

When a user creates a new ML project she/he must provide some information, including: name and general description of the project, the underlying dataset, which algorithm(s) is/are to be used in the training of the base models, configurations of the algorithms, etc.

Multiple algorithms can be used in a single project. This is done when the user wants to build a heterogeneous ensemble, composed of models of different types. In this case, the user must provide the relative proportion of each algorithm (e.g. 50/30/20) as well as the configuration to use for each algorithm. The configuration is dependent on the type of each algorithm. For example, a Decision Tree may be configured in terms of maximum depth or minimum number of instances per leaf (among others), while a neural network may be configured in terms of activation function or the number of layers and respective neurons (among others). This allows for a great range of different ensembles to be built.

When the user requests for a given model to be trained, in the context of a specific ML project, this actually results in the training of multiple base models: one for each block (excluding block replicas) of the dataset. While this is completely transparent to the user, the resulting information is stored in the Base Metrics entity. This table has one line for each base model of each ensemble and contains, among others, data describing the type of algorithm used, the node where the training took place, the resulting performance metrics (e.g. RMSE, MAE, R^2) and the computational cost of training the model (e.g. training time, memory and CPU consumption).

There are also relevant additional data, that are not stored in this data model. Namely, we do not store the meta-data regarding the blocks of each dataset and their location, as these are dynamic and change through time. This information is requested in real-time from the HDFS, when needed, lest we risk using outdated information. The same goes for the location of the base models of a project when there's the need to constitute an ensemble and making predictions.

The architecture of the proposed system, which makes use of this data model, is described in the following Section.

2.2   Architecture

The architecture of the proposed solution is composed of a number of major modules, which are described in this section. The central element is the Storage Layer. This is implemented in the form of an HDFS cluster. Here, large datasets are split into smaller chunks called blocks, of a fixed size (typically 128MB). Multiple data sources can be considered, ranging from files in different formats (e.g. CSV, Parquet, Avro) to databases. The whole process starts with a user uploading a data source into the HDFS, which is then split, distributed and replicated across the cluster.

Once this process finishes, the dataset becomes available to be used in a new ML project. The user can thus create a new project, select the intended dataset, and provide the necessary information already detailed in the previous Section. Once the configuration of the project is complete, the user may start the training of a new ensemble. This process is managed by the Coordination module, which interacts with the Optimization and Metadata modules.

The training of an ensemble takes place in a distributed manner and takes advantage of two aspects: the fact that the blocks that constitute each file are replicated and available in multiple nodes simultaneously, and the principle of data locality. An ensemble is thus trained by training one base model for each block of a file (excluding replicas) which, when combined, constitute the ensemble. There is thus the need to select the most appropriate node where to train each base model for each block, according to criteria such as the state of each node of the cluster where that block is available.

Specifically, the Coordination module needs the following inputs:

Block Location. For a given ML problem, the Coordination module needs to know what blocks are part of the respective dataset, and where each of its replicas are available at the moment of training. Each block is typically available in at least 3 nodes, but this is dependent on the configuration of the HDFS. The number of blocks of a dataset determines the number of base models to be trained (1:1 relationship). This information is obtained from the Metadata module.

Task Cost Prediction. The duration and cost, in terms of computational resources, of the training of a given base model depends on multiple factors, including the size of the dataset (lines and columns) and the type and configuration of the algorithm used. To have an estimate of the cost of each individual task (i.e. the training of each base model), an approach based on meta-learning was developed that uses multiple meta-models, one to predict each intended cost metric, based on metadata collected from a large group of ML problems.

This approach, which provides a fairly good estimate of task complexity/cost is further described in [7,8]. This information is obtained from the Optimization module.

Nodes State. Another relevant aspect to consider when deciding where to train base models is the state of each node in the cluster. The state of a node is characterized by its health (which is calculated and maintained by Hadoop's health checker service), its current load (e.g. memory, CPU, disk) and the size and cost of its tasks queue. This information is provided by the Optimization module.

Task Allocation. In order to implement the actual training of the models of the ensemble, the Coordination module also needs information regarding where each task (base model training) will take place. For instance, when there are three candidate nodes to train a base model for a specific block (thanks to replication), the system will have to choose the best candidate node. The Optimization module carries out a global optimization heuristic, which will provide as output the best allocation of all the necessary tasks across the cluster, based on the elements mentioned in the previous two points (i.e. Tasks cost prediction and nodes state).

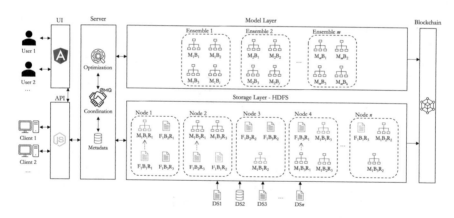

Fig. 2. Depiction of the proposed architecture with a possible sample scenario.

Once the Coordination module receives the tasks allocation for a given ML project, it is ready to start the training process. This process is implemented in ZeroMQ: a brokerless asynchronous messaging library that implements multiple socket communication patterns, useful for implementing distributed systems.

As the distributed training process goes on, each worker node updates the coordinator at regular intervals about its individual progress. This information can thus be accessed in real time by the client applications. As the training of each individual base model ends, the coordinator is notified and the worker node can move on to the next task in its queue, if there is one.

As soon as the training of a base model ends, another process takes place. The model is serialized and is stored in the HDFS. Automatically, the base model will be replicated and distributed across the cluster, according to the replication and balancing factors. This means that soon after the training of a base model, it will be available for making predictions in multiple nodes of the cluster.

This is illustrated in Fig. 2, in which the following nomenclature is used: $F_iB_jR_k$ represents a replica k of block j of file i. The same logic is used for models, which start with the letter M. An arrow between a specific block replica and a model means that that base model was trained from that block, on that node. Replicas of that base model may however exist in other nodes, created through the replication mechanism of Hadoop after the base model was stored in the HDFS.

Thus, when the client requests for predictions from a given ensemble, a similar process takes place to that of training models. First, the Coordination module needs information regarding which of the base models will be used as part of the ensemble. Indeed, depending on the configuration of the ML project, not all the base models may be part of the ensemble. The user may decide to have an ensemble with a number of base models that is smaller than the number of blocks of the dataset. In that case, the Optimization module will decide, according to the performance metrics of the base models and the state of the nodes where they reside, the best group of base models to constitute the ensemble. In the past, we have implemented this process using genetic algorithms, as described in [9]. However, a more efficient approach is now being implemented.

Once the Coordination module receives information about task allocation, it distributes the tasks accordingly by the cluster by means of messages, and awaits for the predictions of each base model. The predictions are then combined to calculate the final prediction by the ensemble. To this end, the mean is used for regression problems, and the mode is used for classification.

Ensembles are thus abstract constructs, depicted in Fig. 2 in the Model Layer, built out of a selection of available base models for a given ML problem.

There are three mode modules in this architecture. The Blockchain module is used to record all the actions of users or clients in the system, including who imported datasets, who requested the training of models, who requested predictions etc. The API module includes all the endpoints necessary for interacting with the services of the system, which in turn communicate with the Coordination module, when applicable, through ZeroMQ messages. This API can be used by external client applications. Finally, there is also a User Interface meant to be used by Human users, which communicates with the previously mentioned API.

The described system is not fully implemented yet. Work is currently undergoing in the Blockchain and Optimization modules. Currently, the Coordinator assigns tasks randomly among available and valid candidate nodes, for each ML project.

3 Validation

This section describes the methodology followed for validating the proposed system, and its results. Given the lack of a physical infrastructure, the distributed system was simulated as a Docker application with 4 containers: 1 acting as the coordinator and 3 acting as workers. Each container is based on the same image, built on top of the ubuntu:bionic image, and in which all the Hadoop ecosystem was installed. To simulate the actual conditions of a real cluster, Hadoop was installed in distributed mode.

The characteristics of the datasets used for testing the system are described in Table 1. Given that the sizer of the datasets is relatively small for what is standard in an Hadoop cluster, HDFS's block size was reduced to 8MB, to force a larger number of blocks and parallelism.

Table 1. Brief characterization of the datasets used for validation.

dataset	rows	columns	size	blocks
city temperatures	2906327	8	140.6 MB	18
mnist	70000	785	127.9 MB	16
sales records	1048575	14	130.9 MB	16

For each dataset, an Ensemble was trained using the proposed system. A heterogeneous Ensemble was used, containing a mixture of three algorithms from the scikit-learn library: random forest (sklearn.ensemble.RandomForestClassifier), decision trees (sklearn.tree.DecisionTreeClassifier) and neural networks (sklearn.neural_network.MLPClassifier). The configuration of algorithms used is detailed in Table 2. String features were transformed using scikit-learn's label encoder.

Table 2. Configurations used to train the heterogeneous Ensembles (defined arbitrarily).

algorithm	parameter	value
decision tree	max_depth	5
	min_samples_split	2
	max_leaf_nodes	5
	ccp_alpha	0
neural network	hidden_layers_size	[5,2]
	activation	relu
	solver	adam
	alpha	0.0001
	learning_rate	constant
	max_iter	200
random forest	nr_estimators	5

In the proposed system, the user can also select the weight of each algorithm in the Ensemble (Fig. 3), which results in a proportional number of models for each algorithm in the Ensemble. For instance, given a dataset split into 18 blocks and the weights of 50, 30 and 20 for the algorithms random forest, decision trees and neural networks, respectively, the Ensemble would have 12 random forest, 3 neural networks and 3 decision trees. For the purpose of these tests, these were the weights used. Moreover, the holdout method was used to evaluate each base model, with 75% of the data used for training and the remaining 25% used for testing. The performance of the model is obtained by computing the average value of a given performance metric (e.g. RMSE, MAE).

In order to compare the results obtained with a baseline, we trained 3 random forests using a commercially available tool (H2O), using the pre-defined configuration of the algorithm, namely 50 trees with a maximum depth of 20. The results are compared with those of the proposed system in Table 3 in terms of the RMSE, R^2 and training time. The performance is equivalent, both in terms of predictive performance or training time. However, it must be stated that the goal at the time is not to have the most accurate models possible but rather to test if all the components of the system are working as intended.

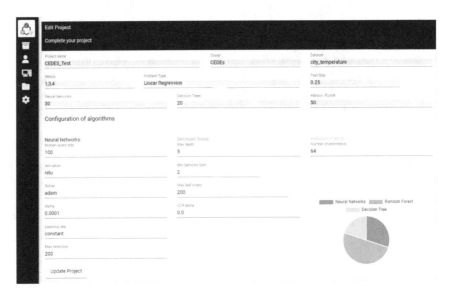

Fig. 3. Prototype of the UI: page for editing the properties of a new ML project.

Table 3. Results of the models trained using the proposed system.

dataset	proposed approach			commercial tool		
	rmse	r2	time (s)	rmse	r2	time (s)
city temperatures	21.40	0.51	120.06	20.13	0.61	128.44
mnist	1.82	0.59	244.50	0.86	0.91	101.23
sales records	63198.12	0.93	17.34	7623.67	0.99	577.5

4   Discussion, Conclusions and Future Work

This paper described the initial steps in the implementation of a distributed learning system that has some key innovative features: it trains models for big distributed datasets following the principle of data locality; it decides in which nodes to carry out the training and predicting tasks according to the state of the cluster; it allows to use heterogeneous Ensembles; and it abstracts all the complexity of distribute learning and associated coordinated tasks from the user. The optimization mechanism is especially important as it will allow for the performance of the ensemble to be improved by adjusting which base models make part of it and with which weight.

The main goal of this paper was to validate the proposed system and test its functionalities rather than to obtain the best possible models. For this reason, we did not invest time in finding the best possible configuration for each ML problem and used the same configuration for all. This may explain the relatively poorer performance of the trained models when compared to those of a commercial tool. Figure 4 shows how the value of the RMSE varies significantly according depending on the base model for the "city temperatures" ensemble, hinting that improvements could be achieved by tweaking the configurations used.

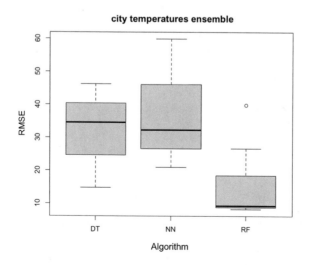

Fig. 4. Distribution of the RMSE for the 18 base models trained for the "city temperatures" dataset, by algorithm.

Moreover, given that we have no dedicated infrastructure yet, we have implemented and tested the system on a Docker instance. In this scenario, we chose to deliberately use a very small block size for HDFS (8MB), in order to increase the number of blocks, so as to validate the communication protocol and overall performance of the system. This leads to very small training sets, which may lead to relatively poorer models, and in turn, lead to generally poor ensembles.

Work now continues on the implementation of the Optimization and Blockchain modules. Moreover, a streaming module based on Kafka is now also being implemented, that will allow data to change over time and new base models to be added to the Ensemble. The Optimization module will be used to determine which models will make part of a given Ensemble in real time, allowing for the Ensemble to evolve over time, as new models are added and others removed. This is only possible due to the decisions taken during the implementation of this system, that allow for this flexibility that is absent in other similar tools. Ultimately, it will allow to better deal with the significant challenges that the field of ML faces today, namely in what concerns learning continuously from large volumes of streaming data.

Acknowledgments. This work was supported by FCT - Fundação para a Ciência e Tecnologia within projects UIDB/04728/2020 and EXPL/CCI-COM/0706/2021.

References

1. Zhou, L., Pan, S., Wang, J., Vasilakos, A.V.: Machine learning on big data: opportunities and challenges. Neurocomputing **237**, 350–361 (2017)
2. Mohammadi, M., Al-Fuqaha, A., Sorour, S., Guizani, M.: Deep learning for IoT big data and streaming analytics: a survey. IEEE Commun. Surv. Tutor. **20**(4), 2923–2960 (2018)
3. Gomes, H.M., Read, J., Bifet, A., Barddal, J.P., Gama, J.: Machine learning for streaming data: state of the art, challenges, and opportunities. ACM SIGKDD Explorations Newsl. **21**(2), 6–22 (2019)
4. Shvachko, K., Kuang, H., Radia, S., Chansler, R.: The Hadoop distributed file system. In: IEEE 26th Symposium on Mass Storage Systems and Technologies (MSST), pp. 1–10. IEEE (2010)
5. Attiya, H.: Concurrency and the principle of data locality. IEEE Distrib. Syst. Online **8**(9), 3 (2007)
6. Dong, X., Yu, Z., Cao, W., Shi, Y., Ma, Q.: A survey on ensemble learning. Front. Comput. Sci. **14**(2), 241–258 (2020)
7. Carneiro, D., Guimarães, M., Silva, F., Novais, P.: A predictive and user-centric approach to machine learning in data streaming scenarios. Neurocomputing **484**, 238–249 (2021)
8. Carneiro, D., Guimarães, M., Carvalho, M., Novais, P.: Using meta-learning to predict performance metrics in machine learning problems. Expert Syst. **40**, e12900 (2021)
9. Ramos, D., Carneiro, D., Novais, P.: Using evolving ensembles to deal with concept drift in streaming scenarios. In: Camacho, D., Rosaci, D., Sarné, G.M.L., Versaci, M. (eds.) IDC 2021. SCI, vol. 1026, pp. 59–68. Springer, Cham (2022). https://doi.org/10.1007/978-3-030-96627-0_6

Enabling Neural Network Edge Computing on a Small Robot Vehicle

Jan Bredereke[(✉)]

City University of Applied Sciences Bremen (HSB),
Flughafenallee 10, 28199 Bremen, Germany
`jan.bredereke@hs-bremen.de`

Abstract. We report on experiences with optimizations that enable neural network edge computing on a small robot vehicle, which serves as a proxy for an autonomous robotic space craft. We realize a visual object recognition task using a neural network on a field-programmable gate array (FPGA) with data processing resources as limited as those of an FPGA suitable for space. We use a quantized neural network nicely matching the properties of an FPGA. The restrictions of the small FPGA require us to sequentialize the processing partially. We employ input frame tiling for this. It allows us to keep the entire neural network on-chip. Furthermore, we split up the visual object recognition task into two stages, using two separate neural networks. The first stage identifies the region of interest approximately, using large and thus few tiles. The second stage looks closely at the single tile containing the region of interest; thus being not that time critical.

1 Introduction

This introduction first motivates our work, then describes our sample application, introduces our solution approach, and finally lists some related work.

Motivation: Neural networks are often used in data centers with powerful and power consuming special hardware. A current research question is how to also make full use of neural networks at the "edge" of the Cloud. That is, close to the sensors and the actors, or even autonomously from data and power supply connections. The CPU of a microcontroller has too scarce data processing resources for this. An FPGA can offer more data processing resources, both in absolute numbers and in relation to its power consumption. An FPGA is very well suited for a highly parallel structure such as that of a neural network. In practice, however, many optimization tasks need to be solved before full use of the potential of the FPGA can be made.

Our particular motivation comes from space craft engineering. On-board computers provide especially scarce data processing resources. Access to the ground segment is usually available intermittently only. Due to space radiation, current off-the-shelf processors would fail soon. Therefore, special processors are used. Their chips feature structural widths of at least 65 nm instead of the denser

L. Braubach et al. (Eds.): IDC 2022, SCI 1089, pp. 33–40, 2023.
https://doi.org/10.1007/978-3-031-29104-3_4

10 nm or less currently in use elsewhere. These special processors compromise on data processing resources to achieve sufficient robustness. An extremely small number of computers of this kind is made. Therefore, they usually are not made with application specific integrated circuits (ASICs), but with programmable standard hardware (FPGAs). Radiation-hard versions of some FPGAs are available, which have suitably larger structural widths.

There is increasing demand for on-board computing resources. Examples are on-board image processing, e.g. for autonomous rovers on other celestial bodies, or for constellations of nano-satellites, each with narrow bandwidth to the ground segment.

The Internet of Things (IoT) is an area with similar challenges. Cost restrictions lead to scarce data processing resources. Many of these devices are battery powered. This further restricts the data processing resources. Thus, challenges and solutions in space craft engineering might be applicable to the IoT, too, and vice versa.

The Sample Application: We employ an autonomous model car as a practical robot vehicle, serving as a proxy for an autonomous robotic space craft. The vehicle is equipped with a camera and a system-on-chip (SoC) Xilinx Zynq-7020. The SoC features an Arm CPU and, in particular, an FPGA Artix-7. The data processing resources of the Artix-7 are quite similar to those of an FPGA suitable for space [HSB+20, Chap. 2.1.3]. The SoC is integrated into a PYNQ-Z1 board. In Fig. 1, the board can be spotted easily due to its pink colour.

Fig. 1. The sample application vehicle with a camera and an FPGA board. (Color figure online)

The task of the vehicle is to visually recognize a person in view, to follow it when it moves, and to obey to driving commands given by simple gestures. It shall do all of this autonomously and in real-time.

Approach for Optimizations: We investigate optimizations that enable neural network edge computing on this kind of small robot vehicle. Many ideas are conceivable. Up to now, we tried out the following:

- Use a pre-trained neural network for inference only on the vehicle.
  Rationale: inference is an order of magnitude cheaper than training.
- Use hardware acceleration for the neural network by an FPGA, with the FPGA choice obeying the restrictions of the application area.
  Rationale: a neural network is inherently concurrent, and an FPGA offers massively parallel execution.

- Reduce hardware resources for the neural network by employing a quantized neural network.
  Rationale: a quantized neural network demands far less bits per node, and recent research showed that the resulting loss of precision per node can be more than compensated by a few more nodes [Blo+18, Umu+17].
- Reduce hardware resources for the neural network by splitting the input image into tiles and process them sequentially.
  Rationale: if the task doesn't fit the chip area available at all, making an area/speed trade-off by partially sequentializing the task can help.
- Reduce the number of tiles required, and thus the number of iterations, by separating locating a person in a frame and determining the person's gesture.
  Rationale: A small object of interest requires many tiles for finding it and evaluating its properties in detail. A two-staged approach can cope with larger, less detailed and thus less numerous tiles for finding. The subsequent gesture recognition needs a single tile only. Our restricted-hardware setting requires the sequentialization by tiling; the expensive part here is finding the "region of interest", not evaluating it.

We report on the details and on our experiences in the remainder of this paper.

Related Work: Several research papers [Man+18, Rof+20, Umu+17, Blo+18, Jok+18, Xu+21] provide valuable work towards our goal, but each achieves partial aspects only.

2 Selection of Suitable Frameworks

Many frameworks exist which promise to support implementing a neural network on an FPGA. When we started our work, we did a little survey on those available at that time. [HSB+20, Chap. 4] Making use of them turned out to be more difficult than expected. In particular, scarce documentation was a problem with several frameworks. Other frameworks did not support our hardware platform, the PYNQ-Z1 board. Two tools did generate code, but it would have required substantial further work to actually make the code run on our board. In this first round, we therefore didn't use any of these frameworks.

Instead, in this first round we manually accelerated the execution of a very simple neural network in a Python environment. The PYNQ-Z1 board provides a user friendly Python development environment. However, execution there is far too slow for our purposes. We manually implemented the neural network on the FPGA and devised a tool chain that feeds data from the Python environment into the FPGA and returns the results back to the Python environment. We used Keras/Tensorflow in the IDE PyCharm to train our neural network, written in Python, in order to obtain the weight parameters for our implementation of the network. We did this offline on standard PC hardware. We then made the inference work on the less powerful PYNQ-Z1 board. We provided the input to our neural network using Python. The PYNQ provides the concept of so-called Python overlays. We wrote the custom driver for such a Python overlay in order to tie the Python code to the AXI bus connecting the CPU and the FPGA of the SoC. We also wrote an IP core in VHDL, consisting of the components of our neural network, suitably connected, interfacing to the AXI bus. We configured the network with the weight parameters from the training. A comparison with a network in Python showed an acceleration of several orders of magnitude. The implementation effort for this first FPGA solution was substantial; however, implementing another network could reuse most of the code except the actual wiring of the network. [HSB+20]

In the second round, we re-evaluated some frameworks. We decided to use the FINN framework [Umu+17], since its documentation had improved considerably at that time. [MKB+21] We eventually stayed with the FINN framework during our further work, which we describe in the next chapter.

3   Prototype Implementations

This section reports how we implemented the optimizations discussed in Sect. 1, and what we learned from this. We applied the optimizations step by step in several rounds. Each subsection reports on one of these iterations.

Tracking a Moving Traffic Sign: In this round, we approached a non-trivial visual object recognition task, and we added our robot vehicle (shown in Fig. 1) as a sample application. [MKB+21] We used the FINN framework [Umu+17] to implement the neural network for visual object recognition on the FPGA automatically. The results are fed back to the CPU of the SoC, where we implemented a simple tracking algorithm in C/C++: the vehicle turns and follows the object recognized. For simplicity, at this point we used one of the pre-trained neural networks that come with the FINN framework. Therefore, the vehicle tracks and follows a particular traffic sign, of which we moved around a decommissioned life-size example in the open air on a public parking space. The resulting system was indeed able to successfully perform this driving task in real-time.

The FPGA hardware is by far too restricted to accommodate a neural network that can process an entire video frame. Therefore, we split each frame into several tiles and fed them into the FPGA sequentially. We used tiles of different size classes, in order to transform both small and large regions with an object into tile-filling images. For simplicity in this round, we implemented the tiling using the OpenCV framework on the CPU. Each frame of size 640×480 pixels was cut into 999 regions of 64×64 pixels and 108 regions of 96×96 pixels. These regions were transformed into 1107 tiles of size 32×32 pixels. The resulting frame rate was 2 fps.

Only the actual inference executed on the FPGA, anything else still executed on the CPU of the SoC. So there was ample room for further optimizations.

Tracking a Moving Person Making Gestures: In our third round, we trained the neural network ourselves, which allowed for a more realistic tracking task. [ACB+21] The task of the vehicle now was to track a person; additionally, this person can command the vehicle by simple arm gestures ("start"/"stop"/"no specific pose", compare Fig. 2). We produced a set of 300 images of persons showing these gestures, and we added 120 random images without a person from a Google web search.

We re-evaluated our choice of a board with a SoC of similar processing capabilities as a radiation-hard FPGA suitable for space. We stuck with our PYNQ-Z1 board, but the Avnet Ultra96-V2 board turned out to be a slightly more powerful alternative still in range. [ACB+21, Chap 3.3]

We re-evaluated our choice of a framework. Many candidates dropped out. This was because they aim at larger FPGAs, have not published their code, provide documentation too scarce, or are not supported anymore. As a result, we stuck with the FINN framework. [ACB+21, Chap. 5]

We considered several publicly available neural network models for visual object recognition. We chose the CNV_1W1A model, because the FINN framework provides additional support for this model. [ACB+21, Chap. 4.5] The model is written using Brevitas, a PyTorch library. Brevitas can

Fig. 2. Arm gestures for commanding the vehicle, at a resolution of 32 × 32 pixels.

export the trained model in the ONNX exchange format, which can be fed directly into the FINN compiler. More details on our tool chain are in our technical report [ACB+21].

We had a first look at alternatives to tiling an image. A Region Proposal Network (RPN) is an interesting idea. A tentative experiment showed the need for additional effort not feasible for us at that time. We also considered a Single Shot Detector (SSD) briefly. [ACB+21, Chap. 6] In hindsight, this would have been feasible only when adding off-chip memory to the FPGA. When the task can be solved on this FPGA by 1000 iterations with a network for a 32 × 32 pixel image, then no network for a full size frame 640 × 480 pixel image can fit on this FPGA to do the task in a single shot. Using the FPGA together with off-chip memory while applying some other kind of iteration might help to realize an SSD network. We did not investigate this further, though. The bandwidth of the off-chip memory link would certainly be a relevant parameter.

Our implementation used 70% of the block RAM and 46% of the look-up-tables of the FPGA. Unfortunately, our implementation had some bugs then, found only after this round, which prevented a successful evaluation of the detection accuracy in a real-life test.

Tiling in Hardware: The thesis of Müller [Mül21], one of our collaborators, improves the efficiency of the tiling by realizing it on the FPGA, too, instead of in software on the CPU of the SoC. Müller's solution allows for a highly flexible configuration of the tiling and scaling at run-time. It uses the PYNQ-Z1 board, too.

Müller integrated his tiling implementation with the implementation of an example CNV neural network from the FINN framework, which was trained with the CIFAR10 data set. All variants of the integrated solution were small enough to fit into the FPGA and its block RAM. Tests proved that the system successfully recognizes differently scaled and differently placed images from the CIFAR10 data set in an input video stream. The hardware solution on the 100 MHz FGPA was up to 4 times faster than a comparable solution in software on the 650 MHz CPU. And Müller identified several aspects of his hardware solution that would allow for further optimization. The development of the solution was more complex than one in software; a large share of this was due to the voluminous AXI4 interface specification. The integration of this hardware tiling into our robot vehicle system needs yet to be done.

Tracking a Moving Person and Only Then Looking for Gestures: In a further round, we separated locating a person in a frame and determining the person's gesture, in order to reduce the number of tiles required; furthermore, we improved the quality of our implementation of the robot vehicle system. [ACB+22] We re-trained the CNV neural network from the previous round, now for detecting a person. For this, we used suitable parts of the huge standard COCO data set. We implemented the inference on the same PYNQ-Z1 board as before. Our implementation used 70% of the block RAM and 44% of the look-up-tables of the FPGA; this should easily allow to add the above tiling in hardware.

When using 24 tiles only, the system achieved 10–12 fps; it could follow a person in up to 2 m distance. When using 396 tiles, the system could follow a person in nearly 6 m distance; however, the frame rate dropped to 2–3 fps. Significant imperfections of the mechanical drive train let the car lose its target more often in the latter case. An overall optimum appeared to be at 138 tiles and 6 fps.

Because of time restrictions, we could not add the recognition of gestures anymore. For a start, it could be implemented in software, since it is used only once per frame. Thus, it is not that time critical.

4   Summary and Outlook

We reported on experiences with optimizations that enable neural network edge computing on a small robot vehicle. The basic approach worked well, doing the cheaper inference only on the vehicle, and employing a field-programmable gate array (FPGA) with its massively parallel resources for a neural network; also, using a quantisized version of a neural network matching the properties of the FPGA well. All of this was expected from recent literature already.

Fitting a realistic visual recognition task into the resources of a small FPGA suitable for space required an area/speed trade-off; that is, processing an input frame in several iterations. We kept the neural network entirely on-chip and avoided the potential memory bandwidth bottleneck of a solution that permanently loads parts of a large neural network into the small FPGA. Instead, we sequentialized processing the input frame by cutting it into several tiles. Tiling is a relatively cheap operation. And Müller proved that tiling can be performed on the very same FPGA, too, thus speeding it up further.

Since we had to sequentialize the processing partially anyway, we split up the visual object recognition task into two stages, using two separate neural networks. The first stage identifies the region of interest approximately, using large and thus few tiles. The second stage looks closely at the single tile containing the region of interest; thus being not that time critical. A single-shot neural network approach would have required another way of sequentialization; and we suspect that it would have become difficult to avoid a memory bandwidth bottleneck there. Of course, the advantage of our two-staged approach partially hinges on the property of our application that there is only one region of interest in a frame at a time.

Future work should integrate hardware tiling into our robot vehicle system. Furthermore, it should investigate the advantages that a neural network brings which is tailored more specific to the application.

References

[ACB+21] Altnickel, P., Catalkaya, T., Bredereke, J., et al.: Gesten- und Objekterkennung durch schwache FPGAs in autonomen Fahrzeugen mittels neuronaler Netze. German. Technical report. HSB (2021)

[ACB+22] Altnickel, P., Cansu, F., Bredereke, J., et al.: Personenerkennung durch schwache FPGAs in autonomen Fahrzeugen mittels Neuronaler Netze. German. Technical report. HSB (2022)

[Blo+18] Blott, M., et al.: FINN-R: an end-to-end deep-learning framework for fast exploration of quantized neural networks. ACM Trans. Reconfigurable Technol. Syst. **11**(3) (2018). https://doi.org/10.1145/3242897

[HSB+20] von Huth, C., Soldin, M., Bredereke, J., et al.: Bericht zum Projekt "Neuronale Netze auf strahlungstoleranten FPGAs für die Raumfahrt". German. Technical report. HSB (2020)

[Jok+18] Jokic, P., et al.: BinaryEye: a 20 kfps streaming camera system on FPGA with real-time on-device image recognition using binary neural networks. In: 2018 IEEE 13th International Symposium on Industrial Embedded Systems (SIES) (2018). https://doi.org/10.1109/SIES.2018.8442108

[Man+18] Manning, J., et al.: Machine-learning space applications on SmallSat platforms with TensorFlow. In: Proceedings of AIAA/USU Conference on Small Satellites (SmallSat) (2018)

[MKB+21] Müller, F., Krekel, N., Bredereke, J., et al.: Applying binarized neural networks on FPGAs to an autonomous driving problem. Technical report. HSB (2021)

[Mül21] Müller, F.: Dynamisches Tiling auf schwachen FPGAs zur Objekterkennung mithilfe kleiner neuronaler Netze. German. Bachelorthesis. HSB (2021)

[Rof+20] Roffe, S., et al.: CASPR: autonomous sensor processing experiment for STP-H7. In: Small Satellite Conference (2020)

[Umu+17] Umuroglu, et al.: FINN: a framework for fast, scalable binarized neural network inference. In: Proceedings of the 2017 ACM/SIGDA International Symposium on Field-Programmable Gate Arrays, FPGA 2017, pp. 65–74. ACM (2017). https://doi.org/10.1145/3020078.3021744

[Xu+21] Xu, X., et al.: DAC-SDC low power object detection challenge for UAV applications. IEEE Trans. Pattern Anal. Mach. Intell. **43**(2), 392–403 (2021). https://doi.org/10.1109/TPAMI.2019.2932429

Using SHAP-Based Interpretability to Understand Risk of Job Changing

Daniel-Costel Bouleanu[1(✉)], Costin Bădică[1], and Kalliopi Kravari[2]

[1] University of Craiova, Craiova, Romania
bouleanu.daniel.y5m@student.ucv.ro, costin.adica@edu.ucv.ro
[2] International Hellenic University, Thessaloniki, Greece
kkravari@ihu.gr

Abstract. We propose a possible solution to the socio-economic problem of understanding the decision of Data Scientists to early change or leave their job. This problem is mainly relevant from a Human Resource perspective, to better understand the factors leading persons working on a Data Science position to change their job in the short term. We consider two Machine Learning prediction models based on Decision Tree and Logistic Regression. For each model we study the importance of the involved features using a Cooperative Game Theory approach based on Shapely value. The experiments were done using a public domain dataset from Kaggle.

1 Introduction

With the increasing competition between businesses on the modern labour market, better understanding and evaluating the emergent risks when hiring new staff became crucial. Such risks span a variety of reasons including for example inconsistent hiring practices, lack of suitable training of HR managers, and failing to get the best fit for the given role, among which of high importance seems to be the risk of the newly hired staff to leave the job too early [10]. Studies performed by specialized companies and researchers based on empirical surveys revealed a multitude of explanations given by new employees that left their job within short time, usually maximum 6 weeks. However, it is of high interest for the companies not only to monitor and understand these explanations, but also to act proactively by predicting with a higher accuracy if a new hire will leave or not his or her new job in the short term in order to mitigate the associated risk.

Recent advances in Artificial Intelligence (AI), Machine Learning (ML) and Cooperative Game Theory (CGT) are providing new tools that can help companies to improve their predictive power as concerning early job leaving by new hires, and thus to better protect themselves against such emerging risks. Historical data regarding the new staff demographic and professional profile, as well as data including the type and size of the company and the geo-economic context can be useful to perform such predictions by using emerging ML algorithms and platforms. Moreover, recent advances in Explainable AI (XAI) can be used to improve companies' ability of better understanding such ML predictions [16].

© The Author(s), under exclusive license to Springer Nature Switzerland AG 2023
L. Braubach et al. (Eds.): IDC 2022, SCI 1089, pp. 41–50, 2023.
https://doi.org/10.1007/978-3-031-29104-3_5

The contribution of this paper is to propose a new possible solution to the socio-economic problem of understanding the decision of Data Scientists to early change or leave their job. This problem is mainly relevant from a Human Resource perspective, to better understand the factors that determine persons working on a Data Science role to early change their job. We consider two Machine Learning models, namely Decision Tree and Logistic Regression, to perform this prediction. For each model we analyze the importance of the involved features using SHAP-based explanations, inspired by Shapely value from Cooperative Game Theory.

2 Related Works

Occupational psychologists concerned with persistent job changing have focused largely on distinguishing between those who are drifting aimlessly and those who are moving between jobs to some purpose. There is, unfortunately, no easy way to distinguish between useful trial and aimless drifting [4]. In this paper we focus on those who are moving between jobs with a purpose, trying to find the aspects that drive their purpose. It is important to note that in the literature, the risk of job changing was also assimilated with job insecurity, which adversely affects job satisfaction, organizational commitment and determines the intention to leave the organization. Those employees who perceived job insecurity reported lower job satisfaction and organizational commitment, as well as higher level of intention to leave the organization (so to change their current job) [15]. Demographics such as age, gender, education, organizational grade have been seen to influence individuals to consider work and career as life priorities and dedicate personal time and energy to their work. Overall, the findings suggest the existence of a relationship between those traits and work involvement [3].

Following nearly two decades of rapid changes in every aspect of the workplace, long term jobs seem to be way less frequent. Factors related to age or gender had been a focus in several studies conducted in Europe, but such inductive research is rare elsewhere [6]. In addition to establishing more clearly the patterns of job changing by gender and education level, understanding it more completely will depend on understanding better the details of both the environment and the person to whom it applies [14]. The study [1] conducted by the National Foundation for American Policy found that 55% of America's billion dollars start-up companies had at least one immigrant in an important position. Another study [7] conducted in United Kingdom showed that firms with a greater share of migrant owners or partners introduced more products and processes, being more innovative. This shows the ongoing effect of globalization, where people are more likely to move to a bigger and more developed city for work. Note that workers in bigger cities earn more than workers in smaller cities and rural areas [13].

The literature is wide and deep, and we only covered a small portion because of space limitation. Some other notable articles are presenting relationships between other factors such as experience – performance [11] and company size – job satisfaction [2]. We have tried to provide a brief overview of the most important factors that can influence one person's decision of changing her job, from both the employee and employer view, so we can compare these approaches and results with our own results.

3   System Development

3.1   Data Collection

Often, when a high-tech company wants to hire a new employee, it goes through an intensive process of interviewing, testing and training to decide what is the best match of the hired person with available positions, thus incurring a significant hiring cost for the company. However, sometimes, even if the employee passes all these checks, he or she might end up not entering the company or leaving it too early to cover incurred hiring costs. In order to minimize such risks, HR department tries to predict how likely are selected new employees to stay in the company. Such predictions can be based on various facts including their skills, results, education, experience, demographics, and past jobs as well as the geo-economic context.

We are using a public domain dataset downloaded from Kaggle [5]. This dataset was proposed to help HR department to better understand and predict the factors that might determine employees to change their job. The dataset refers to an anonymous company that is active in Big Data Science. The dataset is composed of approximately 20000 examples which were split into an 80% part for training and a 20% part for testing, containing 13 features with one more being the target feature.

3.2   Features and Data Pre-processing

Before creating the classification model, we have done a preliminary analysis of the available data set. The features and their semantics are:

- *enrollee_id*: Identifier for each person (number)
- *city*: The city code instead of the city name (number)
- *city_development_index*: Level of development in cities (scaled number)
- *gender*: Gender of candidate (Male, Female, Other)
- *relevant_experience*: Relevant experience of the candidate (Text, Yes/No)
- *enrolled_university*: Type of university enrollment
- *education_level*: Education level of the enrollee
- *major_discipline*: Major of the enrollee
- *experience*: Years of experience of the enrollee (number)
- *company_size*: Number of employees current employer's company
- *company_type*: Type of current company
- *lastnewjob*: Number of years passed since the last job
- *training_hours*: Hours of training (number)
- *target*: If the person is looking for a (new) job or not

We reduced the number of features by excluding those features that could not have any impact on the actual result. For example, the city code should have no importance considering that we have already included the city development index, that clearly provides a higher impact of a given city. We were not able to find any important correlation that could affect our results.

3.3   Model Development

Our approach is based on defining ML classification models to predict if a person is looking for a new job or not. We have developed two such models: i) Decision Tree Classifier that was chosen due to its proven usefulness and relative simplicity for solving decision-related problems, and ii) Logistic Regression for predicting the value of the categorical target variable given a set of independent variables. The decision was made to compare at least two different models, in order to be able to assess if applying SHAP to different classification models will lead to different results. Moreover, these models were chosen to show that even simpler models can give a good insight into this problem when complemented with SHAP explanations.

3.3.1   Decision Tree Classifier

Classifiers based on Decision Tree approach are used successfully in many diverse areas. Their most important feature is that the decision making knowledge it directly related to the input data. A training set can be used to generate a decision tree. The hyper-parameters were determined by experiment, choosing the *entropy* criterion, with the *best* value for the splitter and a *max depth* of 5. Those values prove to give the highest accuracy given our dataset.

3.3.2   Logistic Regression

Logistic Regression is a process of modeling the probability of obtaining a discrete outcome given an input sample. The most common logistic regression is when the dependent (target) variable is dichotomous (with binary outcome, something that can take two states such as true/false, win/loss, etc.). Logistic regression is a useful method for classification problems, where you aim to decide if a new sample fits best into a category [12]. It worth mentioning that Logistic Regression is in fact a particular case of Linear Regression, for which the base functions are modeled using the logistic function. Therefore, all the regularization types available in general to Linear Regression, are also available to Logistic Regression. After our experiments regarding hyperparameterization, the *lgbfs* solver gave both the best and second best accuracy. Having no penalty seems to always increase the accuracy.

4   Experiments and Results

SHAP is considered as state-of-the-art in ML explainability and it is inspired by CGT and Shapley values [9]. While Shapley values measure the contribution of each player to the game outcome, SHAP assumes that the players are represented by the model features, and SHAP values quantify the contribution that each feature brings to the prediction made by the model.

SHAP is computed for each new sample x and each given input feature f of x as an weighted average of all the differences between the prediction of the model obtained for each subset X of features containing $f \in X$ and the model prediction obtained for the subset of features $X \setminus \{f\}$. So we can intuitively observe that $SHAP(x, f)$ quantifies the contribution brought by feature f and sample x to the model prediction. Unfortunately, this direct method for computing SHAP would require training of an exponential number of models, so it is intractable. Instead, approximation methods are used in practice. After model creation, we have computed the SHAP values, based on the predictions of the model, by using the *shap* Python library [8]. Instead of how much was this prediction driven by a feature we actually show how much was this prediction driven by a feature compared to some baseline value for this feature. SHAP values baseline is based on the average of all predictions; i.e. the contribution (SHAP value) of a given feature regards the difference between the actual prediction and the mean prediction.

4.1 General Result over Training Data

We first analyze what explanations can be synthesized from the training data by presenting the importance of the features using a SHAP beeswarm (one-dimensional chart that shows all the information on single axis) graph, for the decision tree classifier. The idea is that a feature is more important if that has larger absolute SHAP value. For determining feature importance, we averaged absolute SHAP values over the whole dataset.

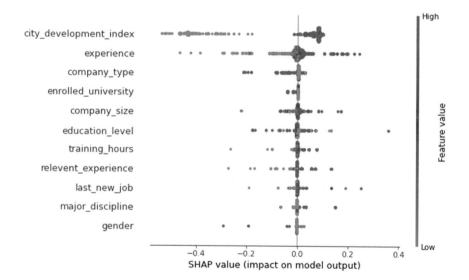

Fig. 1. Distribution of SHAP values given each individual sample

Analyzing Fig. 1 we observe the following:

- The vertical axis represents all the features in the order of their computed importance (how much they influenced the model prediction), while the x-axis is the SHAP value of each feature.
- Darker shade of red means a higher feature value (e.g. more experience), while lighter shade of blue means lower feature value (e.g. less experience).
- Each dot from the graph represent the value of the respective feature(if we are on the experience row, it's experience) for one sample

- Note that the lower the *city_development_index* is (more blue), the least likely people are to search for a new job, while the higher the *city_development_index* is (more towards red), the more likely people are to search for a new job.

4.2 Individual Result over Training Data

The following plots will be done with regard to the value of the positive class, which means a value larger than the base value would result in a higher probability of a true outcome(the result being the value 1 - the wish to change jobs), while a value lower than base value would result in a lower probability of true outcome. Note that if in regard to the negative class, a higher value would result in a higher probability of a negative outcome(the result being the value 0 - no job change) and a lower value would result in a higher probability of the positive outcome.

In Figs. 2 and 3 we analyze the SHAP values of each feature for both models, given an arbitrary data sample.

Fig. 2. SHAP values for a single sample using the Decision Tree Classifier model

Fig. 3. SHAP values for a single sample using the Logistic Regression model

Figures 2 and 3 are interpreted as following:

- The base value is the average of all model output values on the training set.
- The value in bold, **0.74** is the output of our model. In order to get a categorical value(our expected result), the resulted value from the model is compared to the baseline value. If the value is higher then the baseline, this concludes into a positive outcome(i.e. the value 1, which indicates an increased risk of job changing).
- Features colored in red suggest the increase of the SHAP value (dragging the prediction to a value of 1), while features colored in blue suggest the decrease of the value (dragging the prediction to a value of 0).

- The larger a feature bar is, the more important the feature was. For example, note that the *city_development_index* feature pushed the prediction towards 0 by a noticeable margin (from around 0.74 to 0.8, so with an offset of 0.06).
- Note there are also several less significant features that cannot be observed in the Figs. 2 and 3. Their effect on the model prediction is so small that it cannot be observed on the figure.
- We can conclude that this sample has a final output of 0.74, compared to the base value of 0.8356 which would lead to a corresponding prediction of 0.

Note that while *city_development_index* is the most important feature for both DT and LR models, its influence on the prediction is on an opposite rank. This is an interesting way of observing how two different models (Decision Tree Classifier in Fig. 2 and Logistic Regression in Fig. 3) took a very different prediction path, although they were based on the same training data.

4.3 Application Example

We will use this method in order to explain the prediction of an arbitrary example.

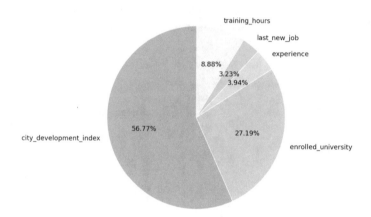

Fig. 4. Features influence on the model

Figure 4 presents the SHAP values obtained for this arbitrary sample, in a percentage format. We conclude that the city development index had the highest SHAP value out of all the features, with a percentage value of 56%. Being a full time student had the second highest SHAP value, of 28%.

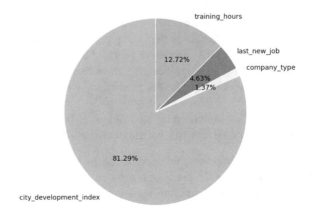

Fig. 5. Negative features

We consider a SHAP value to be negative when it indicates the increase of the risk of job changing. Figure 5 presents negative SHAP values of the sample. We obtained a negative SHAP value of the city development index, being by far the feature that played the biggest role in increasing the risk of changing the job. Another interesting result is that having no training hours leads to a higher risk of changing the job.

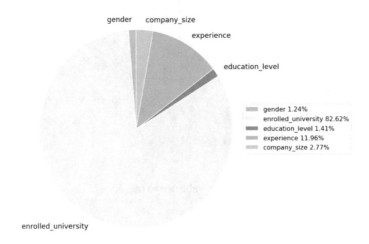

Fig. 6. Positive features

We consider a SHAP value to be positive when it indicates the decrease of the risk of job changing. Figure 6 presents the positive SHAP values of the sample. The positive SHAP values show that being enrolled into university creates a higher risk of changing jobs. Another interesting result is that having almost no experience increases the risk of changing the job. The findings shown in Fig. 6 are consistent with the findings of [4], which claims that a lot of people in their early carrier tend to change jobs a lot.

5   Conclusions and Future Works

In this paper we explored the application of competitive games and Shapley values to explain the predictions of ML classification models for modeling the risk of job changing. More precisely, we considered an example of using SHAP-based explanations to improve transparency of two ML models for predicting if an employee is likely to change his or her current job.

From an HR perspective, the proposed approach can provide some insights into understanding the employees' decisions regarding the changing of their job. Our experiments revealed that the development index of the city plays a huge role in most cases, even more than the employee work experience, which was generally ranked of second importance. This is actually not surprising, as a developed city will always provide more job opportunities than a less developed city.

An even more interesting conclusion is that there is a big gap between the strength of the influence of being in a less developed city onto the decision of not changing the job and the strength of the influence of being in a developed city onto the decision of changing the job. These observations suggest that while the important decision of job changing is determined by various competing factors, the factors determining job change are usually stronger than the factors determining the maintenance of the current job. This study can provide more insights for a company in deciding the right amount of investment for both its current and future employees.

Acknowledgement. Part of this research has received funding from the EU Erasmus+ Programme under grant agreement 2021-1-EL02-KA220-YOU-000028780 with the acronym BRIDGING THE GAP.

References

1. Anderson, S.: Immigrants and billion-dollar companies. National Foundation for American Policy (2019). https://nfap.com/wp-content/uploads/2019/01/2018-BILLION-DOLLAR-STARTUPS.NFAP-Policy-Brief.2018-1.pdf
2. Beer, M.: Organizational size and job satisfaction. Acad. Manag. J. **7**(1), 34–44 (1964)
3. Bozionelos, N.: The big five of personality and work involvement. J. Manag. Psychol. **19**, 68–81 (2004). https://doi.org/10.1108/02683940410520664
4. Cherry, N.: Persistent job changing-is it a problem? J. Occup. Psychol. **49**(4), 203–221 (1976)
5. Kaggle: HR analytics: job change of data scientists predict who will move to a new job (2021). https://www.kaggle.com/arashnic/hr-analytics-job-change-of-data-scientists
6. Lee, C., Huang, G.H., Ashford, S.J.: Job insecurity and the changing workplace: recent developments and the future trends in job insecurity research. Annu. Rev. Organ. Psych. Organ. Behav. **5**(1), 335–359 (2018). https://doi.org/10.1146/annurev-orgpsych-032117-104651
7. Lee, N.: Migrant and ethnic diversity, cities and innovation: firm effects or city effects? J. Econ. Geography **15**(4), 769–796 (2015). https://www.jstor.org/stable/26159673
8. Lundberg, S.: Shap documentation (2018). https://shap.readthedocs.io/en/latest/index.html#
9. Lundberg, S.M., Lee, S.I.: A unified approach to interpreting model predictions. In: Guyon, I., et al. (eds.) Advances in Neural Information Processing Systems, vol. 30, pp. 4765–4774. Curran Associates, Inc. (2017). https://papers.nips.cc/paper/7062-a-unified-approach-to-interpreting-model-predictions.pdf

10. McCrie, R., Lee, S.Z.: Decisions on hiring to meet protective goals. In: McCrie, R., Lee, S.Z. (eds.) Security Operations Management, 4th edn, pp. 71–118. Butterworth-Heinemann (2022). https://doi.org/10.1016/B978-0-12-822371-0.00003-7

11. McDaniel, M.A., Schmidt, F.L., Hunter, J.E.: Job experience correlates of job performance. J. Appl. Psychol. **73**(2), 327 (1988)

12. Murphy, K.P.: Probabilistic Machine Learning: An Introduction. MIT Press, Cambridge (2022)

13. Roca, J.D.L., Puga, D.: Learning by working in big cities. Rev. Econ. Stud. **84**(1), 106–142 (2017)

14. Royalty, A.B.: Job-to-job and job-to-nonemployment turnover by gender and education level. J. Labor Econ. **16**(2), 392–433 (1998). https://doi.org/10.1086/209894

15. Sora, B., Caballer, A., Peiro, J.M.: The consequences of job insecurity for employees: the moderator role of job dependence. Int. Labour Rev. **149**(1), 59–72 (2010)

16. Vultureanu-Albişi, A., Bădică, C.: A survey on effects of adding explanations to recommender systems. Concurr. Comput. Pract. Experience e6834 (2022). https://doi.org/10.1002/cpe.6834

Image-Based Intrusion Detection in Network Traffic

Sergei Golubev and Evgenia Novikova[✉]

St. Petersburg Federal Research Center of Russian Academy of Sciences,
St. Petersburg, Russia
novikova@comsec.spb.ru

Abstract. Recent approaches based on feature transformation to images with further application of the pre-trained deep analysis models have been adopted for different cyber security tasks such as malware detection, intrusion and anomaly detection. The transfer learning (TL) is a technique for solving a new task using experience or knowledge transfer from a solution of the related task. Such approach allows speeding up the process of the problem solution or increasing its performance. This paper reviews existing approaches based on transfer learning with particular focus on data preprocessing step, discusses their advantages and disadvantages. The paper ends up with the proposed approach to feature extraction based on the traffic packets transformation to images, and evaluates its efficiency using Secure Water Treatment data set (SWaT) that models functioning of the modern water treatment facility.

Keywords: Intrusion detection · Image-based features · Transfer learning · Convolutional neural network

1 Introduction

The transfer learning is a technique for solving a new task using experience or knowledge transfer from a solution of the related task. Such approach allows speeding up the process of the problem solution or increasing its performance. The typical examples of the transfer learning is a usage of pre-trained deep models to solve classification and object detection tasks from different subject domains. For example, in [2] the authors apply the Xception model pre-trained on ImageNet data set [14] that contains more than 1,5 mlns of flickr photos with different objects to detect ulcers in endoscopy images. In [4] Inception-ResNet-V2 network is used to extract contextual scene descriptor that allows enhancing the performance of human activity recognition.

L. Nataraj et al. [10] showed that by visualizing malware binaries as grayscale images it is possible to distinguish between samples belonging to different

---

This research is being supported by the grant of RSF #22-21-00724 in St. Petersburg Federal Research Center of the Russian Academy of Sciences.

malware families, moreover, the authors extracted features from the images to construct a classifier and demonstrated quite promising results in malware classification task achieving 98% of the accuracy on a malware database of 9,458 samples with 25 different malware families. The further research in this area showed that transformation of the binaries to images allows constructing efficient analysis models for malware detection and classification. For example, in [1] Alrabaee et al. analyzed the malware author attribution problem, and their experiments showed that the usage of the features extracted from the malware images allows detecting the authorship in case of obfuscated code and different compiler settings. Rong et al. showed that converting raw traffic into RGB images and usage of the pre-trained ResNet-50 model allows detecting known malware with accuracy higher than comparative techniques [13], moreover, the authors demonstrated that the proposed approach is characterized by a stable performance when detecting unseen malware variants.

These findings motivated the authors to analyze current approaches to intrusion detection based on converting network traffic to images, and evaluate if the application of the image-based features and transfer learning techniques allows increasing the performance of the intrusion detection models especially in the case of the unseen attacks. Thus, the authors' *contribution* is as follows:

- analysis of the existing approaches to intrusion detection based on the transfer learning and transformation of the input data to images;
- approach to pre-processing of the network data in PCAP format to images;
- approach to attack detection based on image-based features of the network traffic.

The *novelty* of the approach consists in a novel approach to data preprocessing that is based on transformation of the raw packets to images that allows achieving high accuracy in attack detection.

The paper is structured as follows. The Sect. 2 reviews TL based approaches to intrusion detection that use data transformation to images. Section 3 describes a proposed approach to data preprocessing with particular focus on image construction procedure. Section 4 details on experiments conducted, selected parameters of the image transformation procedure and attack detection performance. Section 5 summarizes the results of the experiments and defines the direction of the future works.

2   Related Works

Though TL and feature construction based on images showed promising results in malware analysis and detection, there are only few works in intrusion detection focusing on image-based features for intrusion detection tasks [3,8,15–17]. Zhao et al. presented one of the first research papers that evaluated the applicability of the TL for intrusion detection [17]. They assessed the efficiency of the TL in case of lack of appropriate labelled data, and proposed framework that enabled extracting features from the raw traffic and mapping them to a new feature space using TL. Their experiments showed that such solution supports higher performance and robustness in case of unknown attacks.

In general, it is possible to outline one widely used approach to traffic transformation to images as feature extraction procedure. It includes following two basic steps:

1. extract numerical and nominal features from the PCAP packets;
2. transform numerical and nominal features to image.

This approach is used in [3, 8, 11, 15, 16]. For example, in [3] authors suggested using a VGG-16 pre-trained model for the network intrusion detection. In order to apply it, the authors transform 41 network features from NSL-KDD data as follows. The features are firstly normalized, and then their number is extended from 41 to 121 due to processing of the nominal parameters in order to generate grey scale image with size 11 * 11. As VGG-16 model requires RGB image as input, the single color channel of the generated image is duplicated for each color channel and reshaped to 224 * 224 * 3. The obtained accuracy for anomaly detection constitutes 89.30% for KDDTest+, and 81.77% for KDDTest-21 in case of binary classification task.

Similar approach is implemented in [11]. The authors use UNSW-NB15 dataset [9] that contains PCAP packets with attack labels. The input data are converted into a 16×16 black and white image in such way that each pixel corresponds to some feature value. For example pixel with coordinates (3, 13) is set to 255 if the packet uses HTTP protocol. To detect attacks, the MobileNetV2 model was used in two modes: binary in order to determine whether it's an attack or not, and multi-class to detect all types of the attacks. The accuracy of the trained model is 97% with the best accuracy obtained for Generic, Fuzzers types of the attacks and normal traffic.

Wu et al. also used converting network features to images in order to detect network attacks [16]. In contrast to [3, 8, 11], the authors firstly trained their one convolutional neural network (CNN) on NSW-NB15 data set and them applied it as pre-trained one on NSL-KDD data set that contains much more different types of the attacks. This enabled authors obtain up to 99.82% of detection rate to KDDTest-21 data set. Interestingly, that the authors considered the transformation of the raw PCAP data to images unnecessary as such preprocessing may lead to loss of important information.

Totally different approach to converting traffic to images is presented in [12]. The underlying idea consists in that texts are classified with high efficiency with CNN when they are transformed to images on the character level, i.e. each character is represented by a pixel. This inspired authors to transform HTTP messages into images and apply convolutional autoencoder to detect anomalies in them. Their experiments showed that suggested approach outperforms traditional unsupervised methods to anomaly detection such as Isolation Forest and one-class support vector machine.

Thus, the analysis of the related research has shown that the essential advantages of network data transformation to image are as follows:

– ability to reveal hidden relationships in attributes and their values;
– ability to be robust in detection unseen attacks;
– ability to extract basic knowledge from limited data sets;

The majority of the research papers focuses on transformation of the numerical and nominal features extracted from the PCAP packets to images. In this paper authors investigate approaches to feature extraction based on direct transformation of the raw PCAP packets to images. The authors were inspired by the approaches developed for malware analysis that operate on binary level and demonstrate high performance in the analysis tasks.

3   The Proposed Approach to Image Based Feature Extraction

The distinctive feature of the proposed approach is a feature extraction step. It consists in transformation of the raw PCAP files to images and includes two steps:

1. every 8-bit vector (or 1 byte) is consequently retrieved from network packet data and mapped to a colour pixel according the rule: 0×0 is black colour, 0xFF is white colour and so on (see Fig. 1)
2. the sequence of pixels are transformed to the gray-scale square image.

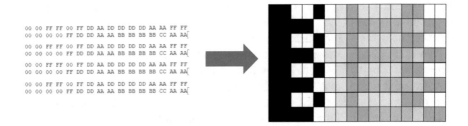

Fig. 1. Process of generating gray-scale image

The choice of image size is an important issue as the PCAP packets could be of different size, and different CNN in particular pre-trained ones have different requirements to the size of input images. In order to solve this problem, we propose to analyze the sizes of test data packets and their distribution. If it is necessary to avoid the problem with cropping images we recommend setting maximum possible packet's size (MPS) and generating all images with size S, where S is the closest integer that has a square root (1). The extra pixels are padded with 0×00 byte value or black colour.

$$S = \begin{cases} \sqrt{\text{Packet\_size}}, & \text{if Packet\_size has square root} \\ floor(\sqrt{\text{Packet\_size}}) + 1, & \text{else} \end{cases} \qquad (1)$$

To analyze the generated images, there are two options. The first option assumes development of a new CNN, its training with further deployment to detect attacks. The second option is to apply TL-based approach by using a pre-trained CNN and fine tuning it using a test data set. In this paper we tested both variants, the performance of two pre-trained models (ResNet34 [6] and MobileNetV3-small [7]) was evaluated, and own CNN model was developed and tested.

4   Experiments and Results

To assess the efficiency of the proposed data preprocessing step, the authors performed a series of the experiments using Secure Water Treatment data set (SWAT) [5]. This data set is generated by a test bed that models a modern water treatment facility with six main technological processes such as raw water intake, desinfection, filtering, purification and etc. The choice of the data set is explained by the fact that it simulates a functioning of the critical infrastructure object. The data from such objects are rarely in the open access, and the application of the transfer learning for this case can solve the problem of the lack of an appropriate data set.

There are several versions of the SWaT dataset, and in the experiments we used SWaT.A6_Dec 2019 dataset, a version of Dec, 2019. This data set includes 3 h of the normal facility functioning and 1 h during which 6 attacks of two types were implemented. The structure of dataset is presented in Table 1. Every attack time period consists of 21 million .pcap network packets. To train the model, we balanced it to keep the ratio: normal packets (70%) and attack packets (30%). Though this operation does not reflect the real distribution of the data, such preprocessing allows achieving better results in attack detection.

Table 1. Structure SWaT data set

Traffic State	Percentage
Data exfiltration attack	9.52%
Disruption of sensor and actuator	6.87%
Normal traffic	83.61%

To define the size of the generated images, we analyzed the sizes of the network packets. Figure 2 shows the distribution of the packets' sizes, and Table 2 gives the statistics on packet lengths. It is clear, that normal and abnormal PCAP packets' have almost similar distribution of their lengths, there is a slight difference in maximum length of the packets. For this reason, we decided to keep the maximum size of the packet which equals to 19034 in this use case, and generated the images using the Eq. (1). Thus, the size of generated images is 138 * 138. The speed of image generation depends on the characteristics of the computational environment. The bigger an image's size the more time it needs to generate a gray-scaled image and vice versa. In the experiments the average speed was 1200 images per second for 138×138 for the following technical environment:

- OS - Windows 10
- Intel Core i7-9700k
- MSI GeForce RT 2070 Super VENTUS 8 Gb
- 32 Gb RAM

The generated images of 138×138 size are shown on Fig. 3. It is possible to see that there is a little difference in these two images which consists in image texture layout.

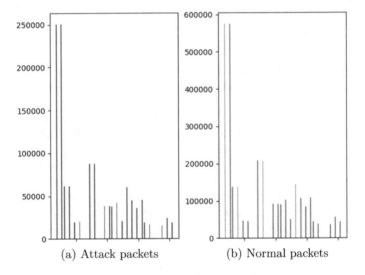

(a) Attack packets   (b) Normal packets

Fig. 2. Packets' sizes distribution

To detect anomalies, we tested two pre-trained models, namely, ResNet34 and MobileNetV3-small, and constructed our own CNN model. The choice of these pre-trained models is explained by the fact that they both show higher classification accuracy on the ImageNet dataset. The light-weight version of the MobileNetV3 was chosen to evaluate the applicability of the approach to resource-constrained environment. The constructed CNN model consisted of the following 10 layers:

Table 2. Packets' sizes statistic

	Normal	**Attack**
Mode	64	64
Median	90	86
80 Percentile	128	128
90 Percentile	148	148
99 Percentile	633	633
Max	19034	14888

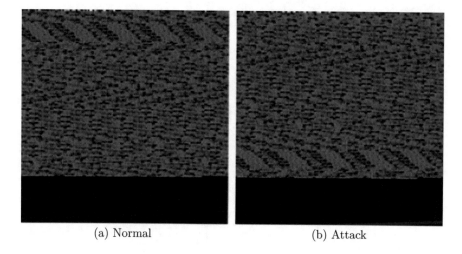

(a) Normal   (b) Attack

Fig. 3. Gray-scale images

- 1 input layer,
- 3 convolutional layers,
- 3 max pooling layers,
- 1 flatten layer,
- 1 dense layer, and
- 1 output layer.

The CNN network was trained 10 epochs, and the intrusion detection accuracy reached 0.99%. Figure 4 shows plots of accuracy and loss during 10 epochs. As we can see after 4 epochs model's accuracy gets its limit (difference between epochs' accuracy is small), thus count of epochs can be reduced to 5–6 epochs. The accuracy of the pre-trained models was significantly lower, the ResNet34 model demonstrated the accuracy that is slightly higher than the random classifier. Table 3 summarizes the experiment results.

Table 3. The accuracy of the attack detection by different analysis models

Model	Accuracy
ResNet34	0.64
MobileNetV3-small	0.54
own CNN	0.99

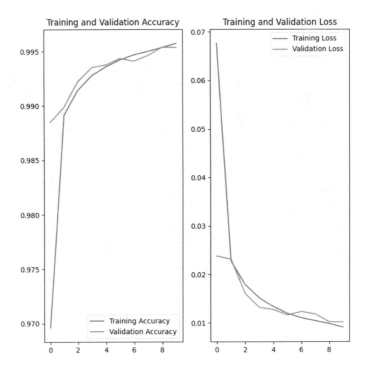

Fig. 4. Accuracy and loss of model

5 Discussion and Conclusions

The convolutional neural networks have shown their efficiency in many applied tasks different from image analysis. They turned out to be efficient in the malware analysis, authorship attribution, the results of the recent research in the intrusion detection have shown that they are able efficiently to detect known attacks and even robust to unseen ones. Application of CNN requires specific transformation of the input data, and one of the possible solutions is to construct images.

This paper analyzed the image-based approaches to feature generation from network traffic. The authors outlined that the most widely used technique assumes extraction of the numerical and nominal data from raw PCAP packets, firstly, and then construction of the images on their basis. However, it has two drawbacks. Firstly, there are additional computations that relate to feature extraction and image construction, and, secondly, the usage of the predefined features may impact on the CNN ability to detect hidden structures by limiting it to detection of the unseen attacks similar to known ones.

The approach proposed in the paper operates with raw PCAP packets, converting them directly to grey-scale images. Such solution allows enhancing the computational performance of image-based techniques as there is no need to pre-process network data in order to retrieve features. A network packet could be directly transformed to gray-scale immediately. To evaluate the efficiency of the approach, the SWaT data set was used, this data set simulates the functioning of the critical infrastructure object, namely, water treatment facility. There is a certain lack in data sets containing attacks on critical infrastructures, and the application of the transfer learning could significantly increase the efficiency of the intrusion detection. However, the implemented experiments showed that the performance of the pre-trained models is slightly higher than the performance of the random classifiers. The probable reason of their failure is that they are trained on the real world images, and have high classification performance on them, thus, they are not good at extracting features from the grey scale images without certain objects. The CNN which was trained by the authors demonstrated quite high accuracy in the attack detection.

However, we also identified certain drawbacks of the approach. Firstly, it is necessary to have information about the maximum packet size in advance, and this parameter may vary from data set to data set. Secondly, the implemented experiments were done on artificially balanced data set, and it is required to assess the efficiency of the approach in the conditions more close to the real world distribution of the abnormal and normal data. These tasks are included in the scope of the future works. Another direction of the future research relates to the analysis of the image-based intrusion detection techniques' ability to handle novel and unseen attacks.

References

1. Alrabaee, S., Karbab, E.M.B., Wang, L., Debbabi, M.: BinEye: towards efficient binary authorship characterization using deep learning. In: Sako, K., Schneider, S., Ryan, P.Y.A. (eds.) ESORICS 2019. LNCS, vol. 11736, pp. 47–67. Springer, Cham (2019). https://doi.org/10.1007/978-3-030-29962-0_3
2. Chollet, F.: Xception: deep learning with depthwise separable convolutions. In: 2017 IEEE Conference on Computer Vision and Pattern Recognition (CVPR), pp. 1800–1807 (2017)
3. Chollet, F.: A transfer learning with deep neural network approach for network intrusion detection. Int. J. Intell. Comput. Res. (IJICR) **12**, 087–1095 (2021)
4. Debnath, B., O'Brient, M., Kumar, S., Behera, A.: Attention-driven body pose encoding for human activity recognition. In: 2020 25th International Conference on Pattern Recognition (ICPR), pp. 5897–5904 (2021). https://doi.org/10.1109/ICPR48806.2021.9412487
5. Goh, J., Adepu, S., Junejo, K.N., Mathur, A.: A Dataset to support research in the design of secure water treatment systems. In: Havarneanu, G., Setola, R.,

Nassopoulos, H., Wolthusen, S. (eds.) CRITIS 2016. LNCS, vol. 10242, pp. 88–99. Springer, Cham (2017). https://doi.org/10.1007/978-3-319-71368-7_8

6. He, K., Zhang, X., Ren, S., Sun, J.: Deep residual learning for image recognition. In: 2016 IEEE Conference on Computer Vision and Pattern Recognition (CVPR), pp. 770–778 (2016). https://doi.org/10.1109/CVPR.2016.90

7. Howard, A., et al.: Searching for mobilenetv3. In: 2019 IEEE/CVF International Conference on Computer Vision (ICCV), pp. 1314–1324 (2019). https://doi.org/10.1109/ICCV.2019.00140

8. Masum, M., Shahriar, H.: TL-NID: deep neural network with transfer learning for network intrusion detection. In: 2020 15th International Conference for Internet Technology and Secured Transactions (ICITST), pp. 1–7 (2020). https://doi.org/10.23919/ICITST51030.2020.9351317

9. Moustafa, N., Slay, J.: UNSW-NB15: a comprehensive data set for network intrusion detection systems (UNSW-NB15 network data set). In: 2015 Military Communications and Information Systems Conference (MilCIS), pp. 1–6 (2015). https://doi.org/10.1109/MilCIS.2015.7348942

10. Nataraj, L., Karthikeyan, S., Jacob, G., Manjunath, B.S.: Malware images: visualization and automatic classification. In: Proceedings of the 8th International Symposium on Visualization for Cyber Security. VizSec 2011. Association for Computing Machinery, New York (2011). https://doi.org/10.1145/2016904.2016908

11. Noever, D.A., Noever, S.E.M.: Image classifiers for network intrusions. CoRR abs/2103.07765 (2021). https://arxiv.org/abs/2103.07765

12. Park, S., Kim, M., Lee, S.: Anomaly detection for http using convolutional autoencoders. IEEE Access 6, 70884–70901 (2018). https://doi.org/10.1109/ACCESS.2018.2881003

13. Rong, C., Gou, G., Cui, M., Xiong, G., Li, Z., Guo, L.: TransNet: unseen malware variants detection using deep transfer learning. In: Park, N., Sun, K., Foresti, S., Butler, K., Saxena, N. (eds.) SecureComm 2020. LNICST, vol. 336, pp. 84–101. Springer, Cham (2020). https://doi.org/10.1007/978-3-030-63095-9_5

14. Russakovsky, O., et al.: ImageNet large scale visual recognition challenge. Int. J. Comput. Vision 115(3), 211–252 (2015). https://doi.org/10.1007/s11263-015-0816-y

15. Wang, W., et al.: Anomaly detection of industrial control systems based on transfer learning. Tsinghua Sci. Technol. 26(6), 821–832 (2021). https://doi.org/10.26599/TST.2020.9010041

16. Wu, P., Guo, H., Buckland, R.: A transfer learning approach for network intrusion detection. In: 2019 IEEE 4th International Conference on Big Data Analytics (ICBDA), pp. 281–285 (2019). https://doi.org/10.1109/ICBDA.2019.8713213

17. Zhao, J., Shetty, S., Pan, J.W., Kamhoua, C., Kwiat, K.: Transfer learning for detecting unknown network attacks. Int. J. Comput. Vision (2019). https://doi.org/10.1186/s13635-019-0084-4

Real-Time Traffic Prediction Using Distributed Deep Learning Based Multivariate Time-Series Models

Hoang-Thong Vo[1,2], Quang-Linh Tran[1,2], Gia-Huy Lam[1,2],
Ngan-Linh Nguyen[1,2], Trong-Hop Do[1,2(✉)], Nguyen Thi Lieu[3],
and Nhu-Ngoc Dao[4]

[1] University of Information Technology, Ho Chi Minh City, Vietnam
{18521462,18520997,18520832,18520989}@gm.uit.edu.vn, hopdt@uit.edu.vn
[2] Vietnam National University, Ho Chi Minh City, Vietnam
[3] Department of Technology, Dong Nai Technology University,
Bien Hoa, Dong Nai, Vietnam
nguyenthilieu@dntu.edu.vn
[4] Department of Computer Science and Engineering, Sejong University,
Seoul, South Korea
nndao@sejong.ac.kr

Abstract. Traffic prediction plays an extremely important role in the intelligent transportation system. The accuracy of traffic prediction helps to reduce traffic jams and helps intelligent transportation system (ITS). However, this problem is also a difficult challenge to solve because of complicated and dynamic spatio-temporal dependencies between different regions in the road network. In this study, we contribute the following suggestions: First, a multivariate approach using feature extraction techniques to increase the performance of the model. Second, we perform a comparative experimental study to evaluate different models, identifying the most effective component. Models are built on distributed and parallel computing platforms.

1 Introduction

Traffic prediction is one of the important tasks of an intelligent transportation system. It has the following main tasks: flow, speed, demand, occupancy, and travel time [6]. In this study, vehicle speed prediction is the main problem we deal with. In an intelligent transportation system, predicting the speed of vehicles can help the city reduce congestion, help transport businesses like Uber, Grab coordinate traffic from places with high traffic density to a place with low traffic density to move faster and save time.

Traffic prediction is a problem with many challenges. In practice, many factors affect traffic prediction. First, traffic data is spatio-temporal, it is always constantly changing in space and time. Second are external factors such as weather conditions, unexpected events or route attributes such as interstate highways,

L. Braubach et al. (Eds.): IDC 2022, SCI 1089, pp. 61–68, 2023.
https://doi.org/10.1007/978-3-031-29104-3_7

U.S. highways, state highways, and county highways. The contribution of this study is two-folds. First, a multivariate approach using feature extraction techniques to enhance the performance of the model. Second, we conduct a comparative experimental analysis to evaluate various models, determining the most effective component. Models are built on a distributed and parallel computing platform.

2   Related Works

Referencing from [7], modern deep learning models are built on distributed and parallel computing platform for big data. A comprehensive and flexible architecture based on distributed computing platform for real-time traffic control was proposed by Amini et al., in which the architecture was based on a systematic analysis of the requirements of the existing traffic control system [8]. Saraswathi et al. presented a system used to predict the number of vehicles on different roads based on the processing technique of streaming data to reduce traffic congestion and visualize the traffic conditions analysis in real-time [9]. Anveshrithaa et al. introduced the proposed model is aspire to predict traffic flow information by integrating Spark and Kafka along with deep neural networks [10]. In the study of data surveys and big data tools, author Jiang et al. have detailed the use of the above tools for traffic estimation and prediction problems [11].

3   Distributed Deep Learning for Time Series Forecasting

3.1   Deep Learning Models for Time Series Forecasting

3.1.1   Long-Short Term Memory - LSTM
The current version of LSTM that BigDL uses is the Vanilla LSTM [1], the original LSTM block with the addition of the forget gate and peephole connections. The architecture consists of two Vanilla LSTM layers, two layers of dropout, and a dense layer (output layer).

3.1.2   Sequence to Sequence
In 2018, Shengdong Du et al. [2] introduced a deep learning framework that serves as a multi-variate time-series forecasting algorithm, called sequence-to-sequence. It can address the dynamics, spatial-temporal and non-linear characteristics of data by applying the encoder-decoder architecture with Long Short-Term Memory components. Follow [2], the sequence-to-sequence is more outperformance than the classic shallow learning and baseline deep learning models which shows that it is capable of dealing with multi-variate time series forecasting with a considerable accuracy.

3.1.3   Temporal Convolutional Networks - TCN

In 2019, Renzhuo Wan et al. introduced a combined model which improves the prediction accuracy and minimizes the multivariate time series data dependence for aperiodic data, is called Multivariate Temporal Convolution Network (M-TCN) [3]. It works as a sequence to sequence with two components: encoder and decoder but using Convolution layers (Residual block, 1D Convolution and Pooling) instead of LSTM components as sequence-to-sequence. Convolution architecture allows a large-scale distributed and parallel computing system, and the inference time gives faster forecasting outcomes than using RNN architecture (RNN not parallel). This is one of the most performance models in addressing multi-variate time-series forecasting problems.

3.2   Distributed Deep Learning Implementation

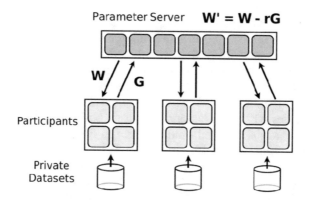

Fig. 1. Distributed deep learning system

Implementing a distributed deep learning algorithm is challenging as the dataset is distributed on a cluster. The deep learning model need to be implemented on every worker node in the cluster. Each worker node in the cluster consumes a partition of the training data. In each iteration, each worker node download the model's parameters from the parameter server. Then, the parameters will be updated based on the consumed training data partition. Then the parameters will be sent to the parameter server. The parameter server will aggregate the parameters sent by all worker nodes and create the most updated parameters of the model, which will be download by all worker nodes in the next iteration. This mechanism is described in Fig. 1.

BigDL is a parallel and distributed artificial intelligence application, supporting users to easily build end-to-end models on large-scale data scale. With a large-scale data set like PEMS-BAY, BigDL is the application of choice for experiencing time-series forecasting algorithms.[1]

---

[1] https://analytics-zoo.readthedocs.io/en/latest/doc/PythonAPI/Chronos/foreca sters.h.

4   Experiment

4.1   Dataset

This section presents information about the Pemsbay dataset and additional data which is used for the experiment.

PEMS-BAY [5] traffic dataset is collected by California Transportation Agencies(Cal Trans) Performance Measurement System (PeMS). 325 sensors were selected in the Bay Area and 6 months of data collected ranging from Jan 1st 2017 to June 30th 2017. With the time window of 5 min, the total number of observed traffic data points is 16,937,179. However, we only use the first 50 sensors for experimenting and instead of 5-min gap, we process for 1-h gap to collect weather data. Consequently, the data which we use for experiment is 217,100 for 50 sensors. The reason why we only choose only 50 sensors is the availability of data about weather, which will be presented in the next paragraph. 1 h is a suitable period of time to see the difference in weather conditions, meanwhile 5 min are too short to capture the big difference in weather conditions, which can make this data useless to solve the traffic prediction problem.

To measure the influence of data about weather. All data about weather of all 50 sensors are collected according to the latitude and longitude of the sensors, which are provided in the dataset. The website visualcrossing[2] provides an API to collect weather data. This is a reliable dataset for collecting weather data because it provides APIs with charging for users beside free APIs for free trials. Data about weather is collected hourly from the first of January 2017 to 30th June 2017. There are 6 attributes in data about weather, which is temperature, humidity, dew, precipitation, wind speed and cloud cover. However, because of the limitation of free APIs, we only collect data about weather of the first 50 sensors of the Pemsbay dataset, so the experiment will be conducted in these sensors.

In addition, we also want to check the assumption that the velocity in each sensor will be affected by the velocity of other nearby sensors, so we also add data about velocity of top 5 nearest sensors to each sensor. There is a dataset provided by the authors of PEMS-BAY, which shows the distance between sensors, we use it to get the top 5 nearest sensors.

Each sensor has 4232 data points for velocity in each hour and a total of 13 attributes, which is a date-time attribute, a velocity attribute, 6 attributes of weather data and 5 attributes of velocity of top 5 nearest sensors.

The Fig. 3 below shows the process of experiment in this problem. We construct 3 datasets to experiment and find the best dataset for solving the problem. The first dataset only has one attribute which is the velocity traffic at the sensor so this is univariable time series problem. The second dataset has 7 attributes which is the velocity traffic at the sensor and 6 attributes about the weather so this is multivariable time series problem. The third dataset in Fig. 2 has 12 attributes which is the velocity traffic at the sensor, 6 attributes about the weather, and the velocity of traffic of top 5 nearest sensors.

---

[2] https://www.visualcrossing.com/weather/weather-data-services.

date	sensor_id	velocity	temp	humidity	dew	precip	windspeed	cloudcover	400045	400394	400122	400479	400838
2017-01-01 00:00:00	400001	71.4	3.9	82.51	1.2	0.0	12.9	66.4	66.8	72.1	68.2	68.4	70.5
2017-01-01 01:00:00	400001	71.0	4.4	79.68	1.2	0.0	11.1	0.5	66.3	72.2	68.6	68.0	70.3
2017-01-01 02:00:00	400001	70.9	5.0	79.31	1.7	0.0	0.0	68.9	68.3	72.2	69.0	67.9	69.9
2017-01-01 03:00:00	400001	71.2	5.7	78.50	2.2	0.0	9.4	92.9	68.4	71.9	69.3	67.2	70.3
2017-01-01 04:00:00	400001	71.2	6.2	75.85	2.2	0.0	7.6	92.9	68.3	71.1	68.9	66.9	70.1

Fig. 2. Illustration of a third dataset with weather related attributes and the remaining attributes being the speed of the nearest sensor IDs calculated based on the closest distance between the point coordinates.

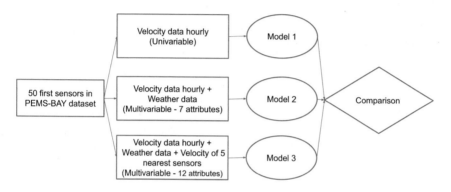

Fig. 3. Experiment procedure

4.2   Experimental Results

The experimental results of the three deep learning models including LSTM, Sequence to Sequence, and Temporal Convolution Network are presented in Table 1, 2, 3, respectively.

Table 1. Results of the LSTM model

	Data Velocity	Data Velocity + Weather	Data Velocity+ Weather+ Top 5
MSE (Mean ± std)	48.26 ± 35.79	48.45 ± 42.86	**44.53 ± 35.05**
RMSE (Mean ± std)	6.46 ± 2.56	6.42 ± 2.69	**6.19 ± 2.49**
MAE (Mean ± std)	**3.27 ± 1.54**	3.46 ± 1.85	3.36 ± 1.81
SMAPE (Mean ± std)	**3.48 ± 1.91**	3.62 ± 2.12	3.51 ± 2.03

In Table 1 and 2, evaluated on the MAE and SMAPE metrics, and the MSE and RMSE metrics, the first and third experimental datasets have a high performance on the LSTM model and Sequence to Sequence model. However, in the Table 3, experimentally experimenting on the first dataset with the TCN model, the RMSE, MAE, and SMAPE evaluation indexes achieved the highest performance. Similar to the previous models, the experimental context of the third dataset always performed on the MSE evaluator with the lowest error.

Table 2. Results of the Sequence to Sequence model

	Data Velocity	Data Velocity + Weather	Data Velocity + Weather + Top 5
MSE (Mean ± std)	47.09 ± 39.09	47.54 ± 33.03	**44.39 ± 34.07**
RMSE (Mean ± std)	6.36 ± 2.61	6.45 ± 2.46	**6.19 ± 2.47**
MAE (Mean ± std)	**3.13 ± 1.60**	3.36 ± 1.61	3.21 ± 1.72
SMAPE (Mean v std)	**3.35 ± 1.91**	3.55 ± 1.93	3.39 ± 1.97

To prove that the weather and speed of the sensor station have a certain influence and help increase the performance of the models, we established the experiment according to the data-centric approach. The results show that evaluating MSE and RMSE metrics with the above approach increases the performance of the model.

Table 3. Results of the Temporal Convolutional Networks model

	Data Velocity	Data Velocity + Weather	Data Velocity + Weather + Top 5
MSE (Mean ± std)	47.84 ± 33.6	50.12 ± 41.50	**47.64 ± 37.25**
RMSE (Mean ± std)	**6.46 ± 2.47**	6.57 ± 2.65	6.60 ± 2.59
MAE (Mean ± std)	**3.228 ± 1.46**	3.47 ± 1.78	3.44 ± 1.87
SMAPE (Mean ± std)	**3.47 ± 1.85**	3.66 ± 2.06	3.63 ± 2.09

In the results, we take a feature engineering approach to add newly enhanced properties to the data. In addition, the way to establish the experiment on the dataset with the choice of the time interval for the training set, validation, and test set is also different from the previous works, so it is difficult for us to compare the previous results studies because previous studies did not establish the same experiment and had the same data approach as our study.

In this experiment, we established and implement the models in a parallel and distributed computing way because the context of the receiving sensor station and the real-time data. By training and inferring data in parallel and distributed computing, they can process faster and forecast data results catch up to the real-time speed of real-life scenarios.

Most traffic prediction problems are usually solved by univariate methods, predicting the speed of vehicles through time series models. We find that such approaches have the limitation that the input to the model is usually small, with only one attribute of time and the output of speed. In order for the models to be more informative for training, we assume that real-life cases, such as weather or nearby traffic sensor stations, influence vehicle speeds. For example, in bad weather such as snowfall, vehicles may travel at a slower speed than they actually are. In addition, at neighboring traffic stations, the speed of vehicles moving from station A to station B also has a continuum with each other or the process of vehicles accelerating or decelerating from point A to point B.

4.3   Comparative Study

Overall, adding data of weather and the velocity of top 5 nearest sensors gives the best performance with mean MSE is 44.39 and RMSE is 6.19. In addition, the model sequence to sequence and LSTM give a better performance than the Temporal Convolutional Networks model. From the above tables of results, we can conclude that multi-variable model is better than univariable. However, when adding data of weather only, the performance of all three models are worse than the univariable model.

5   Conclusion

This study implements several deep learning based multivariable models for traffic prediction problem. In this problem, the dataset is huge and thus the deep learning models need to be implemented and trained in a distributed fashion. In this study, Big-DL framework is used to implement three deep learning models including LSTM, Sequence to Sequence, and Temporal Convolution Network in a distributed system. The PEMS-BAY dataset is used to evaluate and compare the performances of the three models. The experimental results show that the multi-variable model is better than the univariable model for traffic prediction problems. Also, the sequence to sequence model gives the best results among three models.

Acknowledgment. This work was supported by the National Research Foundation of Korea (NRF) grant funded by the Korea government (MSIT) (No. 2021R1G1A1008105).

References

1. Van Houdt, G., Mosquera, C., Nápoles, G.: A review on the long short-term memory model. Artif. Intell. Rev. **53**(8), 5929–5955 (2020)
2. Du, S., Li, T., Horng, S.-J.: Time series forecasting using sequence-to-sequence deep learning framework. In: 9th International Symposium on Parallel Architectures, Algorithms and Programming (PAAP), p. 2018. IEEE (2018)
3. Wan, R., et al.: Multivariate temporal convolutional network: a deep neural networks approach for multivariate time series forecasting. Electronics **8**(8), 876 (2019)

4. Sen, R., Yu, H.-F., Dhillon, I.: Think globally, act locally: a deep neural network approach to high-dimensional time series forecasting. arXiv preprint arXiv:1905.03806 (2019)
5. Li, Y., Yu, R., Shahabi, C., Liu, Y.: Diffusion convolutional recurrent neural network: data-driven traffic forecasting. arXiv preprint arXiv:1707.01926 (2017)
6. Yin, X., et al.: A comprehensive survey on traffic prediction. arXiv preprint arXiv:2004.08555 (2020)
7. Dai, J.J., et al.: BigDL: a distributed deep learning framework for big data. In: Proceedings of the ACM Symposium on Cloud Computing (2019)
8. Amini, S., Gerostathopoulos, I., Prehofer, C.: Big data analytics architecture forreal-time traffic control. In: 2017 5th IEEE International Conference on Models and Technologies for Intelligent Transportation Systems (MT-ITS), pp. 710–715. IEEE(2017)
9. Saraswathi, A., Mummoorthy, A., Anantha Raman, G.R., Porkodi, K.: Real-time traffic moni-toring system using spark. In: 2019 International Conference on Emerging Trends in Science and Engineering (ICESE), vol. 1, pp. 1–6. IEEE (2019)
10. Anveshrithaa, S., Lavanya, K.: Real-time vehicle traffic analysis using long short term memory networks in apache spark. In: 2020 International Conference on Emerging Trends in Information Technology and Engineering (IC-ETITE). pp. 1–5. IEEE (2020)
11. Jiang, W., Luo, J.: Big data for traffic estimation and prediction: a survey of data and tools. arXiv preprint arXiv:2103.11824 (2021)

Machine Learning Model for Flower Image Classification on a Tensor Processing Unit

Anik Biswas$^{(\boxtimes)}$, Julia Garbaruk, and Doina Logofătu

Frankfurt University of Applied Sciences, Frankfurt am Main, Germany
anik.biswas@stud.fra-uas.de, logofatu@fb2.fra-uas.de

Abstract. Identification and categorization of flower images are active research problems in the field of Computer Vision. In the last decade, this problem has been tackled by performing machine learning prediction on the basis of extracted features such as colour or texture extraction. In this project, a novel approach for solving this problem is introduced by integrating and ensembling two efficiently scaled Deep Conventional Neural Network models - EfficientNet and DenseNet. The training experiment would be performed using large number of images from multiple public datasets on multiple complex deep neural network models. To optimize the computational resource and efficiency, the experiment would be run on Tensor Processing Unit (TPU) hardware environment and the efficacy of the same would be assesed in terms of computational power and speed.

Keywords: Image Classification · Computer Vision · Machine Learning · Deep Convolutional Neural Network · EfficientNet DenseNet · TPU

1 Introduction

Flowers have significant cultural, economic and ecological values in our everyday life. Flower classification can be used in various applications like plant species identification, horticulture, medicinal plant etc. As per report [1], there are approximately 325,000 flowering plants discovered in the world. Recognition and identification of flower species is a daunting as well as time consuming challenge even for taxonomists. Therefore, development of a fast and accurate system would be a necessity for effective flower categorization.

The task of flower classification can be accomplished in two different ways. In the first variant, raw flower images are transformed to a convenient format, where some useful features like color, shape or texture can be extracted [3,4]. After feature extraction, traditional machine learning techniques like Support Vector Machine (SVM) classifier can be applied to perform the final classification. In the second technique, the original image is fed directly into a Deep

Neural Network (DNN) like Convolutional Neural Network (CNN) without previous feature extraction. CNN algorithm in particular has been proven to be very efficient in accurate image recognition tasks. It is able to learn hierarchical features automatically for segmentation and classification of images [6]. Our research topic is to design a machine learning model that would classify 104 types of flowers based on their images drawn from five different public datasets. In this study, the proposed machine learning model is built using an ensemble of two Deep Learning Networks namely EfficientNetB7 and DenseNet201. These Neural Networks provide outstanding performance in image classification [7,8]. Ensemble Learning combines predictions from different models and obtains better generalized performance which often is proved to be have superior outcome than the respective individual models [9]. The experiment is carried out in Cloud Tensor Processing Unit (TPU) environment. The motivation behind TPU over conventional CPU or GPU is its speed and high optimization performance in large batches and CNN network [10].

2   Related Work

For several years, various research works have been done on the problem of flower image classification. For example: In [10], a feature based classification technique, using a multiple kernel framework with an SVM classifier was applied to a 103 class flower dataset. It provided a performance accuracy of 88.3%. In recent years, research has focused on the use of DNNs. Xiaoxue et al. [12] proposed a classification model based on Generative Adversarial Network (GAN) and ResNet-101 transfer learning model, and used stochastic gradient descent optimization to update the network weights. Their experiment on Oxford-102 after Data Enhancement resulted a test accuracy of 90.7%. The accuracy was greatly improved in the work by Xia et al. [13]: They used the pre-trained Inception v3 CNN model and transfer learning approach to train the Oxford-102 flower dataset. The model achieved a validation accuracy of 94%.

3   Algorithm and Model Architecture

The two main approaches used for this work are Transfer Learning and Ensemble of two pre-trained Deep CNNs, namely EfficientB7 and Densenet201. This section describes their underlying principles.

3.1   EfficientNet

The need to scale the standard ConvNet architecture stems from the need to have better accuracy for complex problem solving. In the research work by Google [7], a new scaling method for CNN was proposed in which effective compound coefficient was used to scale depth, width and resolution in an efficient way [14]. The architecture of the baseline model EfficientNet B0 consists of different blocks of Mobile Net CNNs connected in sequence. On the basis of B0 other training models can be built by compound scaling of network scaling factors. EfficientNet B7 is the latest development of this family and leverages capability of Google's ImageNet database for model pre-training so that this train model can be subsequently used in further image classification task.

3.2 DenseNet

In this architecture, each layer connects the other layer in a feed-forward fashion. The DenseNet architecture consists of normal Convolution and Pooling Layer cascaded with modules called Dense block [8]. Dense Block connects all layers directly with each other. Each layer receives input from all preceding layers and forwards feature-maps to all subsequent layers. This way, features are combined by concatenating them.

4   Implementation

Transfer Learning is the re-use of knowledge gained from a prior assignment to a new problem for improving classification task. As discussed earlier in our implementation we use pre-trained models EfficientNet-B7 and DenseNet-201 to train on our flower image dataset. Transfer Learning saves training time and computational power while achieving higher accuracy once the model is fine-tuned on the target classification problem.

4.1   Dataset

The original dataset is developed by Kaggle and consists of 104 types of flower images collected from five different public datasets. Each of the file contains the class of the sample in label format and actual pixels in array form. Also, training, testing and validation data is segregated in different folders of the dataset path.

4.2   Tensor Processing Unit (TPU)

The hardware environment used for this task is a Cloud Tensor Processing Unit (TPU). A TPU is an application specific integrated circuit developed by Google with the focus of accelerating Machine Learning training [10].

4.3   Data Access and Visualization

While using TPUs, dataset need to be stored and fetched from Google Cloud Storage Bucket. The *KaggleDatasets* library binds the following datasets to our GCS bucket: *tpu-getting-started* (the Kaggle Competition dataset) and *tf-flower-photo-tfrec* (this dataset consists of multiple public image datasets like Imagenet, iNaturalist etc.). The datasets are stored in dictionaries. Both the internal and external dataset are concatenated to single training, validation and test variables.

The result that we obtain as the final distribution of images in our dataset model after data preparation is the following: 68094 training images, 3712 validation images and 7382 unlabeled test images. A sample visual representation of the flower images with corresponding labels are displayed in Fig 1.

Fig. 1. Sample batch of training image from dataset with labels

4.4   Model Creation and Training

The Deep Neural Network architecture is built with keras Sequential API platform. The base optimizer used is Adam which indeed is a stochastic gradient descent method. Since the problem task is a multi-class image classification, categorical cross-entropy has been chosen as the loss function which would be computed as per Eq. 1.

$$Loss = -\sum_{i=1}^{n} y_i \cdot \log \hat{y_i} \tag{1}$$

where $\hat{y_i}$ is the i-th scalar value of the output of the model, y_i is the target value of the same and n is the output size or number of scalar values of the model output [16]. The model is divided in two parts. In the first part, keras EfficientNetB7 model is used with training weight declared as 'noisy student'. Noisy Student training is a semi-supervised teacher-student self training method used for improving accuracy by adding noise to the student model during training so that the student learns better than the teacher [17]. Similarly keras DenseNet201 model is constructed with ImagNet pre-trained model.

Small addition: In order to improve the robustness of CNN against occlusion, the Random Erasing Data Augmentation technique is used for training our network. In this method, a rectangular region is selected randomly from an image and the pixels with random values are erased [19].

4.5   Model Evaluation

In order to obtain the best outcome from both the EfficientNet and Densenet Models, an probabilistic ensemble average of both the models together has been taken (arithmetic mean). In order to asses performance of the model a confusion matrix has been created which monitors the F1 Score, Precision and Recall.

5   Experimental Results

The algorithm code is run on Kaggle platform in TPU V3 environment and submitted to the Kaggle Competition "Petals to the Metal - Flower Classification on TPU". The evaluation metrics obtained from the test is tabulated below (Table 1).

Table 1. Sensitivity and Specificity Scores of the model

Accuracy	F1 Score	Precision	Recall	Kaggle Score (Out of 1.0)
96.2%	0.955	0.953	0.960	0.954

6   Conclusion

In this project, an efficient Machine Learning architecture is implemented for classifying 104 types of Flower Images. CNN has been proven to be ideal Machine Learning model for complex image classification but building it from scratch, tuning it and scaling it to obtain desired output is a difficult and time-consuming task. The result of this experiment showed that ensembling two high performance pre-trained CNNs namely EfficeintNet and DenseNet leveraged the benefits of Transfer Learning and provided us significantly improved performance in Flower Image detection. Considering the large amount of training dataset and highly deep neural networks used in this project, the computational resource was highly optimized and accelerated by performing the model training in TPU.

References

1. Antonelli, A., et al.: State of the world's plants and fungi 2020. Royal Botanic Gardens, Kew (2020). https://doi.org/10.34885/172
2. Guru, D.S., Sharath Kumar, Y.H., Manjunath, S.: Textural features in flower classification. Math. Comput. Modell. **54**(3–4), 1030–1036 (2011). https://doi.org/10.1016/j.mcm.2010.11.032. ISSN 0895-7177
3. Shaparia, R., Patel, N., Shah, Z.: Flower classification using texture and color features, vol. 2, pp. 113–118 (2017)
4. Mabrouk, A., Najjar, A., Zagrouba, E.: Image flower recognition based on a new method for color feature extraction. In: VISAPP 2014 - Proceedings of the 9th International Conference on Computer Vision Theory and Applications, vol. 2 (2014). https://doi.org/10.5220/0004636302010206
5. Narvekar, C., Rao, M.: Flower classification using CNN and transfer learning in CNN-agriculture perspective. In: 2020 3rd International Conference on Intelligent Sustainable Systems (ICISS), pp. 660–664 (2020). https://doi.org/10.1109/ICISS49785.2020.9316030

6. Alipour, N., Tarkhaneh, O., Awrangjeb, M., Tian, H.: Flower image classification using deep convolutional neural network, pp. 1–4 (2021). https://doi.org/10.1109/ICWR51868.2021.9443129
7. Tan, M., Le, Q.: EfficientNet: rethinking model scaling for convolutional neural networks (2019). https://arxiv.org/abs/1905.11946v5
8. Huang, G., Liu, Z., Van Der Maaten, L.: Densely connected convolutional networks. In: 2017 Proceedings of the IEEE Conference on Computer Vision and Pattern Recognition. IEEE (2017)
9. Ganaie, M., Hu, M., Tanveer, M., Suganthan, P.: Ensemble deep learning: a review (2021). https://arxiv.org/pdf/2104.02395.pdf
10. Jouppi, N., Young, C., Patil, N., Patterson, D.: Motivation for and evaluation of the first tensor processing unit. IEEE Micro **38**(3), 10–19 (2018). https://doi.org/10.1109/MM.2018.032271057
11. Nilsback, M., Zisserman, A.: Automated flower classification over a large number of classes. In: 2008 Sixth Indian Conference on Computer Vision, Graphics & Image Processing, pp. 722–729 (2008). https://doi.org/10.1109/ICVGIP.2008.47
12. Li, X., Lv, R., Yin, Y., Xin, K., Liu, Z., Li, Z.: Flower image classification based on generative adversarial network and transfer learning. In: IOP Conference Series: Earth and Environmental Science, vol. 647, p. 012180 (2021). https://doi.org/10.1088/1755-1315/647/1/012180
13. Xia, X., Xu, C., Nan, B.: Inception-v3 for flower classification. In: 2017 2nd International Conference on Image, Vision and Computing (ICIVC), pp. 783–787 (2017). https://doi.org/10.1109/ICIVC.2017.7984661
14. Putra, T., Rufaida, S., Leu, J.-S.: Enhanced skin condition prediction through machine learning using dynamic training and testing augmentation. IEEE Access 1 (2020). https://doi.org/10.1109/ACCESS.2020.2976045
15. Cloud TPU -Documentation. https://cloud.google.com/tpu
16. Singh, P.P., Kaushik, R., Singh, H., Kumar, N., Rana, P.S.: Convolutional neural networks based plant leaf diseases detection scheme. In: IEEE Globecom Workshops (GC Wkshps), vol. 2019, pp. 1–7 (2019). https://doi.org/10.1109/GCWkshps45667.2019.9024434
17. Xie, Q., Luong, M.-T., Hovy, E., Le, Q.V.: Self-training with noisy student improves imagenet classification. In: IEEE/CVF Conference on Computer Vision and Pattern Recognition (CVPR), vol. 2020, pp. 10684–10695 (2020). https://doi.org/10.1109/CVPR42600.2020.01070
18. He, F.X., Liu, T.L., Tao, D.C.: Control batch size and learning rate to generalize well: theoretical and empirical evidence. In: Proceedings of the Advances in Neural Information Processing Systems, pp. 1143–1152 (2019)
19. Zhong, Z., Zheng, L., Kang, G., Li, S., Yang, Y.: †Cognitive Science Department, Xiamen University, China, University of Technology Sydney (2017). https://arxiv.org/pdf/1708.04896.pdf

Distributed Systems, Agents and Microservices

Distributed Microservices Architecture for Fuzzy Risk-Based Structural Analysis of Resistance to Change in Digital Transformation Practice

Azedine Boulmakoul[1] , Fadoua Khanboubi[1]([✉]) , Bachira Abou El karam[2], and Rabia Marghoubi[2]

[1] LIM Lab. IOS team, FSTM, Hassan II University of Casablanca, Casablanca, Morocco
khanboubi.fadoua@gmail.com
[2] INPT, EVEREST InnovatiVE REsearch on Software, Systems and Data, Rabat, Morocco

Abstract. Modern organizations have to accept to transform themselves in order to respond to customers' imperatives and to withstand the competitive pressure of the market. To do so, they readily embrace digital transformation initiatives and create a climate of trust and serenity for the governance of resistance to change. The objective of this work is to propose a holistic approach for the analysis of resistance to change in the implementation of the digital transformation of smart organizations. Our approach is based on the study of the impact of digital risks on the future of the company, and is illustrated by a simple case study. Our proposal is based on techniques from structural analysis, which refers in particular to Galois lattices. Risk assessment is often tainted by the subjectivity of experts. Thus, we adjust our choices on fuzzy Galois lattices for the structural analysis of risk-driven resistance to change in digital transformation. We also describe a distributed architecture, using a graph-oriented database and some other microservices, that guarantees the monitoring of digital transformation. Our approach is part of a continuous improvement process, which contributes to the best practices of digital transformation development.

Keywords: Fuzzy risk · fuzzy Galois lattice · digital transformation · microservices · distributed architecture · decision-making · organization and management · Resistance to change

1 Introduction

Leveraging today's digital solutions to enhance development processes to improve productivity is a real challenge [1–3]. Nevertheless, the human resources of the company are not always unanimous for all the initiatives declared in the conduct of the digital transformation. This resistance can come from the organization's perimeter managers, as well as from simple employees. Resistance to digital transformation represents an obstacle toresis business decision makers. Committed or planned investments can founder if digital

L. Braubach et al. (Eds.): IDC 2022, SCI 1089, pp. 77–88, 2023.
https://doi.org/10.1007/978-3-031-29104-3_9

transformation initiatives failed. However, employees' mistrust of change does not guarantee the success of a digital transformation project [4–10]. Current research focuses on the problem of resistance to change in organizations. It sheds light on employees' resistance to change during a digital transformation project. Several factors characterize this resistance, which converge together to make employees unfavorable to a planned change in their organizational structures. Employee resistance to change in a digital transformation initiative deserves an analysis to better appreciate the resistance by organizational units and consequently at the overall company level. This work aims at developing a structural analytical approach based on fuzzy Galois lattices [11–15] for the risk analysis of the enterprises' digital transformation while considering th several data do not necessarily present themselves in the form of binary reats raised by the resistance to change. The developed model is then illustrated by a simple example. The paper proceeds as follows: Sect. 2 reviews the literature on fuzzy theory and Galois lattices. Section 3 discusses methodological issues. Section 4 presents the results. The final Section discusses conclusions and future developments.

2   Theoretical Basis on Fuzzy Galois Lattices

In this section we briefly recall the formal fuzzy concepts and their applications. We do not outline the mathematical details of the description of Galois lattices and its fuzzy extensions in particular. The basic algebraic theory is very well described in the work of Ganter et al. [14].

2.1   Fuzzy Formal Concepts Analysis

In the basic model proposed by Wille [13, 14], the Formal Concepts Analysis (FCA) concerns binary relations called formal contexts, but in reality, several data are not in the form of binary contexts, indeed, certain relations between objects and properties can be partially known, imprecise, vague or completely unknown (measurements, observations, judgments…). So FCA has to consider formal contexts of various natures, obtained by measurements, observations, judgments, etc., where the relation between an object and an attribute is no longer Boolean (0 or 1) but can be uncertain, gradual or imprecise [16, 17]. Several approaches have proposed extensions of formal contexts into fuzzy contexts and extensions of the Galois connection to construct the complete lattice of concepts. They are representative of three families: Threshold approaches, single-sided approaches and general approaches [18–21]. Formal fuzzy context analysis is an algebraic method from order theory, which generates concepts from a given set of objects, a set of attributes and a fuzzy relation on them. There is a hierarchy of these concepts, from which a complete lattice can be produced and which corresponds to a Galois lattice. Threshold approaches propose discretization of the fuzzy formal context [22, 23]. The idea is to obtain a binary formal context from a fuzzy context, and to reduce the number of formal concepts in the fuzzy lattice.

2.2   Computing Process

The elaboration of fuzzy contexts, requires as input the elements $[A, \Delta, \Gamma]$ defined above. The calculation path is provided in the Fig. 1. Depending on the analysis objective, the decision-maker selects the fuzzy matrix as input to the process: the matrix K which corresponds to fuzzy context of Employees and resistance to change factors; the matric Ω which corresponds to context of units and resistance to change factors. The decision-maker can have an overall assessment of the resistance to change in the enterprise given by $H(\Delta) = \frac{1}{|\Delta|} \times \sum_i \Phi(\Delta^i)$. Once the fuzzy context is fixed, a threshold binarization technique is positioned to generate the associated concept lattices.

Table 1. K: Fuzzy context of Employees and Resistance to Change factors.

$K = A \times \Gamma$	γ_1		$\cdots$	γ_j	$\cdots$	γ_n	Fuzzy Factor Aggregator for Employees: $\Lambda\left(\Theta^i\right)$
Θ^1							
$\cdots$							
Θ^i				K_{ij}			$\Lambda\left(\Theta^i\right) = \frac{1}{|\Gamma|} \times \sum_j K_{ij}$
$\cdots$							
Θ^m							

Table 2. Ω: Fuzzy context of Units and Resistance to Change factors.

$\Omega = \Delta \times \Gamma$	γ_1	$\cdots$	γ_j	$\cdots$	γ_n	Fuzzy Factor Aggregator for Organizational Units: $\Phi\left(\Delta^i\right)$
Δ^1						
$\cdots$						
Δ^i			Ω_{ij}			$\Phi\left(\Delta^i\right) = \frac{1}{|\Delta^i|} \times \sum_j \Omega_{ij}$
$\cdots$						
Δ^l						
$\odot_{i=1}^{i=l}\left(\Delta^i\right)$	$H(\Delta) = \frac{1}{|\Delta|} \times \sum_i \Phi\left(\Delta^i\right)$					

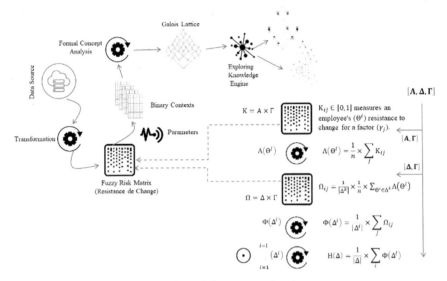

Fig. 1. Fuzzy Formal Concepts exploration process.

3   Method

In this section, we describe a fuzzy model of resistance to change in a digital transformation context for business domain managers of organizational units. Fuzzy set theory has been shown to address uncertainty and imprecision impacting decision making. This resistance to change that manifests itself for digital transformation finds its explanations among the employees of the company for the following reasons, this list is not exhaustive: Fear of losing rank in organization and losing job, Fear of failure when rolling out digital initiatives. Certainly, not all employees experience digital change with the same level of discomfort. They adopt different types of behavior during its implementation and ultimately experience common concerns that they share with their home group. In the following, we give the notations necessary for the understanding of the computation process of the fuzzy matrices intended for the analysis. Let $A = \{\Theta^1, \Theta^2, \ldots, \Theta^m\}$ the set of employees involved in the organizational units. $\Delta = \{\Delta^1, \Delta^2, \ldots, \Delta^l\}$ the set of organizational units, where Θ^i is involved in some $\Delta^k[k = 1..l]$. $\Gamma = \{\gamma_1, \gamma_2, \ldots, \gamma_n\}$, a set of factors of resistance to change in a digital transformation project. K_{ij} measures an employee's (Θ^i) resistance to change for a factor (γ_j). For simplicity we use fuzzy arithmetic mean operator. The literature of fuzzy systems is abundant on fuzzy set aggregators [23]. $|\Delta^k|$ cardinal of Δ^k,

$\Lambda(\Theta^i) = \frac{1}{|\Gamma|} \times \sum_j K_{ij}$, (see Table 1), $\Phi(\Delta^i) = \frac{1}{|\Gamma|} \times \sum_j \Omega_{ij}$, (see Table 2).

$\Omega_{ij} = \frac{1}{|\Delta^i|} \times \sum_{\Theta^\mu \in \Delta^i} K_{\mu j}$, $H(\Delta) = \frac{1}{|\Delta|} \times \sum_i \Phi(\Delta^i)$, (see Table 2). $\odot_{i=1}^{i=l}(\Delta^i)$: Focus on all organizational units of the enterprise (see Table 2), $H(\Delta) = \frac{1}{|\Delta|} \times \sum_i \Phi(\Delta^i)$.

3.1 Illustrative Example: Fuzzy FCA and Risks in Digital Transformation Practice

We propose an illustration of the formal analysis of fuzzy contexts.

Let $A = \{\Theta^1 : Sami, \Theta^2 : Laila, \Theta^3 : Fadoua\}$ the set of employees involved in the organizational units.

$\Delta = \{\Delta^1 : Finance\ Department, \Delta^2 : Marketing\ Department\}$ the set of organizational units, where Θ^1 and Θ^2 are involved in Δ^1; Θ^3 involved in Δ^2.

Γ the set of factors of resistance to change in a digital transformation project.

$$\Gamma = \left\{ \begin{array}{c} \gamma_1 : e-contract\ initiative, \gamma_2 : Mobile\ Web\ Initiative, \gamma_3 : Bitcoin\ Initiative, \\ \gamma_4 : Ecommerce\ WebSite \end{array} \right\}$$

The fuzzy context adjacency matrices are given in Tables 3 and 4.

Table 3. Fuzzy context K: Illustration (Unit × Factors, Employees × Factors).

Unit	$K = A \times \Gamma$	γ_1	γ_2	γ_3	γ_4	$\Lambda(\Theta^i) = \dfrac{1}{\mid\Gamma\mid} \times \sum\limits_j K_{ij}$
Δ^1	Θ^1	0.7	0.3	0	1	0.5
	Θ^2	0.8	0.1	0	0.5	0.28
Δ^2	Θ^3	0.5	0	0	1	0.37

Table 4. Fuzzy context Ω: Illustration (Unit × Factors).

$\Omega = \Delta \times \Gamma$	γ_1	γ_2	γ_3	γ_4	$\Phi(\Delta^i) = \dfrac{1}{\mid\Gamma\mid} \times \sum\limits_j \Omega_{ij}$
Δ^1	0.75	0.2	0	0.75	0.42
Δ^2	0.5	0	0	1	0.37
$\odot_{i=1}^{i=2}(\Delta^i)$	$H(\Delta) = \dfrac{1}{\mid\Delta\mid} \times \sum\limits_i \Phi(\Delta^i) = 0.39$				

Table 5. Binary context of $\Omega = \Delta \times \Gamma$: threshold $= 0.37$.

$\Omega = \Delta \times \Gamma$	γ_1	γ_2	γ_3	γ_4
Δ^1	1	0	0	1
Δ^2	1	0	0	1

Table 6. Binary context of $\Omega = \Delta \times \Gamma$: threshold $= 0.2$.

$\Omega = \Delta \times \Gamma$	γ_1	γ_2	γ_3	γ_4
Δ^1	1	1	0	1
Δ^2	1	0	0	1

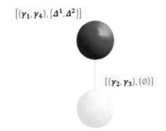

$[\{\gamma_1, \gamma_4\}, \{\Delta^1, \Delta^2\}]$

$[\{\gamma_2, \gamma_3\}, \{\emptyset\}]$

Fig. 2. Galois Lattice, $\Omega = \Delta \times \Gamma$ threshold $= 0.37$.

The closed concepts generated for $\Omega = \Delta \times \Gamma$, with thresholds $0.37, 0.2$ (see Tables 5, 6; Figs. 2, 3) are:

$$\Pi^{0.37} = \{(\gamma_1, \gamma_4 | \Delta^1, \Delta^2)\}, (\{\gamma_2, \gamma_3\} | \emptyset)\},$$
$$\Pi^{0.2} = \{(\gamma_1, \gamma_2 | \Delta^2), (\gamma_2 | \Delta^1), (\{\gamma_3\} | \emptyset)\}.$$

The closed concepts generated for $\Omega = A \times \Gamma$, with thresholds $0.37, 0.1$ (see Tables 7, 8; Figs. 4, 5) are:

$$\Pi^{0.37} = \{(\gamma_1, \gamma_4 | \Theta^1, \Theta^2, \Theta^3)\}, (\{\gamma_2, \gamma_3\} | \emptyset)\},$$
$$\Pi^{0.1} = \{(\gamma_1, \gamma_4 | \Theta^3), (\gamma_2 | \Theta^1, \Theta^2), (\{\gamma_3\} | \emptyset)\}.$$

3.2 Microservices and Distributed Architecture

The design of a software system as well as the construction of its architecture are critical to the success of such systems. Thus, it is essential to adopt the right software architecture and to choose the right development paradigms. Indeed, distributed systems have been evolving for a long time and have proven their ability to ensure best practices in software design and specification. Microservices architecture is an approach to developing an application as a suite of small services, each running in its own process and communicating with lightweight mechanisms [24–26]. These services are built around

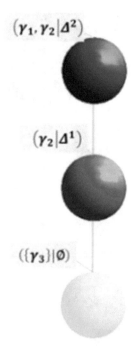

$(\gamma_1, \gamma_2 | \Delta^2)$

$(\gamma_2 | \Delta^1)$

$(\{\gamma_3\} | \emptyset)$

Fig. 3. Galois Lattice, $\Omega = \Delta \times \Gamma$ threshold $= 0.2$.

Table 7. Binary context of $\Omega = A \times \Gamma$: threshold $= 0.37$.

$\Omega = A \times \Gamma$	γ_1	γ_2	γ_3	γ_4
Θ^1	1	0	0	1
Θ^2	1	0	0	1
Θ^3	1	0	0	1

Table 8. Binary context of $\Omega = A \times \Gamma$: threshold $= 0.1$.

$\Omega = A \times \Gamma$	γ_1	γ_2	γ_3	γ_4
Θ^1	1	1	0	1
Θ^2	1	1	0	1
Θ^3	1	0	0	1

business capabilities and can be deployed independently by fully automated deployment machines. The major feature of the microservices architecture is the ability to isolate services. Each microservice is responsible for the management of its own resources and

$$[\{\gamma_1, \gamma_4\}, \{\Theta^1, \Theta^2, \Theta^3\}]$$

$$[\{\gamma_2, \gamma_3\}, \{\emptyset\}]$$

Fig. 4. Galois Lattice, $\Omega = A \times \Gamma$ threshold $= 0.37$.

$$[\{\gamma_1, \gamma_4\}, \{\Theta^3\}]$$

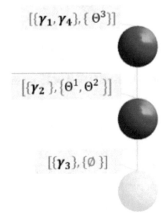

$$[\{\gamma_2\}, \{\Theta^1, \Theta^2\}]$$

$$[\{\gamma_3\}, \{\emptyset\}]$$

Fig. 5. Galois Lattice, $\Omega = A \times \Gamma$ threshold $= 0.1$.

forbids itself direct access to the resources of other services. We have identified different microservices that are driven by the needs manifested in driving a digital transformation subject to change resistance. Figure 6 represents a block diagram of the system. The system has microservices that can provide reliable functions to stakeholders and help decision makers succeed in their commitment. Figure 7 describes the overall architecture of the proposed framework. The proposed architecture is designed on the basis of three layers: the client-oriented analytical layer, the exploration layer and the native execution layer. The client layer encapsulates the microservices functions of access, navigation in the roadmap of the planed digital transformation, choice of the structural analysis method and visualization. The exploration layer integrates the components directly requested by the system's clients. The visualization component offers the functionalities of navigation in graph structures and Galois lattices, etc. The *StructuralETL* microservice takes care of the extraction, transformation and loading of graph data. The Digital Transformation Roadmap component schematizes the procedural guide for conducting the digital transformation. The native execution layer wraps the deep components of computation and elaboration of fuzzy concept lattices, graph analytics and graph mining, procedures for establishing hierarchies of graph *structures*. This layer also concerns the *digital transformation microservice driven by resistance to change*.

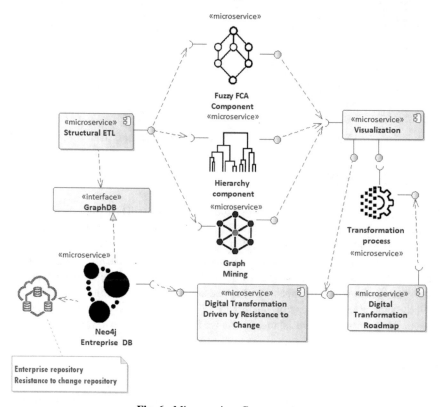

Fig. 6. Microservices Components.

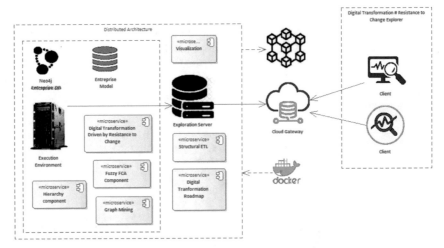

Fig. 7. Distributed microservices deployment ecosystem.

4 Results and Discussion

Figures 2, 3, show the navigation through the concept lattice. The exploration practice is conducted to better determine the optimal grouping of resistance to change factors and organizational units (closed concepts). For instance, the closed concept $(\gamma_1, \gamma_4 | \Delta^1, \Delta^2)$ from $\Pi^{0.37}$, means that the factors of resistance to change (*E-Contract initiative, Ecommerce WebSite initiative*), are important for both units (Finance department, Marketing Department). All units are impacted by resistance to change for digital initiatives; however, the objective is to select the factors to prioritize. This navigational analysis helps to steer the governance of resistance to change in a comprehensive way (Table 9).

5 Conclusion

Digital transformation success when leaders focus on analyzing and understanding employee feedback. The solution that we propose helps to synthesize the climate that reigns among employees. The groupings by factor or by employees give an intelligent global reading to drive a strategy of adherence to digital initiatives. Our approach therefore offers a good upstream strategy to carry out a digital transformation project. The ecosystem proposed in this work concerns the production of a distributed agile and structural decision support framework for decision makers to analyze the resistance to change during a digital transformation project. This platform has been created using a series of microservices aligned with the functional requirements expected by the framework. The proposed ecosystem is in continuous improvement and foresees further enhancements in order to bring more analytical practices to the service of decision makers, to achieve the goals of a digital transformation.

Table 9. Analytics and navigation dashboard.

Global Resistance:	$H(\Delta) = \dfrac{1}{|\Delta|} \times \displaystyle\sum_i \Phi(\Delta^i) = 0.39$		
Unit	Unit's Global Resistance: $\Phi(\Delta^i) = \dfrac{1}{n} \times \displaystyle\sum_j \Omega_{ij}$	Unit's Min Employee Resistance: $\min_j(\Omega_{ij})$	Unit's Max Employee Resistance: $\max_j(\Omega_{ij})$
Δ^1	0.42	0	0.75
Δ^2	0.37	0	1
	Units' Resistance Distribution		
Δ^1	*histogram plot*		
Δ^2	*histogram plot*		
$\odot_{i=1}^{i=2}(\Delta^i)$	*histogram plot*		
	Fuzzy Formal Concepts Navigator		
$\Omega = \Delta \times \Gamma$	**Thresholds**	**Lattices**	
	0.37, 0.2	Lattices plot: **Figures 2, 3**	
$K = A \times \Gamma$	**Thresholds**	**Lattices**	
	0.37, 0.1	Lattices plot: **Figures 4, 5**	

References

1. Scholkmann, A.B.: Resistance to (digital) change. In: Ifenthaler, D., Hofhues, S., Egloffstein, M., Helbig, C. (eds.) Digital Transformation of Learning Organizations, pp. 219–236. Springer, Cham (2021). https://doi.org/10.1007/978-3-030-55878-9_13
2. Trader-Leigh, K.E.: Case study: identifying resistance in managing change. J. Organ. Chang. Manag. **15**(2), 138–155 (2002)
3. Weber, E., Büttgen, M., Bartsch, S.: How to take employees on the digital transformation journey: an experimental study on complementary leadership behaviors in managing organizational change. J. Bus. Res. **143**(2022), 225–238 (2022)
4. Kolagar, M., Parida, V., Sjödin, D.: Ecosystem transformation for digital servitization: a systematic review, integrative framework, and future research agenda. J. Bus. Res. **146**, 176–200 (2022). ISSN 0148-2963
5. Konopik, J., Jahn, C., Schuster, T., Hoßbach, N., Pflaum, A.: Mastering the digital transformation through organizational capabilities: a conceptual framework. Digit. Bus. **2**(2), 100019 (2022)
6. Bovey, W.H., Hede, A.: Resistance to organizational change: the role of defense mechanisms. J. Manag. Psychol. **16**(17) (2001)
7. Kanter, R., Stein, B., Jick, T.: The Challenge of Organizational Change: How Companies Experience It and Leaders Guide It. The Free Press, New York (1992)
8. Khanboubi, F., Boulmakoul, A.: Fuzzy intuitionist approach for resistance to change analysis in a digital transformation process. In: 2021 International Conference on Decision Aid Sciences and Application (DASA), pp. 957–961. IEEE (2021). https://doi.org/10.1109/DASA53625.2021.9682240
9. Khanboubi, F., Boulmakoul, A.: Digital transformation in the banking sector: surveys exploration and analytics. Int. J. Inf. Syst. Change Manage. **11**(2), 93–127 (2020). ISSN 1479-3121

10. Besri, Z., Boulmakoul, A.: An intuitionist fuzzy method for discovering organizational struc-
    tures that support digital transformation. In: Kahraman, C., Cevik Onar, S., Oztaysi, B., Sari,
    I.U., Cebi, S., Tolga, A.C. (eds.) INFUS 2020. AISC, vol. 1197, pp. 331–338. Springer, Cham
    (2021). https://doi.org/10.1007/978-3-030-51156-2_39
11. Boulmakoul, A., Besri, Z.: Performing enterprise organizational structure redesign through
    structural analysis and simplicial complexes framework. Open Oper. Res. J. **7**, 11–24 (2013)
12. Boulmakoul, A., Fazekas, Z., Karim, L., Cherradi, G., Gáspár, P.: Using formal concept
    analysis tools in road environment-type detection. In: Kahraman, C., Cevik Onar, S., Oztaysi,
    B., Sari, I.U., Cebi, S., Tolga, A.C. (eds.) INFUS 2020. AISC, vol. 1197, pp. 1059–1067.
    Springer, Cham (2021). https://doi.org/10.1007/978-3-030-51156-2_123
13. Wille, R.: Concept lattices and conceptual knowledge systems. Comput. Math. Appl. **23**(6–9),
    493–515 (1992). https://doi.org/10.1016/0898-1221(92)90120-7
14. Ganter, B., Wille, R.: Formal Concept Analysis: Mathematical Foundations. Springer, Berlin
    (1999). ISBN 978-3-642-59830-2
15. Boulmakoul, A., Idri, A., Marghoubi, R.: Closed frequent itemsets mining and structuring
    association rules based on Q-analysis. In: 2007 IEEE International Symposium on Signal
    Processing and Information Technology, pp. 519−524, 15−18 December 2007 (2007). https://
    doi.org/10.1109/ISSPIT.2007.4458017. ISBN978-1-4244-1834-3
16. Zadeh, L.A.: The information principle. Inf. Sci. **294**, 540–549 (2015). ISSN 0020-0255
17. Zadeh, L.A.: Fuzzy sets. Inf. Control **8**(3), 338–353 (1965)
18. Brito, A., Barros, L., Laureano, E., Bertato, F., Coniglio, M.: Fuzzy formal concept analysis.
    In: Barreto, G.A., Coelho, R. (eds.) NAFIPS 2018. CCIS, vol. 831, pp. 192–205. Springer,
    Cham (2018). https://doi.org/10.1007/978-3-319-95312-0_17
19. Liu, X., Pedrycz, W.: AFS formal concept and AFS fuzzy formal concept analysis. In: Liu,
    X., Pedrycz, W. (eds.) Axiomatic Fuzzy Set Theory and Its Applications. Studies in Fuzziness
    and Soft Computing, vol. 244, pp. 202–349. Springer, Heidelberg (2009). https://doi.org/10.
    1007/978-3-642-00402-5_8
20. Gutiérrez García, J., Lai, H., Shen, L.: Fuzzy Galois connections on fuzzy sets. Fuzzy Sets
    Syst. **352**, 26–55 (2018). ISSN 0165-0114
21. Mezni, H., Abdeljaoued, T.: A cloud services recommendation system based on fuzzy formal
    concept analysis. Data Knowl. Eng. **116**, 100–123 (2018). ISSN 0169-023X
22. Kwon, S.H.: Threshold selection based on cluster analysis. Pattern Recognit. Lett. **25**(9),
    1045–1050 (2004). ISSN 0167-8655
23. Yager, R.R.: Aggregation operators and fuzzy systems modeling. Fuzzy Sets Syst. **67**(2),
    129–145 (1994). ISSN 0165-0114
24. Khan, O.M.A., Siddiqui, N., Oleson, T.: Embracing Microservices Design. Packt Publishing
    (2021). ISBN-13 978–1801818384
25. Tanenbaum, A.S., Van Steen, M.: Distributed Systems: Principles and Paradigms. Prentice-
    Hall (2007). ISBN 8120334981
26. El Khalyly, B., Belangour, A.: A kubernetes algorithm for scaling virtual objects. In: 2020
    3rd Research of Information Technology and Intelligent Systems (ISRITI) (2020)

Simulating Traffic with Agents, Microservices and REST

Martynas Jagutis, Seán Russell, and Rem Collier[(✉)]

University College Dublin, Dublin, Ireland
martynas.jagutis@ucdconnect.ie, {sean.russell,rem.collier}@ucd.ie

Abstract. Hypermedia Multi-Agent System Simulation is a novel approach to building agent-based simulations in which the environment is modelled as a set of linked hypermedia resources. This paper discusses the implementation of an prototypical simulation system based on this concept and the lessons learnt in the process.

1 Introduction

Agent-Based Modelling (ABM) is a bottom-up approach to studying the behaviour of complex systems that can be modelled as a population of individuals that interact via some shared environment [25]. Hypermedia Multi-Agent System (MAS) Simulation [8] is a novel approach to building ABM simulations in which the environment is modelled as a set of linked hypermedia resources. Each hypermedia resource is a micro-environment that is connected to other related micro-environments by a set of hyperlinks. These micro-environments are implemented as microservices [13,36] that interact using REpresentational State Transfer (REST) [12].

Traffic Simulation Systems (TSS) is a mature area of research with the aim of improving the planning, design, and operation of transportation systems [27]. Approaches to simulation tasks vary in scope and scale from situations where mathematical formulae represent throughput of vehicles to simulations where the individual position of vehicles are represented at each instant. Agent-Based Modelling (ABM) has often been used to simulate the behaviour of individual users of the transport network being simulated [11,16].

This paper presents a proof of concept, in the form of a TSS, for the Hypermedia MAS Simulation approach described in [8]. We do not claim novelty in the features of the simulation, but in the way we construct it through the combination of microservices, hypermedia systems, linked data, multi-agent systems, and affordances in order to create a decentralised simulation that can scale while still delivering support for reusability and extensibility.

The core principle of the Hypermedia MAS Simulation approach is the construction of complex agent-based simulations from suites of loosely-coupled reusable components. Thus a simulation would be composed using a number of sub-simulations, each implemented as instances of **microservices**, that work

L. Braubach et al. (Eds.): IDC 2022, SCI 1089, pp. 89–99, 2023.
https://doi.org/10.1007/978-3-031-29104-3_10

together to achieve global aims. One benefit of this strategy is that such sub-simulations can be implemented using heterogeneous modelling techniques and languages, rather than being restricted to a single language or framework.

Interoperability between sub-simulations can be achieved through the use of **Linked Data** in a manner analogous to its use in the Web of Things (WoT) [15], where the Thing Description standard has been developed [5]. These are machine readable documents that describe the capabilities of the devices that have been deployed, allowing clients to reason about how (or whether) to use them.

The development of similar standards for simulations, could promote the interoperability and reuse of sub-simulations. These descriptions, which we call *Simulation Descriptions*, would specify the nature of the simulation environment as well as describing the inputs and outputs associated with it. Integration of sub-simulations could utilise this description to automatically handle required data transformations when required as the input and output data will contain the semantic meaning. The availability of this semantic information would not be useful without a system capable of making use of it. Using a **Hypermedia MAS** [6] approach would allow agents to navigate the hypermedia space and reason about the simulation, its environments and the entities within them.

The ABM within our proposed Hybrid Simulation framework is central as agents are embodied with the available sub-simulations, where they control the actions of the entities being represented. This presents two distinct challenges, firstly agents may face greater complexity from the integration with multiple sub-simulations, and secondly agents will need to interact with a diverse set of environments.

The increased complexity of agents is congruent with the "Keep It Descriptive, Stupid" principle described in [1], which argues for the simulation of richer, more realistic models. The authors use this principle to argue that the social simulation community needs to adopt BDI style models, tools and programming languages.

Agent interactions in diverse simulation environments can be facilitated through the provision of **affordances**. Affordances are information perceived from the environment that signifies that a particular action may be performed. They allow for a higher level of abstraction in agent-environment interactions, allowing an agent to reason about the actions it can perform instead of having hard coded actions in plans. These affordances could be documented through the hypermedia documents that are published by the system, thus allowing agents to discover and interact with entities in the environment or even other agents regardless of their location.

2   Related Work

The work in this paper stands at the intersection of traffic simulation and multi-agents systems research. To reflect this, related work is first addressed from the perspective of traffic simulation, and then from the perspective of multi-agent systems.

2.1   Traffic Simulation

A large number of commercial and open source systems exist that simulate traffic networks [31]. They can often be fundamentally different in terms of efficiency or features, but generally can be categorised based a number of characteristics.

Traffic flow Behaviour is distinguished based on how individual vehicles are considered. Most systems model the behaviour of vehicles as homogeneous, in that they will behave in predictable ways with little variance in the population [31]. Heterogeneous behaviour of vehicles allow for a greater variety in both the types of vehicles modelled as well as allowing alternate driving styles. This approach is made available by only a small number of systems such as Sumo [19].

Traffic Flow Models are typically described Microscopic, Macroscopic, or Mesoscopic. Microscopic models are based on the simulation of individual vehicles (agents) and include representing the position and velocity of the vehicle on the road network [19,33]. Macroscopic models use statistical approaches to simulate the overall flow of vehicles rather than individuals [26]. Mesoscopic models attempt to benefit from the more individual representation of vehicles in some places while also benefiting from the higher level calculations in others [2,28]. The choice of which traffic flow model to use is based on the intended use and required detail of the simulation system. Macroscopic simulations required less computing resources or time to complete, but may not allow the same level of detail or heterogeneous behaviours of vehicles allowed by microscopic simulations.

The **Representation** of concepts such as position and velocity information also varies. In discrete space models the transport network is represented as a collection of cells that the vehicles can inhabit and move between [23,30]. Continuous models instead represent the same information with greater detail.

2.2   Multi-agent Systems

Recent work in social simulation [4,35] on the use of the Belief-Desire-Intention (BDI) architecture [29] highlights the potential quality improvements that cognitive architectures bring to simulation. An emerging trend in social simulation is outlined that argues that the "Keep It Descriptive, Stupid" (KIDS) principle is more useful than the "Keep It Simple, Stupid" (KISS) principle. Adam and Gaudou argue that the increased computing resources currently available means there is less need to use simplistic agent models in an effort to maximise the performance of the simulation [1]. A better solution is to use richer cognitive agent architectures to facilitate the creation of more nuanced models.

The concept of **Hypermedia Aware Agents** has been proposed by Ciortea et al. through the concept of *Hypermedia MAS*; an approach to building dynamic, open and long-lived MAS [32] that are designed to inter-operate seamlessly with the World Wide Web [6]. This approach was further expanded to outline a vision in which agents are integrated into the hypermedia fabric of the web, and that by doing so, enter into a shared hypermedia environment that is based on the open standards of the Web [7]. In such an environment, devices and physical services can be exposed as first class entities. Hyperlinks and hypermedia controls can be used to discover and interact with those entities or even other agents regardless of their location. Such controls can be published through hypermedia documents that are adapted based on the state of the underlying entities.

Understanding the relationship between **Agents and Microservices Architecture** is critical to the approach advocated in this paper. Microservices have emerged as a key industry approach to building distributed systems that scale. The use of microservices with agents was first explored in bespoke applications built for the IoT domain [20,21]. The *Multi-Agent MicroServices (MAMS)* approach [34] presents an architectural style that can be applied more generally. [24] describes a prototype implemented and deployed using the ASTRA agent programming language [9,10].

The concept of an **Affordance** originates in ecological psychology as a means of representing the relationship between objects in an environment and the potential actions that an agent (human or otherwise) may perform with those objects [14]. Affordances can be viewed as another form of perception of the environment that signifies that a particular action may be performed. They allow for a higher level of abstraction in agent-environment interactions, allowing an agent to reason about the actions it can perform instead of having hard coded actions in plans.

Affordances have been used in some agent based modelling systems, although the implementation can be somewhat varied. Affordances-effect pairs were used in modelling of human behaviour in complex environments [17], affordance fields have been used in path planning simulations [18] and [22] used affordances to identify space occupation actions that could be taken by vehicles in a traffic simulation.

3   A Prototype Traffic Simulation

The vision described in [8] considers simulations to be a composition of two types of interacting microservice: *environment microservices* that host pieces of the simulation environment and *agent-oriented microservices* that implement the intelligent behaviour of the individuals driving the actions of the entities in the simulation.

The environment microservices are considered heterogeneous as different aspects of the simulation environments are implemented by different microservices. While in this example all of the microservices are implemented in Java, this is not a requirement and heterogeneous technologies can be used. The agent-oriented microservices are implemented more homogeneously as they are all using the ASTRA [9] platform. Again, this is not a requirement of the system, rather instead it is a convenience. More importantly, it is not relevant which instance of the agent-oriented microservice is hosting an agent, it's behaviour will not be changed.

3.1   Sketch of Implementation

Figure 1 presents a sketch of the prototype simulation architecture[1]. As can be seen, it consists of three primary environment microservices: the *Traffic Simulator*, which manages the movement of vehicles on the roads; the *Home Simulator*, which represents home and the *Work Simulator*, which represents workplaces. Two secondary environment microservices are also included: the *Clock Service* provides a global discrete time model, and the *Traffic Lights* service implements a traffic light scheduling strategy. The final microservices are a *Management Service*, which is used to configure and manage the prototype and a *Driver Service* that implements the agents. Two driver services exist: one based on ASTRA and one that is pure Java code.

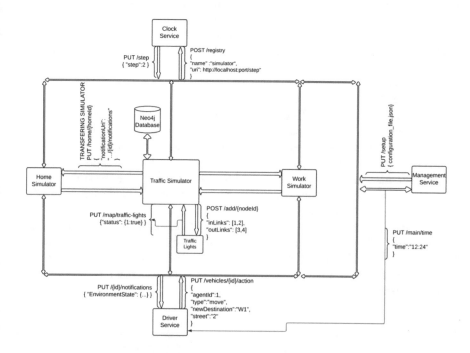

Fig. 1. Overview of Simulation Architecture

The execution of the simulation requires that agents beginning in the home simulator would transition into the traffic simulator at the appropriate time. The agent then navigates through the traffic simulator to the appropriate point and transitions into the work simulator. As highlighted in Fig. 1, the interactions between microservices, which take the from of JSON documents delivered over HTTP, are designed based on the REST architectural style.

---

[1] The source code is available for download at https://gitlab.com/mams-ucd/examples/microservice_traffic_simulator.

Periodic state updates from the environment services are sent to the agents regarding the current state and perception of the body they are connected to. This state update is implemented through a webhook associated with each agent. When joining its first micro-environment, the agent configures this by passing the relevant URI to the micro-environment. Upon each tick of the clock, the micro-environment sends a PUT request to the agent with the body being the updated state. Included within the state update is the current list of valid actions that can be performed. While this is a rather simplistic implementation of affordances, it serves to demonstrate viability of the approach. Agents submit their next action using a PUT request to the relevant micro-simulator. The actions that the agent can perform are: **Plan** used to generate a route to a given destination; **Wait** do nothing; **Enter** enter the intersection (only available if at the end of a street); **Leave** leave an intersection via the street specified; **Accelerate:** increase speed; **Decelerate:** decrease speed; and **Move:** drive along the street at same speed.

Figure 2 illustrates parts of an example interaction between an agent and the traffic simulator. Within the state update, the affordances inform the agent of the actions the can be performed by the vehicle in the next step. In response the agent then posts a document describing the action to be performed, in this case the agent is transitioning the vehicle it controls through a junction while specifying that the intended exit point of the junction is street 434.

In the traffic simulator, streets are implemented as cellular automata; and junctions are implemented as queues. Both simulations operate on a discrete time model which is controlled externally by the clock service. This ensures that the sub-simulations are synchronised. The home/work services are minimal. The only context the agent receives is the current time, and the only action they can perform is the **Plan** action.

Transitioning from the home/work simulation to the traffic simulator is done at the request of the agent through the setting of a plan to travel to a destination (via the traffic simulator). Conversely, the driver automatically transitions to the destination home/work simulator if they enter the intersection that is linked to that simulator.

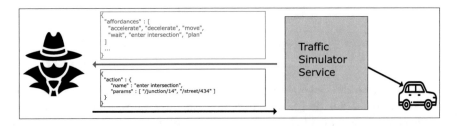

Fig. 2. A representation a subset of the services in the simulation system

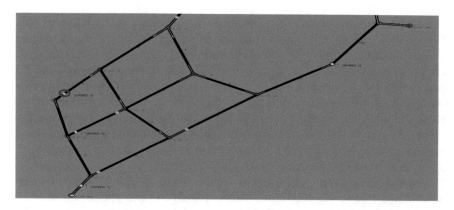

Fig. 3. Graphical representation of the simulation in progress

3.2 Walkthrough

Figure 3 shows a graphical representation of a simulation in progress, with vehicles in transit to their destinations. This simple simulation is composed of 31 streets and 14 intersections as well as a number of homes and workplaces. At the beginning of the simulation all agents are present within a home simulation and at the relevant times they transition into the traffic simulation and make the journey to the intersection of their respective workplace where they are automatically transitioned into that simulation. Similarly, at the appropriate time agents transition back into the traffic simulation and navigate the return journey.

4 Lessons Learnt

During the development of the prototype, the ideas and concepts underlying the vision evolved. Part of this evolution was a natural extension of our initial ideas, part as a consequence of the development and evaluation of the prototype. Some of the main lessons learned are described below.

The **importance of hypermedia links** was demonstrated through the way we used unique identifiers (e.g. junction or street numbers). Using identifiers simplified the implementation, but forced the requirement that the agent must know how to transform the identifier into a URL. While this was not an issue within the traffic micro-environment, which provided a shared context; it was an issue when considering inter-micro-environment interaction. This led to inconsistencies in the representation of the environment and increased the coupling between agents and the environments they inhabited.

The **need for a semantic description of the environment** is motivated by the need to incorporate additional functionality into the simulation. For example, in order to support route planning, a SatNav feature was added. This service was implemented as part of the traffic simulation service because the environment configuration was not accessible to the agents. This enforced a centralised approach to route finding with all agents being required to use that approach. If we had wanted to allow other approaches, it would also have been necessary to implement those approaches as part of the traffic simulation service. In our view, a better approach would be to expose the simulation state (and configuration) using a semantic representation of the traffic network (and following on from that, the wider simulation). Using the appropriate technologies would allow the construction of a shared knowledge graph that agents could explore directly, moving the implementation of functionality, such as route planning, into the agent, allowing for heterogeneous route planning strategies, such as the Q-learning approach advocated in [3].

The most important lesson learnt completing this prototype was to **avoid monolithic micro-simulations where possible**. The traffic simulator is a monolith; it combines the traffic layout, the junction and street simulations, and the route planning service. If the knowledge graph idea above were to be implemented, it would include that too. Monoliths can be bad because they are often difficult to scale and sometimes become bottlenecks in systems. For example, in our prototype, the traffic simulator is a central component that must be navigated when moving between the home and work micro-simulations. The stateful nature of agent-environment interaction would make scaling this design challenging. A better approach is to decompose the monolith into its constituent components (e.g. streets and junctions) that can be implemented as separate micro-environments that are associated via hyperlinks (there would no longer be unique identifiers). The result is a finer grained system that is inherently decentralised and much easier to scale. Further, the street/junction hyperlinks provide a natural basis for implementing a decentralised version of the knowledge graph outlined above. Finally, an additional benefit of this approach is that it would facilitate the integration of heterogeneous road and junction simulators within a single simulation.

Other lessons have been learnt that are more specific to the prototype. For example, the use of the **Plan** action to trigger transition from a home/work simulation into the traffic simulator only works if there is a traffic simulator. In the context of the avoiding monoliths lesson, perhaps a generic **Leave** action would be more appropriate. Also, the use of the **Move** action does not make much sense in the context of the **Accelerate** and **Decelerate** actions. A better model would be to accelerate to the correct speed and then to **Wait** if there is nothing to do. Similarly, the need to perform an **Enter** action at the end of a street does not make sense, it is surely more appropriate to expect that the car will automatically transition into the intersection when it reaches the end of the street.

5   Conclusions

This paper presents an overview of a prototype traffic simulator based on microservices and REST. Overall, the development of the simulator was a positive experience that we believe demonstrates the potential of our approach to building simulations based on Agent-Based Modelling. That said, the prototype has also highlighted some important lessons related to the future direction of our research.

References

1. Adam, C., Gaudou, B.: BDI agents in social simulations: a survey. Knowl. Eng. Rev. **31**(3), 207–238 (2016)
2. Auld, J., Hope, M., Ley, H., Sokolov, V., Xu, B., Zhang, K.: POLARIS: agent-based modeling framework development and implementation for integrated travel demand and network and operations simulations. Transp. Res. Part C: Emerg. Technol. **64** (2016)
3. Beaumont, K., O'Neill, E., Bermeo, N., Collier, R.: Collaborative route finding in semantic mazes. In: Proceedings of the All the Agents Challenge (ATAC 2021) (2021)
4. Bulumulla, C., Singh, D., Padgham, L., Chan, J.: Multi-level simulation of the physical, cognitive and social. Comput. Environ. Urban Syst. **93**, 101756 (2022). https://doi.org/10.1016/j.compenvurbsys.2021.101756
5. Charpenay, V., Käbisch, S.: On modeling the physical world as a collection of things: the W3C thing description ontology. In: Harth, A., et al. (eds.) ESWC 2020. LNCS, vol. 12123, pp. 599–615. Springer, Cham (2020). https://doi.org/10.1007/978-3-030-49461-2_35
6. Ciortea, A., Boissier, O., Ricci, A.: Engineering world-wide multi-agent systems with hypermedia. In: Weyns, D., Mascardi, V., Ricci, A. (eds.) EMAS 2018. LNCS (LNAI), vol. 11375, pp. 285–301. Springer, Cham (2019). https://doi.org/10.1007/978-3-030-25693-7_15
7. Ciortea, A., Mayer, S., Boissier, O., Gandon, F.: Exploiting interaction affordances: on engineering autonomous systems for the web of things. In: Second W3C Workshop on the Web of Things the Open Web to Challenge IoT Fragmentation. Munich, Germany (2019)
8. Collier., R., Russell., S., Golpayegani., F.: Harnessing hypermedia MAS and microservices to deliver web scale agent-based simulations. In: Proceedings of the 17th International Conference on Web Information Systems and Technologies - WEBIST, pp. 404–411. INSTICC, SciTePress (2021). https://doi.org/10.5220/0010711100003058
9. Collier, R.W., Russell, S., Lillis, D.: Reflecting on agent programming with AgentSpeak(L). In: Chen, Q., Torroni, P., Villata, S., Hsu, J., Omicini, A. (eds.) PRIMA 2015. LNCS (LNAI), vol. 9387, pp. 351–366. Springer, Cham (2015). https://doi.org/10.1007/978-3-319-25524-8_22
10. Dhaon, A., Collier, R.W.: Multiple inheritance in AgentSpeak (L)-style programming languages. In: Proceedings of the 4th International Workshop on Programming based on Actors Agents & Decentralized Control, pp. 109–120. ACM (2014)
11. Espié, S., Auberlet, J.M.: ARCHISIM: a behavioral multi-actors traffic simulation model for the study of a traffic system including ITS aspects. Int. J. ITS Res. **5**(n1), p7–16 (2007)
12. Fielding, R.T.: Architectural Styles and the Design of Network-based Software Architectures. Doctoral dissertation, University of California, Irvine (2000)
13. Fowler, M., Lewis, J.: Microservices: a definition of this new architectural term. Technical report (2014). https://martinfowler.com/articles/microservices.html

14. Gibson, J.J.: The Ecological Approach to Visual Perception. Houghton-Mifflin (1979)
15. Guinard, D.D., Trifa, V.M.: Building the Web of Things, vol. 3. Manning Publications, Shelter Island (2016)
16. Horni, A., Nagel, K., Axhausen, K. (eds.): Multi-Agent Transport Simulation MATSim. Ubiquity Press, London (2016). https://doi.org/10.5334/baw
17. Jooa, J., et al.: Agent-based simulation of affordance-based human behaviors in emergency evacuation. Simul. Modell. Pract. Theory 32, 99–115 (2013)
18. Kapadia, M., Singh, S., Hewlett, W., Faloutsos, P.: Egocentric affordance fields in pedestrian steering. In: Proceedings of the 2009 Symposium on Interactive 3D Graphics and Games, I3D 2009, pp. 215–223. Association for Computing Machinery, New York (2009)
19. Krajzewicz, D., Hertkorn, G., Rössel, C., Wagner, P.: SUMO (Simulation of Urban MObility) - an open-source traffic simulation. In: Al-Akaidi, A. (ed.) Proceedings of the 4th Middle East Symposium on Simulation and Modelling (MESM20002), pp. 183–187 (2002)
20. Kravari, K., Bassiliades, N.: A rule-based eCommerce methodology for the IoT using trustworthy intelligent agents and microservices. In: Benzmüller, C., Ricca, F., Parent, X., Roman, D. (eds.) RuleML+RR 2018. LNCS, vol. 11092, pp. 302–309. Springer, Cham (2018). https://doi.org/10.1007/978-3-319-99906-7_22
21. Krivic, P., Skocir, P., Kusek, M., Jezic, G.: Microservices as agents in IoT systems. In: Jezic, G., Kusek, M., Chen-Burger, Y.-H.J., Howlett, R.J., Jain, L.C. (eds.) KES-AMSTA 2017. SIST, vol. 74, pp. 22–31. Springer, Cham (2018). https://doi.org/10.1007/978-3-319-59394-4_3
22. Ksontini, F., Mandiau, R., Guessoum, Z., Espié, S.: Affordance-based agent model for road traffic simulation. Auton. Agent. Multi-Agent Syst. 29(5), 821–849 (2015)
23. Nagel, K., Schreckenberg, M.: Traffic jam dynamics in stochastic cellular automata. Technical report, Los Alamos National Laboratory (1995)
24. O'Neill, E., Lillis, D., O'Hare, G.M.P., Collier, R.W.: Explicit modelling of resources for multi-agent microservices using the CArtAgO framework. In: Proceedings of the 19th International Conference on Autonomous Agents and Multi-Agent Systems (2020)
25. Polhill, J.G., et al.: Crossing the chasm: a 'tube-map' for agent-based social simulation of policy scenarios in spatially-distributed systems. GeoInformatica 23(2), 169–199 (2019)
26. Prevedouros, P.D., Li, H.: Comparison of freeway simulation with INTEGRATION, KRONOS, and KWaves. In: Fourth International Symposium on Highway Capacity, Maui, Hawaii, pp. 96–107 (2000)
27. Pursula, M.: Simulation of traffic systems-an overview. J. Geogr. Inf. Decis. Anal. 3(1), 1–8 (1999)
28. Rakow, C., Kaddoura, I., Nippold, R., Wagner, P.: Investigation of the system-wide effects of intelligent infrastructure concepts with microscopic and mesoscopic traffic simulation. In: 27th ITS World Congress, Hamburg, Germany (2021). https://elib.dlr.de/144810/
29. Rao, A.S., Georgeff, M.P., et al.: BDI agents: from theory to practice. In: ICMAS, vol. 95 (1995)
30. Shang, X.C., Li, X.G., Xie, D.F., Jia, B., Jiang, R., Liu, F.: A data-driven two-lane traffic flow model based on cellular automata. Phys. A 588, 126531 (2022)
31. Ullah, M., Khattak, K., Khan, Z., Khan, M., Minallah, N., Khan, A.: Vehicular traffic simulation software: a systematic comparative analysis. Pak. J. Eng. Technol. 4(1), 66–78 (2021). https://doi.org/10.51846/vol4iss1pp66-78

32. Vachtsevanou, D., Junker, P., Ciortea, A., Mizutani, I., Mayer, S.: Long-lived agents on the web: Continuous acquisition of behaviors in hypermedia environments. In: Companion Proceedings of the Web Conference 2020, pp. 185–189 (2020)
33. Axhausen, K.W., Horni, A., Nagel, K.: The Multi-agent Transport Simulation MATSim. Ubiquity Press (2016)
34. Collier, R.W., O'Neill, E., Lillis, D., O'Hare, G.: MAMS: multi-agent microservices. In: Companion Proceedings of the 2019 World Wide Web Conference, WWW 2019, pp. 655–662. Association for Computing Machinery, New York (2019)
35. Wai, S.Y., Cheah, W.S., Wai, S.K., Khairuddin, M.A.: Towards software engineering perspective for BDI agent. In: 2021 4th International Symposium on Agents, Multi-Agent Systems and Robotics (ISAMSR), pp. 106–110 (2021). https://doi.org/10.1109/ISAMSR53229.2021.9567829
36. Zimmermann, O.: Microservices tenets. Comput. Sci.-Res. Dev. **32**(3–4), 301–310 (2017)

Stock Price Forecasting on BigDL - A Parallel and Distributed Framework

Nhu-Chien Ha[1,2], Quoc-Loc Duong[1,2], and Trong-Hop Do[1,2(✉)]

[1] University of Information Technology, Ho Chi Minh City, Vietnam
{18520527,18521006}@gm.uit.edu.vn, hopdt@uit.edu.vn
[2] Vietnam National University, Ho Chi Minh City, Vietnam

Abstract. In the marketplace, the stock is a noticeable channel of investment. Learning how to invest effectively takes lots of time for new investors. In this scenario, there are various methods, from conventional statistical methods like ARIMA to advance deep learning models such as LSTM, TCN, and Seq2Seq. However, facilitating state-of-the-art models requires many hardware commodities, especially on a big scale. Therefore, the BigDL framework recommends robust APIs to conduct plentiful deep learning models, including time series associated issues, one of the first available frameworks provided on BigDL open source. Below are several stocks, including Apple, Amazon, Google, and Microsoft, which we use to evaluate the new BigDL framework for addressing time series forecasting.

1 Introduction

The stock market is a potential channel to invest and improve the income of individuals in which large companies such as Apple, Amazon, Google, and Microsoft are proof of this success. However, when the number of investors in the marketplace grows, it bears a harsher channel to succeed as a new investor. In addition, economic depression, inflation, and many other aspects are to be considered when an investor decides to put their money into a particular stock. Automatic and supporting methods are born to better assist investors in accumulating property. ARIMA is a typical statistical method and has been applied for many years. However, we need some more effective state-of-the-art approaches such as LSTM, TCN, and Sequence to Sequence models. Ironically, it is harder and slower to implement modern methods due to vital hardware requirements. We expect a parallel and distributed platform to boost our available models in this situation. BigDL is one of the first frameworks to address many practical issues and also consists of time series forecasting, which is a general concept applied to the stock as mentioned above price prediction. This article collects top companies' stock prices in the stock market as mentioned above to evaluate LSTM, TCN, and Sequence to Sequence, which is typically built on Chronos for time series forecasting.

L. Braubach et al. (Eds.): IDC 2022, SCI 1089, pp. 100–110, 2023.
https://doi.org/10.1007/978-3-031-29104-3_11

The BigDL framework [1] is used in building deep learning models based on big data. BigDL enables parallel, distributed execution on clusters like Hadoop, Spark, or K8s. The benefit of using a framework is that it is easy to model the implementation of parallel and distributed computing. Besides, this framework also allows the execution of large and complex data that requires complicated data processing steps, thereby saving time. BigDL includes the DLlib, ORCA, RayOnSpark, Chronos, PPML, and Serving properties. Among them, ORCA and Chronos are mainly used in this paper. ORCA is scalable on Tensorflow and Pytorch for distributed computing and Chronos the AutoML engine for large-scale time series analysis.

2   Related Works

Many authors have researched how to predict stock prices using recurrent networks. In [2], the authors use multi-layer RNN, LSTM, and GRU to forecast Google stock and obtained up to 72% of accuracy. It has become a promising practical application in the finance field. In [3], the authors implemented LSTM to predict a few stocks: Apple, Google, and Tesla, and returned several potential results. This research proved that RNN-LSTM tends to give more accurate results than traditional learning algorithms. In [4], the authors carried out some experiments on recurrent networks as well as Temporal Convolution Network(TCN): Stock2Vec - A Hybrid Deep Learning Framework for Stock Market Prediction with Representation Learning. Moreover, Temporal Convolutional Network indicated the outperforming presence of TCN in most stocks(including Apple and Amazon). These researches are only a tiny part of how time series forecasting is developing on the aforementioned methods. Hence, BigDL sees the importance of making a parallel and distributed implementation and facilitating researchers in studying larger scale available algorithms.

3   Dataset

Dataset is collected directly from Yahoo website[1] which consists of four companies stock prices from late 2011 to 2021 including Amazon, Apple, Google, and Microsoft. There are several key features used to implement stock closing prices which are: *low, high, open, close, volume* and *date*. In particular, *low* is the lowest stock price in a day, *high* is the highest stock price in a day, *open* is the stock price at the opening time in a transaction day, *close* is the stock price at the end of a transaction day, *volume* is total number of matched stock over a transaction day and *date* points to the transaction day.

---

[1] https://finance.yahoo.com.

It has two states for a transaction day which are *Up day* expressing a day with a closing price higher than opening price and *Down day* expressing a day with a closing price lower than the opening price. This price movement is usually depicted by the candlestick chart shown in Fig. 1. In expansion, Fig. 2 shows the trading volume and stock price movement of the four companies in the past.

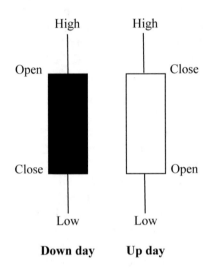

Fig. 1. Candlestick

We conduct a multivariate forecaster for each company. In addition to target data which is *close* price, training data contains some extra features as presented above: *low, high, open* and *volume*. Chronos framework shown in Sect. 4 provides the TSDataset class that allows data initialization for distributed training. Besides, TSDataset class allows data preprocessing such as division (with train/dev/test ratio 8/1/1) and data normalization (the chosen method is Min-MaxScaler).

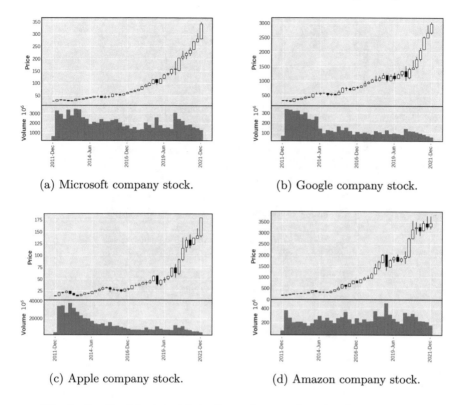

(a) Microsoft company stock.   (b) Google company stock.

(c) Apple company stock.   (d) Amazon company stock.

Fig. 2. Candlestick plot with trading volume and stock price movement.

4 Methodologies

4.1 Frameworks

4.1.1 ORCA

Most Machine Learning models are executed on a single computer. However, executing large-scale data will take a long time, so distributed execution is one way to solve that problem. On the other hand, distributed execution also goes through a rough patch. Therefore, ORCA allows distributed execution on a large cluster (including K8s or Hadoop clusters) through a single node from which distributed execution becomes easy. Initializing ORCA is simply running on the local machine, which automatically executes the distributed execution by providing a Python environment for the nodes (such as a single computer, K8s cluster, or Hadoop cluster). After that, ORCA will automatically provide Apache Spark or Ray as the basic execution engine for data processing or model training. The overview of the architecture of ORCA is shown in Fig. 3.

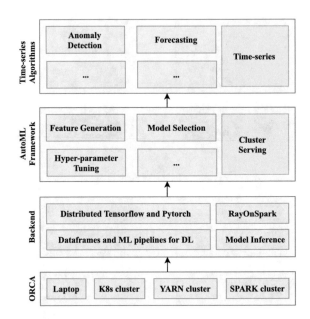

Fig. 3. The overview of the architecture.

4.1.2   CHRONOS

CHRONOS is an application framework for building large-scale time series analysis applications. Chronos provides both deep machine learning models and traditional statistical models for forecasting. The implementation of stock forecasting includes using the *AutoTS pipeline* for data preprocessing, data division, and hyperparameter configuration; model selection using *auto forecasting models*; using *standalone forecasters* for training.

4.2   Models

4.2.1   Long Short Term Memory (LSTM)

LSTM [5] is one of the recurrent neural network (RNN) types that can capture past information and predict the future. One of the advantages of LSTMs is the ability to remember information over a long period compared to recurrent neural networks, which helps forecast stock prices. An LSTM cell consists of input gate, cell state, forget gate, and output gate. It also consists of sigmoid layer, tanh layer and point wise multiplication operation. In an LSTM cell, memorization of previous stages is performed by different gates. Each LSTM node is also responsible for storing the stream of transmitted data. LSTM model is built in BigDL framework for time series prediction task. Figure 4 shows an LSTM node.

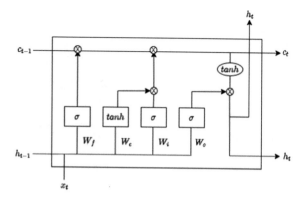

Fig. 4. An LSTM cell.

4.2.2   Temporal Convolution Network (TCN)

TCN [6] shown in Fig. 5 is a computationally efficient way to address several problems related to sequential data, such as anomaly detection and time series forecasting. This model is faster to train in comparison with recurrent networks. To boost TCN performance, this network stacks several *Temporal Block* which is also a pivotal factor to discuss. *Temporal Block* includes two identical chains of Dilated convolution, Batch Normalization, activation ReLU, and Dropout at the end of each chain. The output of the aforementioned chain is added to its input after this input goes through a 1×1 Convolution. Because of the causal computation and special kernels different from regular convolution inside the dilated convolution, TCN captures further dependent information. At the same time, only focus on available data in the past and present daytime to train and make the prediction.

4.2.3   Sequence to Sequence (Seq2Seq)

Seq2Seq [7] available on parallel and distributed environment of BigDL is also LSTM cells connected inside two main blocks, which are encoder and decoder. After the encoder synthesizes a Fixed length vector from time-series input, the decoder receives this context vector to predict the next transaction day. The architecture of the model is shown in Fig. 6.

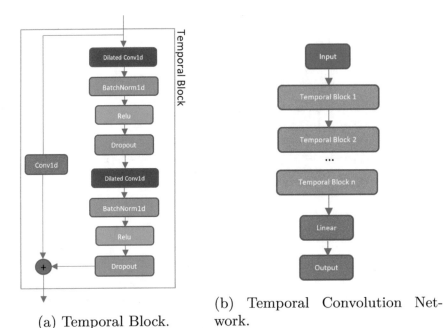

(a) Temporal Block.

(b) Temporal Convolution Network.

Fig. 5. Temporal Convolution Network.

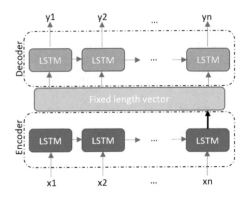

Fig. 6. Sequence to sequence architecture.

5   Experiment and Evaluation

All stock price prediction models are trained with 25 epochs, batch size 32 and optimizer Adam. Figure 7, 8, 9, and 10 show the visualization of predicted price and actual price of four companies Microsoft, Google, Apple, and Amazon, respectively. From Fig. 7, it can be observed that the LSTM and Seq2Seq models, although capturing the upward trend of Microsoft's share price, there is always a difference between the predicted stock price and the actual stock price. That dif-

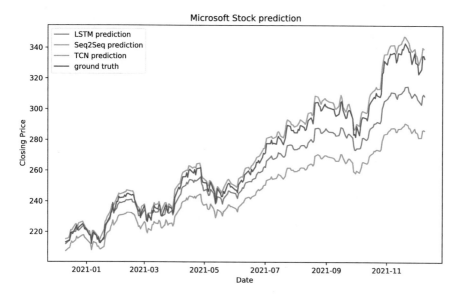

Fig. 7. Microsoft company stock.

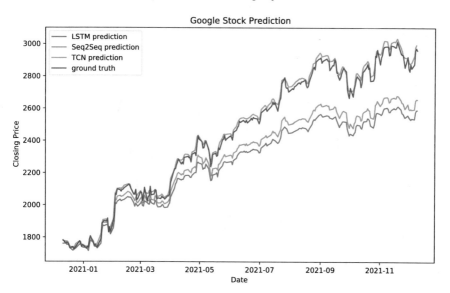

Fig. 8. Google company stock.

ference becomes more significant over time. However, the predictive TCN model captures stock price trends very well, and the difference between the predicted and actual stock prices is relatively small. Figure 8 shows that all three models are good predictors of stock prices between January 2021 and April 2021. As of April 2021, the stock prices predicted by the LSTM and Seq2Seq models are

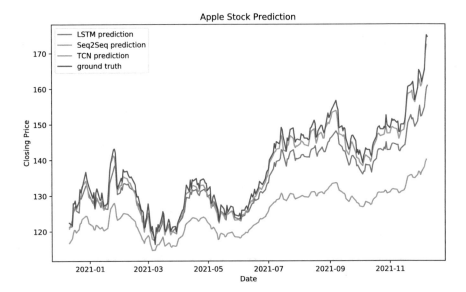

Fig. 9. Apple company stock.

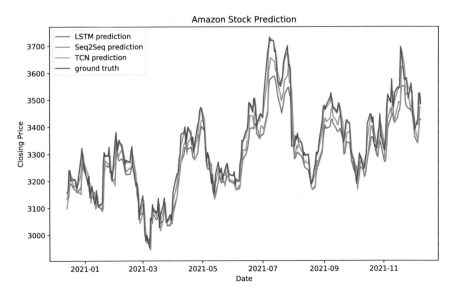

Fig. 10. Amazon company stock.

different. A range from the actual stock price, and that gap grows more prominent over time. However, the TCN model still predicts reasonably well the share price of the Google company. For Fig. 9, the Seq2Seq model predicts Apple's stock price quite severely, with the predicted stock price being much lower than the actual stock price. Besides, the LSTM and TCN models predict pretty well,

Table 1. Measuring the predictive error of the models on each stock.

	Apple		Google		Amazon		Microsoft	
	MSE	SMAPE	MSE	SMAPE	MSE	SMAPE	MSE	SMAPE
LSTM	37.911	0.036	61224.92	0.083	7626.726	0.012	166.955	0.034
Seq2Seq	54.505	0.042	23798.96	0.0502	3860.038	0.028	376.048	0.057
TCN	5.382	0.013	1299.707	0.011	354.101	0.013	14.403	0.012

but as of July 2021, the stock price predicted by the LSTM model and the actual stock price are separated by a slight difference. Figure 10 shows that all three models, LSTM, Seq2Seq, and TCN, are good predictors of Amazon's stock price. All three models capture the company's price trend, and the difference between the predicted stock price and the actual stock price is slight, of which the TCN model is the best predictor.

MSE and SMAPE are two used measures to estimate error for forecasting problems. Table 1 shows the prediction error of the models for each company code. It can be seen that the TCN model predicts meager error rates for the stocks of the companies Google, Apple, and Microsoft. The Seq2Seq model makes predictions with lower error than the other two models, particularly for Amazon's stock ticker.

6 Conclusion and Future Work

Deep learning models such as LSTM, Seq2Seq, and TCN all achieve good results when implemented in the distributed environment of the BigDL framework. The experimental results in this study show that the TCN model gives the best stock price prediction results. Besides, the implementation of predictive models is relatively easy to save time. Training available time series models on BigDL parallel and distributed platforms is worth attempting in the new big data scenario. BigDL also allows developers to build their models serving deeper fine-tuning. Therefore, in the prospect of forecasting in general and in a distributed environment in particular, we can apply many other state-of-the-art models into this big-data platform and experience its convenience, which may not be conducted on single limited hardware.

Acknowledgment. This research was supported by The VNUHCM-University of Information Technology's Scientific Research Support Fund.

References

1. Dai, J.J., et al.: Bigdl: a distributed deep learning framework for big data. In: Proceedings of the ACM Symposium on Cloud Computing, pp. 50–60 (2019)
2. Di Persio, L., Honchar, O.: Recurrent neural networks approach to the financial forecast of Google assets. Int. J. Math. Comput. Simul. **11**, 7–13 (2017)

3. Pawar, K., Jalem, R.S., Tiwari, V.: Stock market price prediction using LSTM RNN. In: Rathore, V.S., Worring, M., Mishra, D.K., Joshi, A., Maheshwari, S. (eds.) Emerging Trends in Expert Applications and Security. AISC, vol. 841, pp. 493–503. Springer, Singapore (2019). https://doi.org/10.1007/978-981-13-2285-3_58
4. Wang, X., et al.: Stock2Vec: a hybrid deep learning framework for stock market prediction with representation learning and temporal convolutional network. ArXiv: abs/2010.01197 (2020)
5. Hochreiter, S., Schmidhuber, J.: Long short-term memory. Neural Comput. **9**(8), 1735–1780 (1997)
6. Misra, V.: Time series forecasting with applications to finance (2021)
7. Du, S., Li, T., Horng, S.J.: Time series forecasting using sequence-to-sequence deep learning framework. In: 2018 9th International Symposium on Parallel Architectures, Algorithms and Programming (PAAP), pp. 171–176. IEEE (2018)

Towards a New Architecture: Multi-agent Based Cloud-Fog-Edge Computing and Digital Twin for Smart Agriculture

Yogeswaranathan Kalyani[✉] and Rem Collier

School of Computer Science, University College Dublin, Dublin, Ireland
kalyani.yogeswaranathan@ucdconnect.ie, rem.collier@ucd.ie

Abstract. The integration of computing paradigms such as Cloud, Fog, and Edge is rapidly adopted in many domains. One such application is Smart Agriculture, which has diverse requirements for applications ranging from Augmented Reality, Satellite, and auto harvesters to a large number of agricultural devices and sensors. Cloud, Fog, and Edge integration solves several problems such as latency and bandwidth issues, security and privacy, and real-time data analytics. However, the combination has a few challenges, including resource allocation, resource scheduling, and task scheduling. In order to overcome these challenges and bring Smart Agriculture to the next levels of productivity and sustainability, based on Cloud, Fog, Edge computing, and Multi-Agent Systems with a concept of Digital Twin, a novel architecture is introduced in this paper. The proposed architecture is expected to be applied in Smart Agricultural farms and fields, opening new aspects of contemporary applications within the Agriculture domain.

Keywords: Cloud-Fog-Edge · Multi Agent Systems · Digital Twin · Smart Agriculture

1 Introduction

The approaches such as Smart Farming, Smart Agriculture, and Precision Agriculture are variations on the same theme. They are typically linked to collecting data from many disparate data sources, including autonomous tractors, harvesters, robots and drones, sensors, and actuators. For example, in crop farming, the agricultural data harvested (such as humidity, temperature, pH and soil conditions) is typically analysed to maximise crop yield whilst minimising costs (e.g. nutrients, pesticides).

Cloud computing helps collect, analyse and store data from the farms and fields. The number of standalone Cloud-based agricultural systems and physical systems is constantly growing. These systems support various monitoring, and

This research is funded under the SFI Strategic Partnership Programme (16/SPP/3296) and is co-funded by Origin Enterprises plc.

L. Braubach et al. (Eds.): IDC 2022, SCI 1089, pp. 111–117, 2023.
https://doi.org/10.1007/978-3-031-29104-3_12

analysis goals [9] by harvesting data from deployed sensors and uploading it to the Cloud for further analysis. A fundamental limitation of this approach is the requirement for pervasive Internet connectivity, which can rarely be achieved on real-world farms. Even when such connectivity does exist, it is often inhibited by high latency and low bandwidth. These real-world concerns limit the potential impact of Cloud-based Smart Agriculture systems.

To overcome these limitations, the last few years have witnessed the extension of Cloud-based approaches like Fog and Edge computing. The main goal of these two computing paradigms is to bring Cloud services and resources to edge devices. Edge computing mainly focuses on the IoT level, whereas Fog computing focuses on an infrastructure level. Handling multiple IoT applications is unsupported in Edge computing and is supported at the Fog level. The location of data collection, processing, and storage mainly happens in network edge and edge devices in Edge computing. In contrast, it occurs near the edge, and core networking in the Fog layer [9]. The authors of [3] explore how these new models address the limitations of the Cloud by providing techniques for moving the computation, data analysis, and storage closer to the network's edge. While the synthesis of Cloud, Fog, and Edge computing are appealing, it is also not trivial, and a number of challenges arise, such as: resource management (including sharing and allocation), task management, and scheduling (workflow, task).

Multi-agent systems are the solution for the challenges mentioned above in the integration of such computing paradigms. The Multi agents incorporate social aspects and human reasoning to solve problems [16]. In this paper, we propose an architecture to integrate all three computing such as Cloud, Fog and Edge and Multi-Agent systems with the Digital Twin model for Smart Agriculture. The agents, which are deployed in the Fog layer, are responsible for: harvesting sensor data; monitoring crop growth through the maintenance of digital twins; and creating, monitoring and scheduling tasks.

2   Related Work

The state of the art in Cloud, Fog and/or Edge Smart Agriculture systems is reviewed in [9]. Reviewed systems solve various problems related to animal management, crop management, greenhouse management, irrigation management, soil management and weather management. The systems deliver a range of benefits (low cost, low latency, saving on bandwidth, reliable data collection) and challenges (Security and privacy, Mobility, Data processing, Power management, hardware cost, internet connectivity). Reviews, focusing on the role of Internet of Things (IoT) in Smart Agriculture [4,6,12] highlight potential application domains, communication protocols, network types, and platforms. Common challenges discussed in these reviews include: Ubiquitous connectivity, data management, hardware and software challenges, authorisation and trust, and scalability.

The role of Multi-Agent Systems (MAS) in Cloud-Fog, Cloud-Edge, or Cloud only applications is motivated by the need to address various challenges, including: task scheduling, resource allocation, resource sharing, and service provisioning. Such systems have typically been suggested for domains such as Smart Cities, Smart Healthcare and Autonomous driving. From a Smart Agriculture perspective, [15] proposes a Cloud-based MAS that implements a round-table consensus-based protocol as part of an adaptive resource management mechanism. [7] proposes a Cloud-based MAS to support knowledge discovery and decision-making for autonomous crop irrigation. [2] is one of a number of systems that present MAS-based agricultural monitoring systems that are implemented using fleets of agent-based Drones [11]. Finally, [5] describes an Agent-based methodology and middleware for constructing IoT applications.

Despite the works mentioned above, we could not find a standard architecture for the combinations in works that have been done in any domain. Although there are several MAS-based models and IoT-based systems available, to the best of our knowledge, no systems currently combine the concepts of Digital Twin, MAS, and Cloud-Fog-Edge. The significant contributions of this research paper are, (1) Propose a novel architecture with a combination of Digital Twin with Multi-Agent Cloud-Fog-Edge; (2) Introduce agents and Digital Twin in the Fog layer; (3) Introduce Mobile Fog zones and Static Fog zones. The objective of agents and Digital Twin in the Fog layer and mobile and Fog zones are clearly described in the following section.

3   Proposed Architecture

As is depicted in Fig. 1, the architecture proposed in this paper is organised over three layers: Cloud, Fog, and Edge. The Cloud plays a number of roles but is primarily the domain of Big Data/Data Analytics. It provides data storage and anonymisation services for farm data, machine learning, and data analysis support. Additionally, the Cloud layer hosts various services for topics such as: yield prediction, disease identification, pest identification, and growth stage estimation. It provides analytical tools that support forecasting outcomes such as expected yield, crop waste, and revenues. The Fog layer represents the Farm Management System. The Multi-Agent System is deployed in this layer and is responsible for overseeing various activities specific to the farm. The Fog is further decomposed into two sub-strata: static Fog nodes are hosted in the farm itself, while mobile Fog nodes are associated with farm equipment, such as Tractors or Combine Harvesters and any other available hardware, such as Drones or Agricultural Robots. The key idea is that the static Fog nodes host local data storage /compute services for in-situ farm data analysis. In contrast, the mobile Fog nodes allow for the extension of the Fog to areas of the farm with sporadic or no Internet connectivity. Finally, the Edge consists of any agricultural (or other) devices hosting sensors (in-situ soil moisture sensor, tractor-mounted NDVI) or actuators deployed on the farm, be it in fields or part of farm equipment such as Tractors or Combine Harvesters. Edge devices are connected directly to static Fog services where internet connectivity permits. If such connectivity is unavailable, then the data provided by these devices is harvested on-demand or opportunistically through data collection services hosted on mobile Fog nodes.

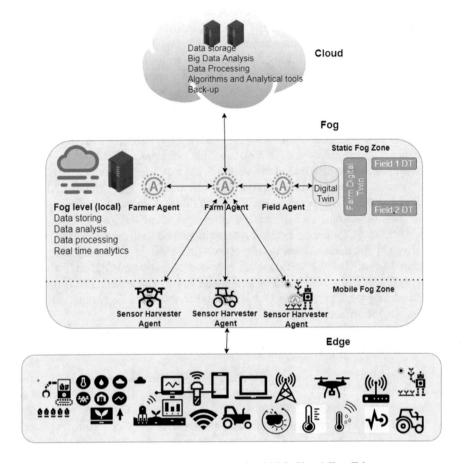

Fig. 1. Proposed Architecture for MAS-Cloud-Fog-Edge

In addition to data collection, the Fog is responsible for overseeing other support activities such as: soil analysis, crop monitoring, pest/disease identification, or irrigation. Example tasks include: growth monitoring, lodging, nutrient analysis, information on the condition of the land, disease and pest monitoring, and irrigation needs. Where related to areas of the farm that lack Internet connectivity, tasks may be allocated to dedicated agents deployed to mobile Fog nodes. These agents are responsible for the achievement of their assigned tasks. This may require the support of the farmer (for example, directing the farmer via Augmented Reality to the key locations for extracting soil samples for a field [8]), or the farmer may play little or no role in the task (e.g. opportunistic sensing of data [10]), with the agent instead having full responsibility for its completion.

Managing tasks is not the only role for agents, as it includes three additional types of agent: the *Farm agent*, *Field agents*, and the *Farmer agent*. The *Farm agent's* primary responsibility is managing the farm and overseeing the use of farm resources as optimally as possible in the achievement of tasks whilst accounting for the needs/preferences of the farmer, who is represented indirectly by the *Farmer agent*. *Field agents* are created for each field on the farm to maximise the profit for the field, typically by maximising yield while minimising cost. To achieve this, each *Field agent* maintains an associated Digital Twin [13] of their field. The Digital Twin is underpinned by sensor data gathered by the system and is used to model the performance of the field against predicted performance based on historical data. The agent uses this model to create and maintain an associated Field Management Plan encoded as tasks, including sensing activities (for updating/ calibrating the model) or interventions, such as irrigation or spraying fertiliser or pesticide. Collectively the plan determines what tasks should be completed and when. The divergence between the predicted and actual models triggers reassessment of the field management plan by the relevant *Field agent*, potentially resulting in recommended alterations to the plan. Setting and revising the plan is a collaborative activity that occurs between the *Field agent(s)*, the *Farm agent* and the farmer (via the *Farmer agent*).

A central component of our approach is the Digital Twin of fields. It provides the primary context for activities of the *Field agents* and underpins the recommendations that agents make. Raw sensor data is harvested from Edge devices deployed across the farm by a specific type of task agent known as a *Sensor Harvester*. Third-party data sources supplement it (e.g., weather, satellite, or even financial data) and are used in conjunction with Cloud-based Data Analytics services to provide insights that the agents can apply. Some insights help improve the accuracy of the Digital Twin; others facilitate decision-making activities, helping to identify when a Field Management Plan may need to be altered or how it should be altered. These Cloud-based services are also used in the formation of these plans. Using the Digital Twin model in this architecture helps farmers get information on the fields and overall farm behavior. It also forecasts yield prediction, growth stage, nutrient information, and weather-report.

To ensure that our architecture is future-proofed, we have designed it to be agnostic to implementing the Digital Twin. To cater for this, our architecture is also designed to allow the incorporation of new tasks. To achieve this, task descriptions are encoded using a flexible format, and new task types can be supported by creating new task-specific agents. Our vision aligns with the recently proposed idea of the Computing Continuum [1] which is concerned with developing digital infrastructure that is used by complex workflows typically combining real-time data generation and processing and computation [14]. Additionally, this architecture can increase sensors (soil moisture probes, NDVI sensors, weather stations), and facilitate upgrading equipment such as guidance in tractors and variable spray booms/map-based variable treatment.

4   Conclusions and Future Work

This paper presents a novel approach to integrating MAS and Cloud-Fog-Edge with Digital Twin in Smart Agriculture. In the whole process of this architecture, there is a high possibility of combining several technologies and increasing the number of edge and fog devices (scalability) to achieve the objective of this model. For example, Blockchain makes it possible to monitor crop growth until handover to suppliers, Artificial intelligence and Robotics to help in the fields/farms, and Drones to aid in task achievements in the farm monitoring process. Smart Agriculture will benefit from these new technologies with the proposed architecture such as production increase, water-saving, better quality, cost reduction, early pest and disease identification, and better sustainability.

In future work, we plan to focus on these areas: (1) Implement the use cases for Smart Agriculture using the proposed architecture; (2) Investigate the design constraints and requirements in integrating MAS and the computing paradigms with the Digital Twin model; (3) Further research on identifying challenges in combining new computing paradigms and MAS and possible solutions; We believe that other domains have enough scope to adopt the proposed MAS-based Cloud-Fog-Edge computing with Digital Twin architecture with additional changes based on the domain's requirements.

References

1. Antoniu, G., Valduriez, P., Hoppe, H.C., Krüger, J.: Towards integrated hardware/software ecosystems for the edge-cloud-HPC continuum (2021)
2. Cavaliere, D., Loia, V., Senatore, S.: Towards a layered agent-modeling of IoT devices to precision agriculture. In: 2020 IEEE International Conference on Fuzzy Systems (FUZZ-IEEE), pp. 1–8. IEEE (2020)
3. Escamilla-Ambrosio, P.J., Rodríguez-Mota, A., Aguirre-Anaya, E., Acosta-Bermejo, R., Salinas-Rosales, M.: Distributing computing in the internet of things: cloud, fog and edge computing overview. In: Maldonado, Y., Trujillo, L., Schütze, O., Riccardi, A., Vasile, M. (eds.) NEO 2016. SCI, vol. 731, pp. 87–115. Springer, Cham (2018). https://doi.org/10.1007/978-3-319-64063-1_4
4. Farooq, M.S., Riaz, S., Abid, A., Umer, T., Zikria, Y.B.: Role of IoT technology in agriculture: a systematic literature review. Electronics 9(2), 319 (2020)
5. Fortino, G., Russo, W., Savaglio, C., Shen, W., Zhou, M.: Agent-oriented cooperative smart objects: from IoT system design to implementation. IEEE Trans. Syst. Man Cybern.: Syst. 48(11), 1939–1956 (2017)
6. Glaroudis, D., Iossifides, A., Chatzimisios, P.: Survey, comparison and research challenges of IoT application protocols for smart farming. Comput. Netw. 168, 107037 (2020)
7. González-Briones, A., Castellanos-Garzón, J.A., Mezquita-Martín, Y., Prieto, J., Corchado, J.M.: A multi-agent system framework for autonomous crop irrigation. In: 2019 2nd International Conference on Computer Applications & Information Security (ICCAIS), pp. 1–6. IEEE (2019)
8. Huuskonen, J., Oksanen, T.: Soil sampling with drones and augmented reality in precision agriculture. Comput. Electron. Agric. 154, 25–35 (2018)
9. Kalyani, Y., Collier, R.: A systematic survey on the role of cloud, fog, and edge computing combination in smart agriculture. Sensors 21(17), 5922 (2021)
10. Liang, Q., Cheng, X., Huang, S.C., Chen, D.: Opportunistic sensing in wireless sensor networks: theory and application. IEEE Trans. Comput. 63(8), 2002–2010 (2013)

11. Mariani, S., Picone, M., Ricci, A.: About digital twins, agents, and multiagent systems: a cross-fertilisation journey. arXiv preprint arXiv:2206.03253 (2022)
12. Navarro, E., Costa, N., Pereira, A.: A systematic review of IoT solutions for smart farming. Sensors **20**(15), 4231 (2020)
13. Ricci, A., Croatti, A., Mariani, S., Montagna, S., Picone, M.: Web of digital twins. ACM Trans. Internet Technol. (TOIT) (2022)
14. Rosendo, D., Silva, P., Simonin, M., Costan, A., Antoniu, G.: E2Clab: exploring the computing continuum through repeatable, replicable and reproducible edge-to-cloud experiments. In: 2020 IEEE International Conference on Cluster Computing (CLUSTER), pp. 176–186. IEEE (2020)
15. Skobelev, P., Larukchin, V., Mayorov, I., Simonova, E., Yalovenko, O.: Smart farming – open multi-agent platform and eco-system of smart services for precision farming. In: Demazeau, Y., Matson, E., Corchado, J.M., De la Prieta, F. (eds.) PAAMS 2019. LNCS (LNAI), vol. 11523, pp. 212–224. Springer, Cham (2019). https://doi.org/10.1007/978-3-030-24209-1_18
16. Villarrubia, G., Paz, J.F.D., Iglesia, D.H., Bajo, J.: Combining multi-agent systems and wireless sensor networks for monitoring crop irrigation. Sensors **17**(8), 1775 (2017)

A Survey on the Use of the Multi-agent Paradigm in Coordination of Connected and Autonomous Vehicles

Giacomo Cabri[✉], Letizia Leonardi, and Enzo Rotonda

Laboratorio CINI Smart Cities and Communities,
University of Modena and Reggio Emilia, Modena, Italy
{giacomo.cabri,letizia.leonardi,enzo.rotonda}@unimore.it

Abstract. In this paper we present our work of searching, filtering and selecting relevant papers about the exploitation of the multi-agent paradigm in self-driving vehicles, along with the results of our survey.

1 Introduction

One of the fields that is likely to be most affected by the exploitation of autonomous software components and in which there can be investments in the next years is certainly the automotive one [1]. In fact, for several years, cars are full of sensors capable of receiving information from the surrounding environment and act accordingly.

In this paper, in particular, we focus on automotive sector to analyze how the use of intelligent systems is evolving in the self-driving car sector, specifically referred to the approaches based on the multi-agent paradigm for the coordination of autonomous vehicles [2].

1.1 Adopted Methodology

In our work, the research question is the following: RQ-Which are the multi-agent based approaches that are exploited for the coordination of autonomous vehicles?
The question clarifies also the Research scope that is the coordination of Connected and Autonomous Vehicles (CAVs).

We proceed with the *search phase*, during which we choose the databases for our search and then execute the needed queries to retrieve all necessary material. After the search phase, during the *selection phase*, we select only the most relevant documents on the basis of keywords and abstracts. Then, we read the whole text to verify that the leading theme was the one requested. Table 1 shows, respectively, the number of papers that matched the query, the number of papers that passed the filter step, the number of papers that also passed the second screening phase and, therefore, are considered relevant.

© The Author(s), under exclusive license to Springer Nature Switzerland AG 2023
L. Braubach et al. (Eds.): IDC 2022, SCI 1089, pp. 118–124, 2023.
https://doi.org/10.1007/978-3-031-29104-3_13

Table 1. Sources and selected papers.

Database	Results	Filtered	Relevant
Scopus (https://www.scopus.com/)	53	37	33
Web Of science (https://www.webofscience.com)	18	9	6
MoReThesis (https://morethesis.unimore.it/)	5	2	2
Others (https://scholar.google.com/, https://ieeexplore.ieee.org, https://www.mdpi.com, https://www.researchgate.net)	19	10	6
Total	95	58	47

1.2 Exploited Agent Approaches

The multi-agent based approaches that we have found in the surveyed papers are: *Multi-agent reinforcement learning (MARL)* [3]; *Consensus-Based Auction Algorithm (CBAA)* [4]; *Fuzzy Logic* [5]; *Deep Learning (DL)* [6]; *Mixed Integer Linear Program (MILP)* [7].

In our work, we have found that there are also combinations of above explained approaches. Moreover, there are some works that exploit agents but do not adopt a specific approach.

2 Classification of the Papers

In Table 2, we can find the classification based on the most used approaches. In particular, we observe that multiple approaches are often used to try to solve a problem, while for 17 papers it has not been possible to individuate a specific used approach.

From Table 2, we can observe that the most used approach to address the problems arising from the coordination of CAVs in multi-agent systems is MARL, sometimes combined with deep learning or fuzzy logic. In fact, among the 47 relevant papers we analyzed, globally 13 use MARL. Instead, the most individually used approach is the CBAA.

In Table 3, we report the distribution of the approaches per year. It is easy to understand that, with the passage of time, some approaches have been neglected or at least not given so much coverage. For example, we can see that some approaches such as MILP and Fuzzy logic have not undergone the same push of others. We can see how, in the last two years, the most popular approaches are: the MARL, the CBAA and the MARL + deep learning. Moreover, some works use agents without a specific approach.

Table 2. Documents per approach.

Approach	N. of documents	Papers
MARL	5	[8–12]
CBAA	7	[13–19]
Fuzzy	4	[20–23]
Deep learning	1	[24]
MILP	5	[25–29]
MARL + Deep	7	[30–36]
MARL + Fuzzy	1	[37]
Not specified	17	[38–54]

Table 3. Documents per approach and per year

Year	MARL	CBAA	Fuzzy	DL	MILP	MARL+DEEP	MARL+Fuzzy	Not specified	TOTAL
2005	–	–	–	–	–	–	–	1	1
2006	–	–	–	–	–	–	–	1	1
2007	–	–	1	–	–	–	–	–	1
2008	–	–	–	–	–	–	–	1	1
2009	–	–	–	–	–	–	–	1	1
2010	–	–	–	–	–	–	–	1	1
2012	–	1	–	–	–	–	–	1	2
2013	–	–	–	–	1	–	–	–	1
2014	–	–	1	–	–	–	–	–	1
2015	–	–	–	–	–	–	–	1	1
2018	–	–	1	–	1	–	–	–	2
2019	–	1	–	–	1	1	1	1	5
2020	1	1	–	–	1	2	–	3	8
2021	4	4	1	1	1	4	–	6	21

3   Conclusions

In an increasingly interconnected and smart world, cars of the future will be completely autonomous and able to act without a human intervention. Their coordination is one of the challenges that must be faced. In this paper, we have shown how research in the field of autonomous vehicle coordination has evolved, focusing on the various used approaches and how they have evolved over time. The multi-agent paradigm seems to be the best candidate to solve the related problems. In the immediate future, we could have completely hyper-connected cities, where there will be cars that will communicate with traffic lights, with road signs, with public parking lots and with other vehicles to safely maneuver thanks to CAVs.

References

1. Brenner, W., Herrmann, A.: An overview of technology, benefits and impact of automated and autonomous driving on the automotive industry. Digital Marketplaces Unleashed, 427–442 (2018)
2. Bertogna, M., Burgio, P., Cabri, G., Capodieci, N.: Adaptive coordination in autonomous driving: motivations and perspectives. In: 2017 IEEE 26th International Conference on Enabling Technologies: Infrastructure for Collaborative Enterprises (WETICE), pp. 15–17. IEEE (2017)
3. Busoniu, L., Babuska, R., De Schutter, B.: A comprehensive survey of multiagent reinforcement learning. IEEE Trans. Syst. Man, Cybern. Part C (Appl. Rev.) **38**(2), 156–172 (2008)
4. Choi, H.-L., Brunet, L., How, J.P.: Consensus-based decentralized auctions for robust task allocation. IEEE Trans. Rob. **25**(4), 912–926 (2009)
5. Yager, R.R., Zadeh, L.A.: An Introduction to Fuzzy Logic Applications in Intelligent Systems, vol. 165. Springer Science & Business Media, Cham (2012)
6. Gupta, J.K., Egorov, M., Kochenderfer, M.: Cooperative multi-agent control using deep reinforcement learning. In: Sukthankar, G., Rodriguez-Aguilar, J.A. (eds.) AAMAS 2017. LNCS (LNAI), vol. 10642, pp. 66–83. Springer, Cham (2017). https://doi.org/10.1007/978-3-319-71682-4_5
7. Earl, M.G., D'Andrea, R.: Modeling and control of a multi-agent system using mixed integer linear programming. In: Proceedings of the 41st IEEE Conference on Decision and Control, 2002, vol. 1, pp. 107–111. IEEE (2002)
8. Dresner, K., Stone, P.: Multiagent traffic management: opportunities for multiagent learning. In: Tuyls, K., Hoen, P.J., Verbeeck, K., Sen, S. (eds.) LAMAS 2005. LNCS (LNAI), vol. 3898, pp. 129–138. Springer, Heidelberg (2006). https://doi.org/10.1007/11691839_7
9. Parvini, M., Javan, M.R., Mokari, N., Arand, B.A., Jorswieck, E.A.: AoI aware radio resource management of autonomous platoons via multi agent reinforcement learning. In: 17th International Symposium on Wireless Communication Systems (ISWCS), vol. 2021, pp. 1–6. IEEE (2021)
10. Peake, A., McCalmon, J., Raiford, B., Liu, T., Alqahtani, S.: Multi-agent reinforcement learning for cooperative adaptive cruise control. In: 2020 IEEE 32nd International Conference on Tools with Artificial Intelligence (ICTAI), pp. 15–22. IEEE (2020)
11. Skrynnik, A., Yakovleva, A., Davydov, V., Yakovlev, K., Panov, A.I.: Hybrid policy learning for multi-agent pathfinding. IEEE Access **9**, 126034–126047 (2021)
12. Troullinos, D., Chalkiadakis, G., Papamichail, I., Papageorgiou, M.: Collaborative multiagent decision making for lane-free autonomous driving. In: Proceedings of the 20th International Conference on Autonomous Agents and MultiAgent Systems, pp. 1335–1343 (2021)
13. Daoud, A., Balbo, F., Gianessi, P., Picard, G.: A generic multi-agent model for resource allocation strategies in online on-demand transport with autonomous vehicles. In: Proceedings of the 20th International Conference on Autonomous Agents and Multiagent Systems (AAMAS 2021), p. 3 (2021)
14. Hunt, S., Meng, Q., Hinde, C.J.: An extension of the consensus-based bundle algorithm for multi-agent tasks with task based requirements. In: 2012 11th International Conference on Machine Learning and Applications, vol. 2, pp. 451–456. IEEE (2012)
15. Molinari, F., Katriniok, A., Raisch, J.: Real-time distributed automation of road intersections. IFAC-PapersOnLine **53**(2), 2606–2613 (2020)
16. Rizvi, S.R., Zehra, S., Mukkamala, R., Olariu, S.: ASAP: an agent-assisted smart auction-based parking system in internet of things. In: Proceedings of the 9th ACM Symposium on Design and Analysis of Intelligent Vehicular Networks and Applications, pp. 1–8 (2019)

17. Wu, J., Wang, Y., Shen, Z., Wang, L., Du, H., Yin, C.: Distributed multilane merging for connected autonomous vehicle platooning. Sci. Chin. Inf. Sci. **64**(11), 1–16 (2021). https://doi.org/10.1007/s11432-020-3107-7

18. Cabri, G., Gherardini, L., Montangero, M., Muzzini, F.: About auction strategies for intersection management when human-driven and autonomous vehicles coexist. Multimedia Tools Appl. **80**(10), 15921–15936 (2021)

19. Gambelli, M., Mariani, S., Cabri, G., Zambonelli, F.: Combining coordination strategies for autonomous vehicles in intersections networks. In: The 14th International Symposium on Intelligent Distributed Computing, IDC 2021, Online, Italy, 16-18 September 2021. IEEE (2021)

20. Bi, Y., Srinivasan, D., Lu, X., Sun, Z., Zeng, W.: Type-2 fuzzy multi-intersection traffic signal control with differential evolution optimization. Expert Syst. Appl. **41**(16), 7338–7349 (2014)

21. Ikidid, A., El Fazziki, A., Sadgal, M.: A fuzzy logic supported multi-agent system for urban traffic and priority link control. J. Univ. Comput. Sci. **27**(10), 1026–1045 (2021)

22. Hsu, P.L., Liu, H.H.: The multi-agent system with an adaptive-fuzzy algorithm for flow control of traffic networks. In: 2007 IEEE International Conference on Systems, Man and Cybernetics, pp. 3300–3304. IEEE (2007)

23. Tan, M.K., et al.: Decentralized traffic signal control for grid traffic network using genetic algorithm. In: 2019 IEEE 6th International Conference on Engineering Technologies and Applied Sciences (ICETAS), pp. 1–6. IEEE (2019)

24. Ma, H., Sun, Y., Li, J., Tomizuka, M.: Multi-agent driving behavior prediction across different scenarios with self-supervised domain knowledge. In: IEEE International Intelligent Transportation Systems Conference (ITSC), pp. 3122–3129. IEEE (2021)

25. Ashtiani, F., Fayazi, S. A., Vahidi, A.: Multi-intersection traffic management for autonomous vehicles via distributed mixed integer linear programming. In: Annual American Control Conference (ACC), pp. 6341–6346. IEEE (2018)

26. Falsone, A., Molinari, F., Prandini, M.: Uncertain multi-agent MILPs: a data-driven decentralized solution with probabilistic feasibility guarantees. In: Learning for Dynamics and Control, pp. 1000–1009. PMLR (2020)

27. Okoso, A., Otaki, K., Nishi, T.: Multi-agent path finding with priority for cooperative automated valet parking. In: IEEE Intelligent Transportation Systems Conference (ITSC), pp. 2135–2140. IEEE (2019)

28. Okoso, A., Okumura, B., Otaki, K., Nishi, T.: Network-flow-problem-based approach to multi-agent path finding for connected autonomous vehicles. In: IEEE International Intelligent Transportation Systems Conference (ITSC), pp. 1946–1953. IEEE (2021)

29. Yu, J., LaValle, S.M.: Multi-agent path planning and network flow. In: Frazzoli, E., Lozano-Perez, T., Roy, N., Rus, D. (eds.) Algorithmic Foundations of Robotics X. STAR, vol. 86, pp. 157–173. Springer, Heidelberg (2013). https://doi.org/10.1007/978-3-642-36279-8_10

30. Chen, J., Yuan, B., Tomizuka, M.: Model-free deep reinforcement learning for urban autonomous driving. In: IEEE intelligent transportation systems conference (ITSC), pp. 2765–2771. IEEE (2019)

31. Chen, S., Dong, J., Ha, P., Li, Y., Labi, S.: Graph neural network and reinforcement learning for multi-agent cooperative control of connected autonomous vehicles. Comput.-Aided Civil Infrastruct. Eng. **36**(7), 838–857 (2021)

32. Palanisamy, P.: Multi-agent connected autonomous driving using deep reinforcement learning. In: 2020 International Joint Conference on Neural Networks (IJCNN), pp. 1–7. IEEE (2020)

33. Qiu, X., Li, X., Wang, J., Wang, Y., Shen, Y.: A DRL based distributed formation control scheme with stream-based collision avoidance. In: 2021 IEEE International Conference on Autonomous Systems (ICAS), pp. 1–5. IEEE (2021)

34. Shi, H., Zhou, Y., Wu, K., Wang, X., Lin, Y., Ran, B.: Connected automated vehicle cooperative control with a deep reinforcement learning approach in a mixed traffic environment. Transp. Res. Part C: Emerg. Technol. **133**, 103421 (2021)
35. Yuan, Q., Fu, X., Li, Z., Luo, G., Li, J., Yang, F.: GraphComm: efficient graph convolutional communication for multiagent cooperation. IEEE Internet Things J. **8**(22), 16359–16369 (2021)
36. Zhang, Y., Qian, Y., Yao, Y., Hu, H., Xu, Y.: Learning to cooperate: application of deep reinforcement learning for online AGV path finding. In: Proceedings of the 19th International Conference on Autonomous Agents and Multiagent Systems, pp. 2077–2079 (2020)
37. Han, W., Zhang, B., Wang, Q., Luo, J., Ran, W., Xu, Y.: A multi-agent based intelligent training system for unmanned surface vehicles. Appl. Sci. **9**(6), 1089 (2019)
38. Mahdavi, A., Carvalho, M.: Distributed coordination of autonomous guided vehicles in multi-agent systems with shared resources. In: SoutheastCon, pp. 1–7. IEEE (2019)
39. Al-Nuaimi, M., Wibowo, S., Qu, H., Aitken, J., Veres, S.: Hybrid verification technique for decision-making of self-driving vehicles. J. Sens. Actuator Netw. **10**(3), 42 (2021)
40. Bevly, D., et al.: Lane change and merge maneuvers for connected and automated vehicles: a survey. IEEE Trans. Intell. Veh. **1**(1), 105–120 (2016)
41. Dresner, K., Stone, P.: A multiagent approach to autonomous intersection management. J. Artif. Intell. Res. **31**, 591–656 (2008)
42. Gindullina, E., Mortag, S., Dudin, M., Badia, L.: Multi-agent navigation of a multi-storey parking garage via game theory. In: IEEE 22nd International Symposium on a World of Wireless, Mobile and Multimedia Networks (WoWMoM), pp. 280–285. IEEE (2021)
43. González, C.L., Zapotecatl, J.L., Gershenson, C., Alberola, J.M., Julian, V.: A robustness approach to the distributed management of traffic intersections. J. Ambient. Intell. Humaniz. Comput. **11**(11), 4501–4512 (2020)
44. Dresner, K., Stone, P.: multiagent traffic management: a reservation-based intersection control mechanism. In: Autonomous Agents and Multiagent Systems, International Joint Conference on, vol. 3, pp. 530–537. IEEE Computer Society (2004)
45. Khayatian, M. et al.: Cooperative driving of connected autonomous vehicles using responsibility-sensitive safety (RSS) rules. In: Proceedings of the ACM/IEEE 12th International Conference on Cyber-Physical Systems, pp. 11–20 (2021)
46. Makarem, L., Gillet, D.: Fluent coordination of autonomous vehicles at intersections. In: IEEE International Conference on Systems, Man, and Cybernetics (SMC), pp. 2557–2562. IEEE (2012)
47. Liu, S., Feng, Y., Wu, G.: Reservation-based network traffic management strategy for connected and automated vehicles: a multiagent system approach. In: IEEE International Intelligent Transportation Systems Conference (ITSC), pp. 2150–2155. IEEE (2021)
48. Mandiau, R., Champion, A., Auberlet, J.-M., Espié, S., Kolski, C.: Behaviour based on decision matrices for a coordination between agents in a urban traffic simulation. Appl. Intell. **28**(2), 121–138 (2008)
49. Mateo, R.M.A., Lee, Y., Lee, J.: Collision detection for ubiquitous parking management based on multi-agent system. In: Håkansson, A., Nguyen, N.T., Hartung, R.L., Howlett, R.J., Jain, L.C. (eds.) KES-AMSTA 2009. LNCS (LNAI), vol. 5559, pp. 570–578. Springer, Heidelberg (2009). https://doi.org/10.1007/978-3-642-01665-3_57
50. Rodriguez, M., et al.: A gradient-based approach for coordinating smart vehicles and traffic lights at intersections. IEEE Control Syst. Lett. **5**(6), 2144–2149 (2020)
51. Shao, Y., Rios-Torres, J.: Traffic prediction for merging coordination control in mixed traffic scenarios. In: Dynamic Systems and Control Conference, vol. 84287, p. V002T23A002. American Society of Mechanical Engineers (2020)
52. Sharon, G., Stern, R., Felner, A., Sturtevant, N.R.: Conflict-based search for optimal multi-agent pathfinding. Artif. Intell. **219**, 40–66 (2015)

53. Yu, C.Y., Wu, M.H., He, X.S.: Vehicle swarm motion coordination through independent local-reactive agents. In: Advanced Materials Research, vol. 108, pp. 619–624 (2010). Trans Tech Publ
54. Zong, F., He, Z., Zeng, M., Liu, Y.: Dynamic lane changing trajectory planning for CAV: a multi-agent model with path preplanning. Transportmetrica B: Transport Dynamics, pp. 1–27 (2021)

Proof-of-Concept (PoC) Biometric-Based Decentralized Digital Identifiers

Anatoliy Lut$^{(\boxtimes)}$ (iD)

Shanghai, China
`lut.anatoliy@gmail.com`

Abstract. This article describes decentralized digital identifiers based on proof-of-concept (PoC) biometrics and the application-level implementation of these identifiers for user authentication through blockchain with a smart contract. To establish PoC, user credentials are calculated from the biometric data of the user's own face, which is cryptographically encrypted (PKI-less infrastructure), stored, and decrypted when necessary at both on-chain and off-chain level. Data recording, storage, and access at the on-chain level are provided by smart-contract functions.

Keywords: Authentication · Blockchain · Biometric · Decentralized · Identity · PoC

1 Introduction

As we know, the blockchain madness began with publication, revealing to the world the potential use of cryptography as a purely peer-to-peer version of electronic cash, which would allow online payments to be sent directly from one party to another without going through a financial institution [1]. Today, this potential and the technologies described in the publication have been transformed into new projects (e.g., Ethereum, Hyperledger Fabric, Azure Blockchain as a service).

When we use an identity provided via third-party authentication protocols (for example, OAuth 2.0 protocol), we encounter certain problems related to the risk of losing the possibility to use the account; or the account may be blocked or lack of confidence in the new systems that the user is trying to log into: we do not want to trust them with sensitive information or literally sacrifice confidentiality.

These problems can be solved with using the blockchain-based authentication system described in this article, it is possible to store only a certain set of data without disclosing sensitive information in an unencrypted form, where only the owner will be able to use the digital identity, since it can be authenticated only based on his or her biometric data. This article aims to fill the gap in the security of and trust in existing user authentication systems by combining a smart contract, biometrics, and the PKI-less infrastructure encryption and decryption of user data with the immutability provided by a blockchain network.

A. Lut—Independent Researcher.

L. Braubach et al. (Eds.): IDC 2022, SCI 1089, pp. 125–130, 2023.
https://doi.org/10.1007/978-3-031-29104-3_14

Important. Experiment results are recorded as blockchain transactions in *.json file, and Python source codes are recorded as *.py files. User photos, Figures, IBV, CBV, VC shares, and VC-reconstructed results, including Table 1 as result of biometric feature matching are available here: https://github.com/zerotoolz/pba.

2   Concept Overview and Methods

This article comprises the collaboration of several algorithms, such as

- face detection, user-image extraction
- biometric feature data extraction
- visual cryptography
- hashing
- smart contract and blockchain for account registration and access to it
- biometric data matching

The purpose of combining these methods is to create a proof of concept (PoC), which makes it possible to implement decentralized digital identifiers based on biometric data in various projects, so as to propose a decentralized model where the user cannot be censored from large structures, to create a trustful and immutable credentials-authentication environment. At the same time, this approach does not require users to sacrifice their anonymity, which distinguishes it from existing authentication methods (generally, the OAuth 2.0 protocol), the basic work principle of which is the need to give access to user's personal information, for example, e-mail, age, or country.

To solve the biometrics and programming tasks, the Python3x and Solidity languages were chosen, as were the most common open source libraries, NumPy, Python Imaging Library (PIL), Argparse, Random, hashlib, scikit-learn, and OpenCV 3.4.2.

2.1   Face Detection, User-Image Extraction

OpenCV library provides 17 Haar Feature-based Cascade Classifiers by default; in the current paper, only haarcascade_frontalface_default.xml was used as the main technology for face detection. Based on the AdaBoost multiple facial detection algorithm, Haar classifiers are suitable for detecting regions such as the eyes, nose, or/and mouth of a hypothetical face. Firstly, it detects the user image from a camera, then identifies and encodes it into a gray image. Next, it loads the classifier to decide if picture contains a face or not. If so, picture continues and paints a square box on the appropriate part of image, after which it saves the result as another image. The usage-detection process is represented in Listing 1: Face detection, user-image extraction.

2.2   Biometric Feature Data Extraction and Cryptography

By default, OpenCV offers three of the most popular and affordable face-recognition algorithms: Eigen Face, Fisher Face and Local Binary Pattern Histograms (LBPH). The choice of LBPH was based on the function of this algorithm, which allows the comparison of the input face image to the image of a user who requires authentication. This is basically a 1 × 1 comparison; also, LBPH has succeeded in recognition at the expense of performance [3].

The main idea of the local binary pattern (LBP) operator is to accumulate the local pattern value in a processing image by comparing every image pixel with its neighbor pixel. The LBP operator uses the center-pixel value as the threshold value and compares it with the value of eight pixels nearby. If the nearby-pixel value is higher than or equal to the pixel at center, the value of the pixel location is indicated as 1 and as 0 if not. Finally, the LBP operator obtains the value of a binary number for each pixel in the image. For example, in the window of 3 * 3, the LBP operator takes the center-pixel value as the threshold value of the window, compares it with the value of eight gray neighbor pixels. If the neighbor-pixel value is equal to or bigger than the center-pixel value, the value of the neighbor-pixel position is set as 1, otherwise as 0, with a central pixel value and the intensity of the neighbor pixel.

After feature extraction Listing 2 the next step is to extract the histogram (Listing 3, which contains the biometric-data vector only, and was created on the basis of the digital image from the dataset. It is saved as *.npy array.

For the encryption (Listing 4) and decryption (Listing 5) of user images with PKI-less infrastructure, it is necessary to divide user image into two parts using visual cryptography (VC), which encodes an original image into several shares. Each share appears with unique noise (white noise). Here, a conventional share is a transparent copy, in which each pixel is transparent or not. According to the algorithm, we can combine the shares, then the secret image can be recognized without any high-level or high-cost computation.

An example of VC is shown in Fig. 2. It is important to note that VC works well with encryption, splitting image and decryption without PKI. In the VC example, black is 1, white is 0. By splitting an image into two halves and adding the white noise, the VC keeps biometric vectors anonymous [2]. Listing 4. Visual cryptography's encoding process. During the decryption process, VC applies an inverse transformation to two ciphered shares, called share1 and share2 to recover the original image with Listing 5. Visual cryptography's decoding process. To ensure that the 1/2 initial biometric vector (IBV) and the 2/2 IBV files are original and unchanged, a SHA3-256 deterministic hash function was implemented for the purposes of this paper (Listing 6).

2.3 Visual Cryptography Output-Image Smoothing and Biometric Feature Matching

Once the VC algorithm works (Listing 4 and Listing 5), there will be white noise in the decoded image that needs to be optimized to improve the recognition, extraction, comparison, and reduction of the threshold value (Listing 7). For optimization, the filter built into OpenCV uses the cv2.GaussianBlur function, with the result saved in Listing 5, Line 18. LBHP methods and the vector array of extracted biometric data, used in this paper in the user-registration section, need to be matched with the candidate's biometric-data vector for candidate authentication. These vectors can be compared with each other by using Euclidean distance. To complete the feature-matching process, one must use the threshold (Listing 8), which can be defined according to Table 1.

2.4 Blockchain Network and Smart Contract

To test the PoC, the Kovan testnet was chosen. Kovan is a new testnet for Ethereum and uses the Parity's Proof-of-Authority consensus engine, with benefits regarding the following: rapid deployment and components iteration, and no cpu-intensive usage. A smart contract here is special Solidity executable code, that contains all need data and functions. A smart contract allows the user to deploy any transaction without an intermediary third party. These transactions are easily traced and cannot be tampered with or reversed. The smart contract contains all details of the terms and executes them automatically [4]. In this case, the smart contract contains (Listing 9), but is not limited to, the following five functions:

- CreateAccount, create next credentials:

  - Login. Cannot be shorter than 2 bytes or longer than 32 bytes. Additionally, the user cannot register two identical logins.
  - IBV1CloudLink: link to 1/2 IBV at cloud storage.
  - IBV1CloudHash: 1/2 IBV SHA3-256 hash.
  - IBV2LocalHash: 2/2 IBV SHA3-256 hash.

- GetIBV1CloudLink: returns access link to 1/2 IBV at cloud storage.
- GetIBV1CloudHash: returns 1/2 IBV SHA3-256 hash.
- GetIBV2LocalHash: returns 2/2 IBV SHA3-256 hash.
- GetUserBlockchainAddress: returns user blockchain address.

Communication between the smart contract and Kovan testnet was established via Injected Web3 connection type.

3 Biometrically Based Decentralized Digital Identifiers

The proposed blockchain-based concept of a biometric identifier is a new type of user authentication. This method of user authentication is able to ensure that the specified user exists and that his or her data has not changed since registration. Additionally, this concept meets the following requirements:

- The user can trust the system he or she wants to log into, since no sensitive data in an unprotected form is transferred to a third party.
- The account exists in an independent, uncensored user-authentication system, where any account can exist independently without the risk of blocking.

To verify the correctness of the concept, an experiment was conducted regarding digital identifiers and the entire concept. Figure 4 shows how the proposed PoC works as an authentication infrastructure for a third-party service provider.

3.1 User Registration and Authentication

For user registration, it is necessary to create a digital identity (based on facial biometric data) with Listing 1. For face detection, haarcascade_frontalface_default is used (Line 2), then the algorithm converts the received face image into grayscale and saves it in a.png format (Lines 5–20). Next, the saved image must be split into two shares according to the VC algorithm (Listing 4). To guarantee the immutability of the shares, one must calculate their hash sum (Listing 6). The next step is to save the 2/2 IBV share (Listing 4, Line 37) in cloud storage, such as Google Drive etc., and to obtain a direct link to access the 2/2 IBV share. The 1/2 IBV share (Listing 4, Line 36) can be stored on the user's device (for example, on a SIM card). One can specify any login, but the following rule applies: the login cannot be shorter than 2 bytes or longer than 32 bytes. Of course, this rule can be changed, but the user needs to remember the login. The final step of user registration is to save the data (1/2 IBV hash, login, 2/2 IBV link, 2/2 IBV hash) to the blockchain using the smart contract (Listing 9 CreateAccount function, Lines 8–17).

3.2 User Authentication

First of all, it is necessary to obtain the candidate biometric vector (CBV) with Listing 1. To detect the user's face, the default frontal-face cascade classifier (Line 2) is used; next, the algorithm converts the received face image into grayscale and saves it in a.png format (Lines 5–20). Subsequently, LBPH (Listing 2) should be extracted from the saved image (Line 7–20), after which it is necessary to save the extracted 8-bit unsigned integer (uint8) array by using cv2.face.LBPHFaceRecognizer_create in.xml format, with the possibility of further face recognition (Lines 21–22). Next, the biometric vector (CBV; Listing 3) must be extracted from the saved result (Listing 2, Line 22), after which an authentication request containing the following data must be sent: login, CBV, and 1/2 IBV. After the authentication request has been received, it is necessary to check the presence of the specified login on the blockchain, using the smart contract (Listing 9, Lines 27–29). If there is a match (i.e., if the answer is $\neq$ 0x0000000000000000000000000000000000000000), one requests 1/2 IBV hash, 2/2 IBV link, 2/2 IBV hash (Listing 9, Lines 18–26). The next step is to request the 2/2 IBV share in the cloud storage with a direct link to access the 2/2 IBV share. After that, the hash sum of the 2/2 IBV share received from the smart contract and the hash sum of the 2/2 IBV share obtained from cloud storage need to be compared. The same process applies to the 1/2 IBV hash sum from the smart contract and the 1/2 IBV hash sum from the authentication request. Only three results are possible:

1. Both shares are not identical: cancel process.
2. Only one share is identical: cancel process.
3. Both shares are identical: proceed to next step.

The next step is to restore the IBV from two parts, 1/2 IBV and 2/2 IBV, using VC decryption (Listing 5). The obtained IBV result must be smoothed (Listing 6), then the biometric vector must be extracted (Listing 2 and Listing 3) and compared with the CBV measurement method (the Euclidean distances in Listing 8). The result of this operation is determined by answering the following question: "Is the user authenticated? Yes or No." In this case, the threshold is 1.514.

4 Limitations and Conclusions

The concept has not been implemented in accordance with accepted standards, the associated smart contract is not optimized for the economical use of Gas and not optimized for a Web3.js provider. Is no implemented storing IBV data in decentralized storage, for example InterPlanetary File System (IPFS) or in devices/systems (eSIM, SIM cards) is more convenient for everyday use.

In same time the PoC experiment confirms the possible implementation of biometric-based decentralized digital identifiers as an authentication basis for third-party systems that are trusted and immutable in mixed off-chain and on-chain implementation. Future research might focus on improving the algorithms used to recognize bio-metric data; adding Web3.js support; using other types of blockchain networks, using a SIM card as IBV storage.

References

1. Nakamoto, S.: Bitcoin: A Peer-to-Peer Electronic Cash System (2008)
2. IEEE Standard for Biometric Open Protocol. IEEE 2410 (2017)
3. Özdil, A., Özbilen, M.M.: A survey on comparison of face recognition algorithms. In: 2014 IEEE 8th International Conference on Application of Information and Communication Technologies (AICT) (2014)
4. Pham, H.L., Tran, T.H., Nakashima, Y.: A secure remote healthcare system for hospital using blockchain smart contract. In: 2018 IEEE Globecom Workshops (GC Workshops) (2018)

Text and Research

Correlation Between Researchers' Centrality and H-Index: A Case Study

V. Carchiolo$^{(\boxtimes)}$, M. Grassia, M. Malgeri, and G. Mangioni

Dip. Ingegneria Elettrica, Elettronica e Informatica,
Università degli Studi di Catania, Catania, Italy
vincenza.carchiolo@dieei.unict.it

Abstract. The quality of research' work is often very hard to assess, While controversial, the use of indices to measure the performance of the researchers and of institutions is widely accepted in scientific communities. In this paper, we build and study the co-authorship network of the MAT/05 "academic discipline" (as defined by Italian law), and investigate if the centrality measures from Network Science are enough to predict the h-index of a researcher, when fed to classic Machine Learning algorithms. Our results show that some models and combinations of features work better than others, and also that the partial network (e.g., without authors that do not belong to MAT/05) may not be enough to predict accurate h-index values.

1 Introduction

Finding a metric to compare scientific performance is an important problem since the '80s due to the strong impact in funding and/or earning of researchers in public and private institutions. Subramanyam [1] discussed the increasingly collaborative endeavor of Scientific research highlighting that the nature and the average number of collaborations is quite different from one discipline to another, and depend either upon specific characteristics of the discipline, such as the nature of the research problem, the research environment and external factors such as demographic factors and country's body of law. They concluded that *Bibliometric* methods offer a convenient and non-reactive tool for studying collaborations among researchers. One of the most famous, while controversial, bibliometric index is the Herfindahl-Hirschman index (h-index), introduced by Hirsch [2], which is largely used nowadays in many countries to rank, evaluate and compare researchers and academic institutions. In particular, Hirsch proposed the h-index as an answer to the following question: *"Among the rest of"* – researchers – *how does one quantify the cumulative impact and relevance of an individual's scientific research output?*.

The collaboration among researchers naturally produces common scientific publications, which can be easily represented as co-authorship networks. Such networks, of course, exhibit *topological* properties and can be analyzed using the Network Science tools. While one may argue that the study of collaboration through co-authorship analysis provides a partial representation, co-authorship

© The Author(s), under exclusive license to Springer Nature Switzerland AG 2023
L. Braubach et al. (Eds.): IDC 2022, SCI 1089, pp. 133–143, 2023.
https://doi.org/10.1007/978-3-031-29104-3_15

is still a good way to measure collaboration [3] among scientists. In fact, co-authorship analysis is important due to bibliometric importance in studying collaborative activities and research collaborations [4]. Jeong et Alii [5] affirmed that informal communication, cultural proximity, academic excellence, external fund inspiration, and technology development levels play significant roles in the determination of specific collaboration modes. In this context, node/link centrality is one of the most important metric that permits to asses the importance of a connection and/or a node helping to explain the performance of each single user. In the literature, there are several studies of specific groups of researchers aiming at understanding the reasons behind certain phenomena: for instance, Moschini et Alii [6] compare the distribution of Elsevier's Scopus subject areas of authors' documents, their bibliographical references and their citing documents using the h-index as a measure of interdisciplinarity, by analyzing 120 researchers from the Italian Institute of Technology and the National Institute for Nuclear Physics, while in [7,8] authors studied a preliminary dataset from three subsets of academics belonging to the same academic discipline (that authors suppose share topics and habits). Hu et Alii [9] studied the correlation between network centrality and research topics using papers published in the five top ranked journals from 1980 to 2016; they found that some of the most connected authors acts as hubs in a *small-world* network.

The dataset studied in this paper contains the information of the mathematics (MAT/05) subset of Italian Academic researchers of the Ministry of Education [10], enriched with data extracted querying Elsevier's Scopus database. Scopus is one of the largest database of scientific publication. A specialization, or scientific disciplinary sector, is a grouping of researchers imposed by law according to common topics used in Italy to hire researchers in public universities.

This paper, along the lines of co-authorship networks analysis, studies the links between bibliometrics performance indices, mainly h-index, and centrality measures. In particular, such analysis is meant to investigate if the network topology (i.e., the collaboration patterns) influences the scientific productivity, maybe even the quality. Specifically, we build three networks which are characterized by their "depth", i.e., the number of hops used to build the network itself, which translates to how many collaborators of collaborators include in the network, even if they do not belong to the same research field. The paper is organized as in the following: Sect. 2 introduces the MAT/05 dataset, in Sect. 3 we discuss the use of centrality measures and the setup of several datasets. Section 4, that is the core of contribution, presents and discusses the use of a machine learning approach to predict the h-index using centralities.

2   MAT/05 Data Source and Co-authorship Network

As discussed in the previous section, we aim at studying the correlation, if any, between some performance indices and the structural and topological properties of the researcher's network, focusing on centrality measures. In previous works [7,8] authors made a preliminary study but it was based on a small dataset that contains only a subset of Italian academic belonging to three specific *academic disciplines*. We crafted a semi-automatic procedure that creates a dataset of one academic discipline: the procedure starts from Italian academics that belong to a given academic discipline (the *seed*) and searches depth-first all coauthors, performing disambiguation, consistency check and cleaning. In this paper, we analyze the co-authorship network built from the MAT/05 (Mathematical analysis) academic discipline. The resulting data are organized into three datasets and has been extracted from Scopus Database according to Elsevier's access policies [11,12] in the period May, 2021 - March, 2022:

1. *depth-zero*: this is the dataset containing only Italian researchers from the MAT/05 academic discipline and their co-authorships. This implies that the collaborations with scientists not belonging to the MAT/05 disciplinary sector are not included. depth-zero = {vertices = 679, edges = 1990};
2. *depth-one*: starting from depth-zero we added all direct co-authors even if they do not belong to the MAT/05 group. Of course, depth-one includes depth-zero as a subset. depth-one = {vertices = 7891, edges = 13764};
3. *depth-two*: starting from depth-one we included also the co-authors of coauthors ('two' means that we stopped the search at the second hop), therefore depth-one is a subset of depth-two. depth-two = {vertices = 338475, edges = 1043750}.

The resulting dataset contains the weighted co-authorship networks annotated with all the metrics related to the authors, e.g. h-index, documents count, period of publications and several other information about each document, e.g., authors list, date of publication, type of publication, etc. The main reasons why we choose the Italian Academic researches in our study arev (i) studying the impact of external constraints on the scientific results. In this case the main constraint is the grouping into academic discipline by law, (ii) limiting the size of the dataset to facilitate the discovering of "hidden" effects embedded in the results. Moreover, we know the Italian academy, therefore we have a better knowledge of group's internal dynamics.

Note that in our experiments we use the Greatest Connected Component (GCC, also known as Largest Connected Component) of each network in order to filter out isolated nodes and smaller groups. We refer to the datasets and results using the name DBMat05<weighted>-<depth>, where: <weighted> is W if analysis takes into account weights and <depth> = $\{0, 1, 2\}$. For example, DBMat05-0 means unweighted depth-zero dataset whilst DBMat05W-1 means weighted depth-one dataset.

3 Centrality analysis

In this Section, we describe the details of centrality measures calculated on the co-authorship networks described in Sect. 2 and used to enrich networks. Since it is possible to weight edges with the number of common publications between any couple of authors, we analyzed both the weighted and unweighted version of the networks. As previously discussed, the first step of our analysis is to extract the GCC of the three instances of the dataset (depth-zero, depth-one and depth-two)) of the Italian researcher belonging to MAT/05. Table 1 shows the size of the GCC for each network. As expected, they are very different, both in the number of vertices and edges.

Table 1. Network Structure: Number of vertices

Network	$\#GCC$	$\#GCC \cap$ MAT/05
depth-zero	611	611
depth-one	7 712	644
depth-two	338 462	671

Once removed the isolated vertices, we calculated centrality measures [13] using graph-tool [14] package. Degree centrality (D), that refers to the total number of direct connections between node i and other $g-1$ nodes in a network with g nodes. The higher the degree centrality of a node is, the more nodes this node has contact with, and the more important this node is in the network. Local clustering coefficient (L), as defined in [15], is a measure of the degree to which nodes in a graph tend to cluster together. It gives an indication of the embeddedness of single nodes. Closeness centrality (C) [16] is a measure of centrality calculated as the reciprocal of the sum of the length of the shortest paths between the node and all other nodes in the graph, thus, it measures how close the i node is to all the other nodes in the network. Betweenness (B) [17] quantifies the number of times a node acts as a bridge along the shortest path between two other nodes. Eigenvector centrality (E) [18] is a measure of the influence of a node in a network. It assigns relative scores to all nodes in the network based on the concept that connections to high-scoring nodes contribute more to the score of the node in question than equal connections to low-scoring nodes. PageRank centrality (P) [19] is a generalization of Eigenvector centrality in which is used a scaling factor. After the calculation of the centrality measures, the selection of the nodes of the networks related to the set of researchers belonging to the MAT/05 set was carried out. Table 1 shows the number of 679 researchers, belonging to the set MAT/05, that also belong to the GCC. It was, therefore, decided to limit our analysis to the 611 researchers in common among the three networks in order to make the results comparable. Finally, the value of the h-index, as extracted from MAT/05 dataset, was added to each of the

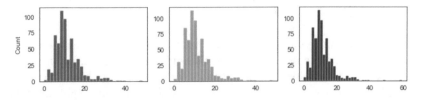

Fig. 1. h-index distribution in the depth-zero (red), depth-one (green), and depth-two (blue) networks.

remaining 611 nodes. Therefore all the analysis and predictions of the h-index discussed later in the Sect. 4 exclude the "lone wolves" researchers.

As result of our computation, we obtained six different centrality sets with 611 vertices where each node contains the h-index, ranging over $[0, 48]$, and the six centrality measures calculated for both weighted and unweighted networks for a total of 13 features.

We first report the summary statistics of the node centrality measures, and plot their distributions on the six different networks discussed above. Finally, we discuss the correlation among the features in each dataset, that is very useful to evaluate possible pre-processing useful to predict the target variable, h-index, as discussed in the next section. We show the distribution of all centrality measures in Fig. 2, where each sub-figure contains the plot of one of the centrality in the different datasets in order to highlight the different trends, and summarize the such using the minimum, maximum and average value in Tables $1, 2, 3, 4, 5, 6, 7, 8$ and 7.

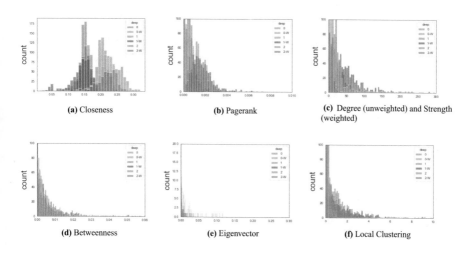

(a) Closeness   **(b)** Pagerank   **(c)** Degree (unweighted) and Strength (weighted)

(d) Betweenness   **(e)** Eigenvector   **(f)** Local Clustering

Fig. 2. Centrality measures distribution.

By analyzing Fig. 2 and Tables from $2, 3, 4, 5, 6$ and 7, we notice that the distributions of the closeness shift towards smaller values as the depth grows, which is somehow expected due to the normalization factor, but the same phenomenon also happens when comparing the unweighted and weighted networks at the same depth. On the contrary, the values of betweenness, degree and local clustering coefficient frequency decreases when compared to the weighted version. On the other hand, PageRank distribution is quite different. In fact, the smaller values of PageRank can be observed for the DBMat05-1 and DBMat05W-1 networks, while they are larger but with less frequency in the case of DBMat05W-2 e DBMat05W-1. Eigenvector centrality is characterized by very scattered values, which is particularly true in deeper networks, and also ranges over a larger interval with respect to the betweenness, PageRank, degree, and local clustering coefficient that are very close to the minimum one. Also, let us note that the average value of eigenvector centrality and of local clustering coefficient in the different datasets differs of several orders of magnitude. Finally, in Fig. 3 we display the correlations among the different node centrality metrics. We use the Pearson's Correlation Coefficient, that is, the measure of the linear relationship between two random variables - X and Y. In particular, the correlation is positive among all centrality indices (except for the eigenvector centrality) and h-index with different values in function of the depth of the network and the kind of centrality index.

Table 2. Closeness

Dataset	min	max	mean
DBMat05-0	9.54E-02	3.20E-01	2.32E-01
DBMat05-1	1,58E-01	2,69E-01	2,04E-01
DBMat05-2	8.12E-02	1.96E-01	1,52E-01
DBMat05W-0	1.31E-01	2.81E-01	2.18E-01
DBMat05W-1	2.07E-02	1.91E-01	1,38E-01
DBMat05W-2	8.12E-02	1.96E-01	1,52E-01

Table 3. Pagerank

Dataset	min	max	mean
DBMat05-0	4.11E-04	7.19E-03	1.63E-03
DBMat05-1	1.00E-06	9.80E-05	1.30E-05
DBMat05-2	2.90E-05	2,21E-02	8.20E-04
DBMat05W-0	4.30E-05	2.08E-02	7.82E-04
DBMat05W-1	2.86E-04	6.28E-03	1.64E-03
DBMat05W-2	1.00E-06	9.80E-05	1.37E-05

Table 4. Degree

Dataset	min	max	mean
DBMat05-0	1.0	3.70E+01	6.44E+00
DBMat05-1	1.0	4.29E+02	2.42E+01
DBMat05-2	1.0	1.48E+03	6.72E+01
DBMat05W-0	1.0	4.25E+02	2.41E+01
DBMat05W-1	1.0	1.21E+02	2.46E+01
DBMat05W-2	1.0	1.54E+03	6.73E+01

Table 5. Local CLustering

Dataset	min	max	mean
DBMat05-0	0.0	2.46E-01	2.24E-02
DBMat05-1	0.0	2.30E-02	6.20E-05
DBMat05-2	0.0	7.07E-01	1.28E-03
DBMat05W-0	0.0	6.87E-01	6.14E-03
DBMat05W-1	0.0	6.18E-01	4.05E-03
DBMat05W-2	0.0	1.33E-04	3.49E-07

Table 6. Eigenvector

Dataset	min	max	mean
DBMat05-0	1.0	3.70E+01	6.44E+00
DBMat05-1	1.0	4.29E+02	2.42E+01
DBMat05-2	1.0	1.48E+03	6.72E+01
DBMat05W-0	1.0	4.25E+02	2.41E+01
DBMat05W-1	1.0	1.21E+02	2.46E+01
DBMat05W-2	1.0	1.54E+03	6.73E+01

Table 7. Betweeness

Dataset	min	max	mean
DBMat05-0	0.0	9.10E-02	5.64E-03
DBMat05-1	0.0	3.76E-02	4.73E-04
DBMat05-2	0.0	1.48E-01	6.38E-03
DBMat05W-0	0.0	1.42E-01	5.53E-03
DBMat05W-1	0.0	8.64E-02	7.27E-03
DBMat05W-2	0.0	4.00E-02	5.35E-04

4   Predicting the h-index from the Centrality Metrics

In this Section, we present the analysis of the role of the network centrality to determinate the h-index using a classical machine learning approach. That is, we train various regressor models and see what centrality measures are most important for an accurate prediction. The scale of the centrality measures are very different from we normalized them as required by the machine learning algorithms to run efficiently. We should note that, among the bibliometric indices, we selected the h-index as target variable due to its importance and diffusion, and also because it is widely used in institutions to evaluate the work of a researcher.

We use various machine algorithms for this study. In particular, we employ: Support Vector Regression (SVR) models, a Kernel Ridge Regressor (KRR), and a Multi-Layer Perceptron (MLP). We refer the Reader to Table 8 for a summary of the parameters used. SVR algorithms use a kernel to perform dimensionality augmentation. That is, they map the (lower) dimensional input data into a high dimensional point. Moreover, they have the peculiarity that the training depends only on a subset of the training data, because the cost function ignores samples whose prediction is close to their target. Note that we use both linear and Radial Base Function (RBF) kernels with several regulation parameters. Regarding the Linear kernel, we use a linear Hyperplane, which is similar to a Linear regressor when the kernel is based on a Radial Base Function (RBF) that has a Gaussian distribution. KRR combines ridge regression with the so-called "kernel trick". In the case of a RBF kernel, it learns a RBF function in the space induced by the respective kernel and by the data. The function learned by KRR is similar to SVR, but is trained using a different loss function. Specifically, while the KRR uses the Mean Squared Error loss, SVR uses epsilon-insensitive loss, both combined with a regularization parameter named $L2$. Finally, we also use a Multi-Layer Perceptron (MLP) using a Limited-memory BFGS (LM-BFGS) optimization algorithm. It is a quasi-Newton method that approximates the Broyden-Fletcher-Goldfarb-Shanno (BFGS) algorithm using a limited amount of computer memory, which has a number of non-linear layers (called hidden layers) between the input and output ones. In this study, we use two hidden layers, which are enough to approximate many functions while keeping the model simple. Let us note that we find the best parameters of each model using a grid search.

The metrics used to evaluate the performance of a model are: Mean Absolute Error (MAE), Mean Squared Error (MSE), Mean Absolute Relative Error (MRE) and R^2. MAE is the expected value of the absolute error loss or -norm loss, MSE is the expected value of the squared (quadratic) error or loss, MRE is a relative measure that is computed cutting the null value in order to avoid non-significant values and, finally, R^2 is the proportion of the variation in the dependent variable that is predictable from the independent variable(s). It normally ranges over $[0, 1]$, and can be more (intuitively) informative than MAE and MSE in regression analysis evaluation, as the former can be expressed as a percentage, whereas the latter measures have arbitrary ranges.

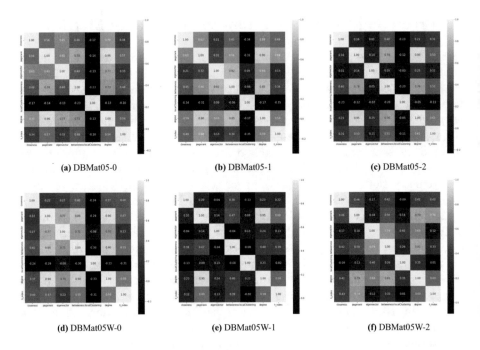

(a) DBMat05-0   **(b)** DBMat05-1   **(c)** DBMat05-2

(d) DBMat05W-0   **(e)** DBMat05W-1   **(f)** DBMat05W-2

Fig. 3. The correlaction among the node centrality measures and the authors' h-index.

Table 8. Hyper-parameters of the regression model.

Models	Hyper-parameters
SVR linear kernel	$C : [1, 10^1, 10^2, 10^3]$,
	$\gamma :$ np.logspace$(-2, 2, 10)$
SVR BRF kernel	$C : [1, 10^1, 10^2, 10^3]$,
	$\gamma :$ np.logspace$(-2, 2, 10)$
Kernel Ridge BRF	$\alpha : [1, 10^{-1}, 10^{-2}, 10^{-3}]$,
	$\gamma :$ np.logspace$(-2, 2, 10)$
MLP Regressor	$\alpha : [10^2, 10^1, 1, 10^{-1}, 10-2]$,
	$solver : ['lbfgs']$,
	hidden_layer_sizes $: [(x, y) : x \in (12, 10, 8) \ and \ y \in (6, 4, 2)$

In order to evaluate the best feature combination, we compare the regression algorithms with all combination of the six centrality measures. Therefore, we have used an exhaustive search (grid) with 63 different combinations of features. In the following, we refer to combination with C_i where i ranges from 0 to 62. C_1 to C_5 are the combination with only one feature, whereas, C_{62} is the combination containing all features.

Table 9. Bst results of regression by through classic ML approaches

DataSets	MAE	MRE	MSE	R^2
DBMat05-0	**3.4217**	**0.3556**	22.0360	**0.3283**
DBMat05-1	**2.8222**	**0.2710**	15.1379	0.5312
DBMat05-2	2.7647	**0.2933**	13.5948	0.5327
DBMat05W-0	3.4878	0.3294	21.3972	0.3537
DBMat05W-1	2.4077	**0.2258**	**9.8447**	**0.6893**
DBMat05W-2	**2.4040**	**0.2218**	**10.0470**	0.6579

Table 10. Best results of regression through MLP

DataSets	MAE	MRE	MSE	R^2
DBMat05-0	3.5224	0.3955	**21.4990**	0.3074
DBMat05-1	2.8308	0.2914	**14.1052**	**0.5549**
DBMat05-2	**2.6998**	**0.2936**	**12.8042**	**0.5959**
DBMat05W-0	**3.4262**	0.3682	**20.2073**	**0.3623**
DBMat05W-1	**2.3998**	0.2379	12.8042	0.6793
DBMat05W-2	2.4136	0.2342	10.1213	**0.6848**

The Tables 9 and 10 highlight the best results (obtained using classic ML approaches or the MLP regressor) in terms of each of the four indices. Table 11 shows the combination of features used to achieve the best score in terms of R^2 related to ML and MLP. Let us note that MLP performs better in 4 out of 6 cases and typically uses a lower number of features to get the best result.

The Fig. 4 shows the best value of MAE, MSE, MRE an R^2 (y-axis) obtained with every feature combination (x-axis) for each dataset (using different color for each dataset). It is possible to appreciate that, usually, the best result are obtained for DBMat05W-1 and DBMat05W-2 using only three features as it is also visible in numerical form in Table 11.

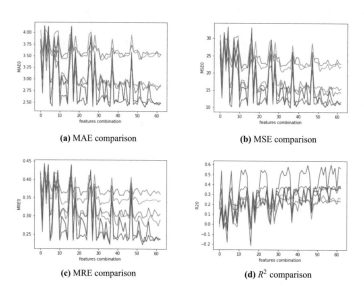

(a) MAE comparison

(b) MSE comparison

(c) MRE comparison

(d) R^2 comparison

Fig. 4. Comparison between performance indices and combination.

Table 11. Features used by the best model. Capital letters represent the centrality used as features. Bold highlights the best result.

DataSets	ML Model	R^2	Features	MLP Hidden Layer	R^2	Features
DBMat05-0	KRR	**0.3283**	$\mathscr{P} - \mathscr{E}$	s=(8, 2)	0.3074	$\mathscr{P} - \mathscr{E} - \mathscr{B}$
DBMat05W-0	KRR	0.3537	$\mathscr{P} - \mathscr{E} - \mathscr{L}$	s=(12, 6)	**0.3623**	$\mathscr{P} - \mathscr{E} - \mathscr{L}$
DBMat05-1	SVR	0.5312	$\mathscr{C} - \mathscr{E} - \mathscr{D}$	s =(12, 6)	**0.5549**	$\mathscr{B} - \mathscr{D}$
DBMat05W-1	KRR	**0.6893**	$\mathscr{P} - \mathscr{L} - \mathscr{D}$	s=(10, 4)	0.6793	$\mathscr{P} - \mathscr{D} - \mathscr{L}$
DBMat05-2	KRR	0.5327	$\mathscr{C} - \mathscr{E} - \mathscr{L} - \mathscr{D}$	s = (8, 4)	**0.5959**	$\mathscr{C} - \mathscr{E} - \mathscr{L}$
DBMat05W-2	SVR	0.6579	$\mathscr{C} - \mathscr{E} - \mathscr{D}$	s =(10, 4)	**0.6848**	$\mathscr{C} - \mathscr{D}$

5   Conclusions

We analyze the co-authorship network of the Italian researchers belonging to the MAT/05 *academic discipline*, as defined by the Italian law, and predict the h-index of each researcher using as features the centrality metrics. In particular, we build six networks: the unweighted and weighted (i.e., with the number of co-authored papers as edge weight) networks that include only the researchers that belong to such academic discipline, all their co-authors (1-hop), and also the co-authors of their co-authors (2-hops). As shown in Sect. 4, ML models are able to predict the h-index correctly often using only 3 features, MLP performs better with respect to the R^2 score, which is more informative about the actual performance than the other metrics, and also needs less input features for each node (author). Moreover, ML approaches often achieve best result with the PageRank and Eigenvector centrality, the best set of features to feed to the MLP is different in each of the networks. Future works may involve studying other academic disciplines and how as the difference shape of coauthirship networks [20] impacts on h-index prediction. Moreover, in order to improve the predictions further leveraging the full network topology through Geometric Deep Learning approaches [21].

References

1. Subramanyam, K.: Bibliometric studies of research collaboration: a review. J. Inf. Sci. **6**(1), 33–38 (1983)
2. Hirsch, J.E.: An index to quantify an individual's scientific research output. Proc. Natl. Acad. Sci. **102**(46), 16569–16572 (2005)
3. Lundberg, J., Tomson, G., Lundkvist, I., Skr, J., Brommels, M.: Collaboration uncovered: exploring the adequacy of measuring university-industry collaboration through co-authorship and funding. Scientometrics **69**(3), 575–589 (2006)
4. Acedo, F.J., Barroso, C., Casanueva, C., Galán, J.L.: Co-authorship in management and organizational studies: an empirical and network analysis*. J. Manag. Stud. **43**(5), 957–983 (2006)
5. Jeong, S., Choi, J.Y., Kim, J.: The determinants of research collaboration modes: exploring the effects of research and researcher characteristics on co-authorship. Scientometrics **89**, 967–983 (2011)

6. Moschini, U., Fenialdi, E., Daraio, C., Ruocco, G., Molinari, E.: A comparison of three multidisciplinarity indices based on the diversity of scopus subject areas of authors' documents, their bibliography and their citing papers. Scientometrics **125**(2), 1145–1158 (2020)

7. Carchiolo, V., Grassia, M., Malgeri, M., Mangioni, G.: Preliminary characterization of Italian academic scholars by their bibliometrics. In: Camacho, D., Rosaci, D., Sarne, G.M.L., Versaci, M. (eds.) Intelligent Distributed Computing XIV. IDC 2021. Studies in Computational Intelligence, vol. 1026. Springer, Cham (2022). https://doi.org/10.1007/978-3-030-96627-0_31

8. Carchiolo, V., Grassia, M., Malgeri, M., Mangioni, G.: Analysis of the co-authorship sub-networks of Italian academic researchers. Stud. Comput. Intell. **1015**, 321–327 (2022)

9. Hu, X., Li, O.Z., Pei, S.: Of stars and galaxies - co-authorship network and research. China J. Account. Res. **13**(1), 1–30 (2020)

10. Ministero dell'Università e della Ricerca : Professori e Ricercatori. Accessed May 2021

11. Elsevier B.V.: Elsevier Developer - API Service Agreement. https://dev.elsevier.com/academic_research_scopus.html. Accessed July 2021

12. Elsevier B.V.: Elsevier Developer - Academic Research. https://dev.elsevier.com/api_service_agreement.html. Accessed July 2021

13. Das, K., Samanta, S., Pal, M.: Study on centrality measures in social networks: a survey. Soc. Netw. Anal. Min. **8**(1), 1–11 (2018). https://doi.org/10.1007/s13278-018-0493-2

14. Peixoto, T.P.: Graph-tool - efficient network analysis. Accessed April 2022

15. Watts, D.J., Strogatz, S.H.: Collective dynamics of small-world networks. Nature **393**(6684), 440–442 (1998)

16. Opsahl, T., Agneessens, F., Skvoretz, J.: Node centrality in weighted networks: generalizing degree and shortest paths. Soc. Netw. **32**(3), 245–251 (2010)

17. Brandes, U.: A faster algorithm for betweenness centrality. J. Math. Soc. **25**(2), 163–177 (2001)

18. Langville, A.N., Meyer, C.D.: A survey of eigenvector methods for web information retrieval. SIAM Rev. **47**(1), 135–161 (2005)

19. Page, L., Brin, S., Motwani, R., Winograd, T.: The PageRank citation ranking: bringing order to the web. In: WWW 1999 (1999)

20. Carchiolo, V., Grassia, M., Malgeri, M., Mangioni, G.: Co-authorship networks analysis to discover collaboration patterns among Italian researchers. Future Internet **14**(6) (2022)

21. Carchiolo, V., Cavallo, C., Grassia, M., Malgeri, M., Mangioni, G.: Link prediction in time varying social networks. Information **13**(3) (2022)

Towards an Online Multilingual Tool for Automated Conceptual Database Design

Drazen Brdjanin[1]($\boxtimes$), Mladen Grumic[2], Goran Banjac[1], Milan Miscevic[3], Igor Dujlovic[1], Aleksandar Kelec[1], Nikola Obradovic[1], Danijela Banjac[1], Dragana Volas[4], and Slavko Maric[1]

[1] Faculty of Electrical Engineering, University of Banja Luka, Patre 5, 78000 Banja Luka, Bosnia and Herzegovina
{drazen.brdjanin,goran.banjac,igor.dujlovic,aleksandar.kelec,
nikola.obradovic,danijela.banjac,slavko.maric}@etf.unibl.org
[2] Syrmia, Kralja Petra I Karadjordjevica 14,
78000 Banja Luka, Bosnia and Herzegovina
mladen.grumic@syrmia.com
[3] AUTO1 Group, Bergmannstraße 72, 10961 Berlin, Germany
[4] codecentric, Mese Selimovica 57, 78000 Banja Luka, Bosnia and Herzegovina

Abstract. The paper presents an early prototype of an online multilingual tool named TexToData, which is aimed at automatic conceptual database design. TexToData is the first online web-based tool enabling automatic conversion of a natural language text into the target conceptual database model represented by a UML class diagram. It implements the entire process through the orchestration of web services, whereby some core functionalities are carried out by external services. The tool usage is illustrated with examples of automatic generation of a conceptual database model from the text represented in different natural languages.

1 Introduction

Data modeling is an important part of information system design, and data models constitute the essence of any information system. The process of data modeling is not straightforward, and typically requires many iterations before the final model is obtained. Therefore, automatic generation of data models is very appealing and has been the subject of research for many years.

Since Chen's eleven rules [4] for the translation of requirements specified in a *natural language* (NL) into an E-R diagram, a lot of research [15] has been done in the field of *natural language processing* (NLP) to extract knowledge from requirements specifications and automate conceptual database design. However, there is still no tool able to automatically convert NL text into the corresponding *conceptual database model* (CDM).

L. Braubach et al. (Eds.): IDC 2022, SCI 1089, pp. 144–153, 2023.
https://doi.org/10.1007/978-3-031-29104-3_16

The existing NLP-based tools typically support one single source NL (mainly English, Spanish or German) and do not provide multilingual support. Finally, there is also no online NLP-based solution enabling the automated CDM design. To fill this research gap, we started a project aimed at developing an online tool for automated CDM design with multilingual support. In this paper, we present an early prototype of the corresponding tool named TexToData, which provides the required functionality including the following characteristics: online web-based tool, multilingual support, UML[1]-based representation (class diagram) of the generated CDM, automatic layouting of the generated diagram including editing and formatting functionalities, and XMI[2]-based export in order to enable model portability.

The paper is structured as follows. After this introduction, the second section shortly presents the related work. The third section presents the TexToData tool, while the fourth section provides some illustrative examples of its usage. The final section concludes the paper.

2 Related Work

According to [17], the existing (semi)automated approaches to CDM design can be classified as: (1) linguistics-based, (2) pattern-based, (3) case-based, (4) ontology-based, and (5) multiple approaches.

Our tool, as well as the majority of the existing tools, belongs to the linguistics-based category. These approaches use NLP techniques to convert NL text into the CDM. The development of these approaches started with Chen's rules, which have been further enhanced and extended (e.g. [10,12,13]). To overcome some ambiguities inherent to NL text, some studies tried to put constraints regarding the supported vocabularies and sentence structure (e.g. [1]), or to use formal languages (e.g. Z). In some studies (e.g. [8]), the linguistics-based approach is supported by a linguistic dictionary (e.g. WordNet++ [6]), which delivers semantic links between concepts (synonyms, antonyms, etc.), as well as syntactical and morphological information.

The NLP toolkits are typically employed to carry out common NLP tasks (tokenisation, PoS tagging, etc.), whereby general-purpose toolkits (such as NLTK [2] and Stanford CoreNLP [11]) are usually used. There are also several online services (such as TextRazor[3] and Bitext[4]) providing necessary NLP functionalities. In our approach we employ the TextRazor service.

---

[1] Unified Modeling Language.
[2] XML Metadata Interchange.
[3] http://www.textrazor.com.
[4] http://www.bitext.com.

The most important tools applying the linguistic approach are: LIDA [13], COLOR-X [8], CM-Builder [9], ER-Converter [12], and ACDM [7]. The main representatives of other categories are: pattern-based APSARA [14], case-based CABSYDD [5], ontology-based OMDDE [16], and HBT [17] belonging to the multiple category.

Currently, NLs are commonly used for requirements specifications and most approaches to (semi)automated CDM design are based on NLP. However, their effectiveness and limitations are usually deeply related to the source NL since they depend on the grammar complexity and the lexicon scope. This is the main reason why NLP-based approaches are modestly used for NLs with complex morphology and why some non-NLP-based alternatives have been proposed, such as approaches taking *models* (i.e. graphically represented require-ments) as the basis for automated CDM design. Among a number of model-driven proposals, there is only one single online tool named AMADEOS[5] [3], which enables automatic derivation of an initial CDM from a set of business process models. Apart from the functionalities publicly provided to the end-user, AMADEOS also exposes services for automatic model-driven CDM derivation. In our approach, we employ some AMADEOS services that are necessary to complete the whole CDM design process in the TexToData tool, such as diagram layouting and model export.

3   TexToData Tool

This section presents the implemented TexToData[6] tool. TexToData is the first online web-based tool aimed at automatic conversion of NL text into the CDM represented by the UML class diagram. It implements the entire process of CDM synthesis through orchestration of internal and external web services. The external services are employed to carry out the core functionalities, such as text translation from the source NL into English and vice-versa (in order to enable multilingual support), and NLP of English text. Figure 1 shows the architecture of the TexToData tool.

The *client web application* allows users to upload a source text. When the entire synthesis process is finished, the client application receives the JSON[7] response and visualizes[8] the class diagram in the browser. The visualized diagram is editable so users can additionally improve it. It is also possible to export the model in the XMI format.

The *server-side* is implemented as a set of web services. The *Orchestrator* service orchestrates the whole process, whereby each activity is implemented by the corresponding service. In a positive usage scenario (Fig. 2), the orchestrator receives a source text, and returns the automatically generated CDM. The source text is firstly sent to the *TextTranslator* service which detects the

---

[5] http://m-lab.etf.unibl.org:8080/amadeos.

[6] http://m-lab.etf.unibl.org:8080/Textodata.

[7] JavaScript Object Notation.

[8] The implementation is based on jsUML2 (http://www.jrromero.net/tools/jsUML2).

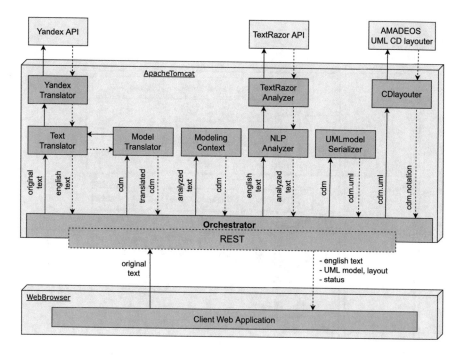

Fig. 1. TextToData architecture

source NL and, in case the source NL is not English (Fig. 2(a)), forwards the text to the external translation service through the corresponding adapter – currently we employ the *Yandex* service via the *YandexTranslator* adapter. The orchestrator further sends the English text to the *NLPAnalyzer* service responsible for NLP, which employs the external NLP service via the corresponding adapter – currently we employ the *TextRazor* service via the *TextRazorAnalyzer* adapter (Fig. 2(b)). After NLP is finished, the analyzed text is sent to the *ModelingContext* service which generates an internal representation of the CDM (Fig. 2(c)). If the source NL is not English (Fig. 2(d)), then the CDM is sent to the *ModelTranslator* service, which further employs the *TextTranslator* service to translate each model element back into the source NL. The CDM is further sent to the *UMLmodelSerializer* service which serializes the generated class diagram in the XMI format (Fig. 2(e)). After the serialization, the model is sent to the *CDlayouter* service, which employs the corresponding *AMADEOS layouting service* and returns a layout of the class diagram. Finally, the model and the diagram are merged into a single JSON object, and returned to the client.

The core of the tool is the *ModelingContext* service, which is responsible for the CDM generation. It takes English text (already analyzed by the NLP service) split into sentences as input, where each input word is tagged with the corresponding part of speech and dependency to other words. Based on the implemented strategy for the model generation, it produces a CDM containing a

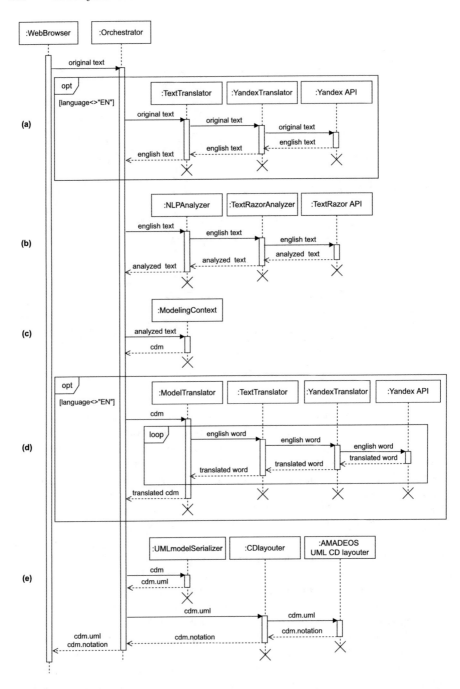

Fig. 2. Positive usage scenario of TexToData: (a) translation of text from source language to English, (b) NLP analysis of English text, (c) generation of CDM in English, (d) translation of CDM from English back to source language, and (e) generation of corresponding UML model and diagram

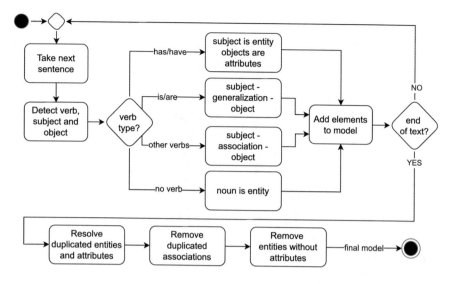

Fig. 3. From English text to CDM

set of classes with attributes, and a set of their relationships (associations and generalizations). Currently, we apply a simple strategy for the CDM generation, with rules split into two logical phases, as shown in Fig. 3. In the first phase, sentences are analyzed separately, adding detected classes, attributes[9] and relationships to the initial model. The second phase is responsible for cleaning up the initial model, where duplicate classes, relationships, as well as classes without attributes are removed, producing the final model as output.

4 Illustrative Examples

This section illustrates the usage of the TexToData tool. Firstly, we provide a screenshot of the tool in action (Fig. 4), then we provide two examples of CDMs automatically derived from textual specifications in different NLs.

The first example (Fig. 5) provides a simple textual specification in English, and the corresponding CDM automatically generated by the TexToData tool (after some minor layout improvements).

In order to illustrate the multilingual support, we provide another example (Fig. 6), where we use a textual specification in Italian that is equivalent[10] to the sample English text used in the previous example.

---

[9] Currently, only attribute names are generated, while other properties (such as data type, multiplicity, etc.) will be a part of our future work.

[10] This textual specification represent the result of a direct translation of the sample source English text by the Google Translate service (https://translate.google.com). It is possible that CDMs of different structure will be generated for the same source text represented in some other NL, due to some translation inconsistencies.

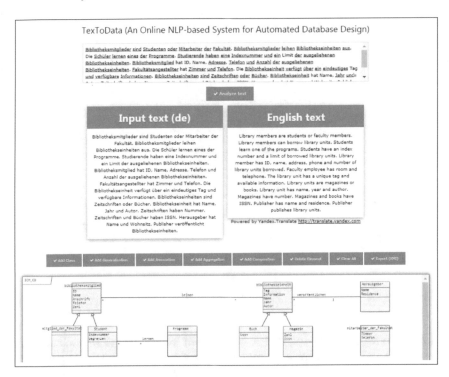

Fig. 4. Screenshot of TexToData in action

Although the presented examples are quite simple, they are very illustrative and show that the proposed approach and implemented tool enable automatic CDM derivation from textual specifications represented in different NLs, even from NLs with very complex morphology (such as Slavic and some other languages). This further implies that is not necessary to develop and employ different NLP services for different NLs, but only one single NLP service for the English language is enough.

Since we currently apply a very simple strategy for automatic CDM derivation from the English text, we are not able to provide some valuable evaluation results related to the effectiveness of the implemented tool, through the assessment of the correctness and completeness of the automatically generated models from textual specifications in different NLs, given the different translation and NLP services, as well as different source languages. A more extensive evaluation of the approach and the implemented tool will be a part of our future work.

Library members are students or faculty employees. Library members borrow library units. Students study one of the programs. Students have index number and limit of library units borrowed. Library member has id, name, address, telephone and number of library units borrowed. Faculty employee has room and phone. Library unit has unique tag and available information. Library units are magazines or books. Library unit has name, year and author. Magazines have number. Magazines and books have ISSN. Publisher has name and residence. Publisher publishes library units.

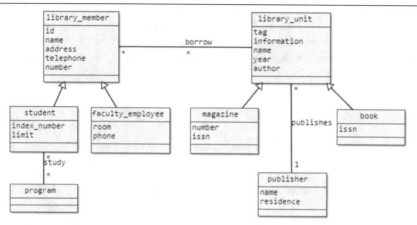

Fig. 5. Sample source text in English (top) and corresponding CDM (bottom)

I membri della biblioteca sono studenti o dipendenti di facoltà. I membri della biblioteca prendono in prestito unità della biblioteca. Gli studenti studiano uno dei programmi. Gli studenti hanno il numero di indice e il limite delle unità di biblioteca prese in prestito. Il membro della biblioteca ha id, nome, indirizzo, telefono e numero di unità della biblioteca prese in prestito. L'impiegato della facoltà ha stanza e telefono. L'unità della libreria ha un tag univoco e le informazioni disponibili. Le unità della biblioteca sono riviste o libri. L'unità della biblioteca ha nome, anno e autore. Le riviste hanno il numero. Riviste e libri hanno ISSN. L'editore ha nome e residenza. L'editore pubblica le unità di libreria.

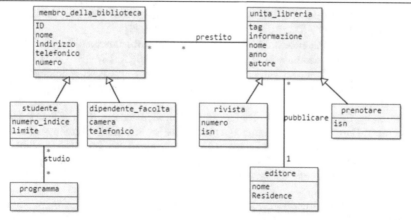

Fig. 6. Textual specification in Italian (top) and corresponding CDM (bottom)

5   Conclusion

In this paper we presented TexToData – the first online web-based tool with multilingual support, which enables automatic conversion of NL text into the CDM represented by the UML class diagram. TexToData generates the CDM through an orchestration of web services, whereby some functionalities are performed by external services that are publicly available and free of charge.

The presented examples of automatically generated models based on the textual specifications represented in different NLs imply that the implemented tool has a potential for further development and usage both in academia and industry.

References

1. Ambriola, V., Gervasi, V.: On the systematic analysis of natural language requirements with CIRCE. Autom. Softw. Eng. **13**(1), 107–168 (2006)
2. Bird, S., Loper, E.: NLTK: the natural language toolkit. In: Proceedings of the ACL Interactive Poster and Demonstration Sessions, pp. 214–217. ACL (2004)
3. Brdjanin, D., Vukotic, A., Banjac, D., Banjac, G., Maric, S.: Automatic derivation of the initial conceptual database model from a set of business process models. Comput. Sci. Inf. Syst. **19**(1), 455–493 (2022)
4. Chen, P.: English sentence structure and entity-relationship diagrams. Inf. Sci. **29**(2–3), 127–149 (1983)
5. Choobineh, J., Lo, A.W.: CABSYDD: case-based system for database design. J. Manag. Inf. Syst. **21**(3), 281–314 (2004)
6. Dehne, F., Steuten, A., van de Riet, R.: WordNet++: a lexicon for the Color-X-method. Data Knowl. Eng. **38**(1), 3–29 (2001)
7. Du, S., Metzler, D.P.: An automated multi-component approach to extracting entity relationships from database requirement specification documents. In: Kop, C., Fliedl, G., Mayr, H.C., Métais, E. (eds.) NLDB 2006. LNCS, vol. 3999, pp. 1–11. Springer, Heidelberg (2006). https://doi.org/10.1007/11765448_1
8. Fellbaum, C., Miller, G.: Color-X: Using Knowledge From WordNet for Conceptual Modeling, pp. 353–377. MIT Press, Cambridge (1998)
9. Harmain, H., Gaizauskas, R.: CM-Builder: a natural language-based CASE tool for object-oriented analysis. Autom. Softw. Eng. **10**(2), 157–181 (2003)
10. Hartmann, S., Link, S.: English sentence structures and EER modeling. In: Proceedings of the 4th Asia-Pacific Conference on Conceptual Modelling, vol. 67, pp. 27–35 (2007)
11. Manning, C., Surdeanu, M., Bauer, J., Finkel, J., Bethard, S., McClosky, D.: The stanford CoreNLP natural language processing toolkit. In: Proceedings of the 52nd Annual Meeting of the ACL: System Demonstrations, pp. 55–60. ACL (2014)
12. Omar, N., Hanna, P., McKevitt, P.: Heuristics-based entity-relationship modelling through natural language processing. In: Proceedings of the AICS 2004, pp. 302–313 (2004)
13. Overmyer, S.P., Benoit, L., Owen, R.: Conceptual modeling through linguistic analysis using LIDA. In: Proceedings of the ICSE 2001, pp. 401–410. IEEE (2001)
14. Purao, S.: APSARA: a tool to automate system design via intelligent pattern retrieval and synthesis. SIGMIS Database **29**(4), 45–57 (1998)

15. Song, I.-Y., Zhu, Y., Ceong, H., Thonggoom, O.: Methodologies for semi-automated conceptual data modeling from requirements. In: Johannesson, P., Lee, M.L., Liddle, S.W., Opdahl, A.L., López, Ó.P. (eds.) ER 2015. LNCS, vol. 9381, pp. 18–31. Springer, Cham (2015). https://doi.org/10.1007/978-3-319-25264-3_2
16. Sugumaran, V., Storey, V.C.: Ontologies for conceptual modeling: their creation, use, and management. Data Knowl. Eng. **42**(3), 251–271 (2002)
17. Thonggoom, O.: Semi-automatic conceptual data modelling using entity and relationship instance repositories. PhD Thesis, Drexel University (2011)

ProExamit: Development of Cross-Platform, Fully Automated and Expandable in Terms of Detectable Violations Proctoring

Kirill Krinkin, Valeriia Dopira[✉], Mark Zaslavskiy, Maxim Subbotin, Nikita Iyerussalimov, Sergey Sorokumov, Yaroslav Gosudarkin, Sergey Glazunov, and Dmitry Ivanov

Saint Petersburg Electrotechnical University "LETI", Saint Petersburg, Russia
kirill@krinkin.com

Abstract. The work describes the development of a novel proctoring platform, ProExamit, aimed to be an automatic solution for the problem of exam during online learning. The comparison of popular commercial proctoring services showed that existing solutions do not provide fully automated cross-platform service with the ability to be extended with new detected violation types. In order to create a new solution, an architecture of extendable and cross-platform proctoring solution was implemented in the ProExamit web service which is a fully automated proctoring system for online exams. Video processing and violation detection are implemented as separate software modules for camera and screencast. There are three types of critical violations on webcam: student not detected, student not looking at the monitor, and an unknown user detected. The advantage of the developed proctoring is the algorithm for detecting websites and applications opened by the student on the video. The proctoring ProExamit described in this paper is used at the Department of Software Engineering and Computer Applications at Saint Petersburg Electrotechnical University "LETI" by 13 teachers and more than 350 students for a year.

1 Introduction

In today's world, online education grows, influenced, of course, by the COVID-19 pandemic. However, the demand for education has increased the teacher shortages [1–3]. Organizing online learning is complicated in general but also because of the many ways that students can use to cheat. In order to address this problem, educational organizations use conventional solutions - proctoring systems. Proctoring could help to automatize the process of exam.

The goal is to develop a fully automated proctoring system for exams that is also cross-platform and expandable in terms of detectable violations.

The rest of the paper is organized as follows. Section 2 compares the existing proctoring platforms. Section 3 describes the proctoring architecture, possible

L. Braubach et al. (Eds.): IDC 2022, SCI 1089, pp. 154–159, 2023.
https://doi.org/10.1007/978-3-031-29104-3_17

violation types in ProExamit. Section 4 contains conclusions and directions for further research respectively.

2   State of the Art

In order to review existing platforms according to the stated goal (e.g. full automation) the desired use case should be described. The use case implies two roles for users - teachers and students. Usage of a proctoring platform contains authentication, starting and stopping proctoring session. Students interact with a proctoring platform using their web browser on a computer equipped with web camera and microphone. A Proctoring platform performs recording of students' actions using screencast, web camera and microphone data. After a session is completed (finished and processed by a platform) teachers can access session's raw data and violations report. All violations would be found by processing the video with machine learning. A user interface is to be provided to teachers to view the violations found.

In order to determine existing trends and general problems in the domain area, a search of proctoring platforms corresponding to the use case above was conducted. The comparison is looked at the most popular proctoring platforms. The comparison was made according to the following criteria:

1, 2, 3. Screen, web camera and audio recording: These criteria represent the data received from the student for further processing to find violations.

4. Student identification by photo Comparing the student in webcam and in the picture that was uploaded [4].

5. Absence of a face in the video Absence of the student's face in the camera image. Considered undesirable during the exam, as there is no way to monitor what the student is doing at this time.

6. Checking for unauthorized persons on web camera Checking for extraneous faces. All faces that do not pass the comparison with the photo are marked as extraneous.

7. Program context detection: By analyzing video of student's recorded screen will be determined if student is using the computer to dishonestly take the exam. Two approaches are used: (1) detection of unwanted applications based on screencast of students (Proctortrack, ProctorExam, Examus), (2) detection of applications using data directly from the OS of student's computer (ProctorEdu). The advantage of the first approach is that does not need access to the student's OS, which means that it is not necessary to install software on PC. But because of using ML algorithms accuracy might be reduced.

8. Gaze direction recognition: By analyzing video from webcam the task of determining angles of gaze direction and angles of the student's head rotation is solved.

9. Audio track recognition from a microphone: On exam, student has to be alone in the room.

10. Data processing machine: This criterion affects whether or not the student needs to install additional software.

11. Session grade generation time: There are two approaches: post-processing and real-time. The advantage of the first approach is proctoring consumes computing and network resources more balanced, while if the data is processed in real-time, the peak consumption of the computation increases server capacity, which limits the system bandwidth.

The comparison of analyzed proctoring services is listed in Table 1.

Table 1. Comparison of proctoring services

Criteria	Proctortrack	ProctorExam	Examus	ProctorEdu	Proctoru	Proctorio
1. Screen recording	no	no	yes	no	no	no
2. Web camera recording	yes	yes	yes	yes	yes	yes
3. Audio recording	yes	yes	yes	no	yes	no
4. Student identification	yes	yes	yes	yes	yes	yes
5. Absence of face	yes	yes	yes	yes	yes	yes
6. Unauthorized persons checking	yes	no	yes	yes	no	yes
7. Program context detection	yes	no	yes	no	no	no
8. Gaze direction	no	no	yes	yes	no	yes
9. Audio recognition	no	no	no	no	no	no
10. Processing machine	student	student	student	remote	student	student
11. Session grade	real-time	real-time	both	after exam	real-time	real-time

Among the reviewed analogs in Table 1, Examus and ProctorEdu have the most features implemented. Most of the analogs require the installation of external software on the students' computers. Five out of six analogs generate and send real-time statistics of violations to the teacher.

Thus, there are no analogs in which all the required violation detectors have been implemented. Analogs do not satisfy the goal of this article, in particular, there are no fully cross-platform solutions among reviewed systems. The review showed that a proctoring platform should be able to detect and process voice along with other data. Also, it is might be more convenient for teachers to have access to video streams. A full violation report will be received after the user completes the exam. Proctoring will be automated the meaning of a fully automated search for violations during the exam.

3   Architecture

Because of the State of the art conclusions, it is obvious that for achieving full cross-platform experience proctoring platform should be web application with a client-server architecture. The client part is aimed for the student to record a video and for the teacher to view the session analysis results. The server part is responsible for interaction with the database and between the video handlers.

The web application (Fig. 1) uses Node JS as the backend and React JS as the frontend. The WebRTC media server [5] is Kurento [6]. The task processing service XQueue [7] is responsible for managing the video session queue. Video handlers service is used for preliminary detection of cheating moments.

Recorded data is stored in the MongoDB database. It allows the creation of its own data storage schema, which is relevant for a system with many fields for data about users, exams, and violations. Usually, a group of students writes one exam at one time, so a room system was invented. The exam room database consists of exam date, subject name, teacher name, approximate exam duration, and the link to be opened after the exam starts if it is needed. The second database stores information about each student's exam.

3.1   Violations

There are three types of seriousness of violations:

1. *Danger* is a serious violation in which there is a very high chance that the student has cheated during the exam. The color is red. Violations are:

   - student NOT detected
   - student NOT looking on monitor
   - student logged into messengers

2. *Warning* is about a possible violation, which is worth looking at the teacher. The chance of cheating is still high. The color is yellow. Violations are:

   - Unknown persons detected
   - Students accessed unwanted websites during the exam

3. *Allowed* applications. The list includes those allowed by the teacher and also those that do not pose a threat of cheating. The color is blue.

Teachers can view the video with the ability to transit to the moment with violations. Below screencast and web camera videos, there are infraction scales (Fig. 2). The entire scale is a timecode of the video, on which the violations of students are highlighted in blocks. The block's color described above will help the teacher understand the type of violation.

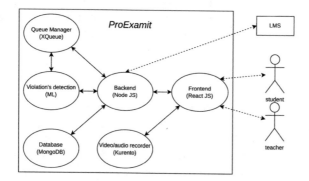

Fig. 1. Architecture

Fig. 2. Scale

On the right of the video, there are cards of violations. They contain brief information and two buttons: "load" and "image". ProExamit allows discovering applications opened by a student by analyzing application icons and keywords on screencast.

3.2   Video Handlers

The violation detection service is implemented using handlers. The handler is a python module that is designed to process the received task from the task queue for the presence of violations in the video stream. The results of handlers are a frame-by-frame report with found violations. Consecutive time intervals at which the same violations are recognized are combined into a single interval.

The novelty of ProExamit is the development of an algorithm for finding a list of suspicious applications that the user opened during the exam. An array of lexemes is recognized on the frame from the screen (using OCR frameworks). The resulting array is searched for keywords that are used in the interfaces of unwanted programs. The dictionary for each program contains a list of keywords that it uses. The handler returns an array of applications and their timecodes.

Videos are recorded in *webm* format. The standard video recording rate is 30 frames per second. In ProExamit the average frame processing time of webcam video is 0.096 s, and the maximum processing time is less than 0.2 s. The average time of screencast frame processing is 2.87 s, and the maximum time does not exceed 5 s. Such a long processing time is due to the dependency on the amount of text in the frame, the frame size, and the number of splitting areas of the frame.

In order to reduce the processing time, an option of skipping N frames between frames being processed was added. The accuracy is not affected since student violations take seconds instead of several frames.

4   Conclusion and Future Work

Proctoring helps to organize distance learning and automatize the process of exams. An architecture of extendable and cross-platform proctoring solution was implemented in the ProExamit. Video processing and violation detection are implemented as separate software modules for webcamera and screencast. These modules are the software component responsible for detecting violations on video using computer vision and machine learning. Developed proctoring

ProExamit is automated in terms of a fully automated search for violations during the exam.

The proctoring ProExamit described in this paper is used at the Department of Software Engineering and Computer Applications at Saint Petersburg Electrotechnical University "LETI". ProExamit used: 13 teachers, 382 students, 3559 exams were held. ProExamit allows making sure the integrity of tests and exams. It has reduced thousands of teachers' hours and automated the detection of resources used by students. Proctoring ProExamit detailed is on this page [8].

It is planned to interact with Learning Management System (LMS). The connection of proctoring to the LMS via Learning Tools Interaperability (LTI) protocol will allow seamless integration of the product into the learning process of interested organizations. Resource monitoring and intelligent resource allocation will help avoid errors or server crashes at times with a large number of users.

References

1. Carver-Thomas, D., Leung, M., Burns, D.: California Teachers and COVID-19: How the Pandemic is Impacting the Teacher Workforce. Learning Policy Institute (2021)
2. Vlasova, E.Z., Barakhsanova, E.A., Goncharova, S.V., Ilina, T.S., Aksyutin, P.A.: Teacher education in higher education systems during pandemic and the synergy of digital technology. Propositos y representaciones 8(3), 30 (2020)
3. Ghildial, P.: How COVID deepened America's teacher shortages. BBC, 21 December 2021. https://www.bbc.com/news/world-us-canada-59687947
4. Bruce, V., Young, A.: Understanding face recognition. Br. J. Psychol. 77(3), 305–327 (1986)
5. Lopez, L., et al.: Kurento: the WebRTC modular media server. In: Proceedings of the 24th ACM international conference on Multimedia, pp. 1187–1191, October 2016)
6. Spoiala, C.C., Calinciuc, A., Turcu, C.O., Filote, C.: Performance comparison of a WebRTC server on docker versus virtual machine. In: 2016 International Conference on Development and Application Systems (DAS), pp. 295–298. IEEE, May 2016
7. XQueue. (2022). [Source code]. https://github.com/openedx/xqueue
8. PROEXAMIT: FULLY AUTOMATED PROCTORING. Monitoring of certification events. https://online.osll.ru/proexamit_eng

A Novel Two-Stages Information Extraction Algorithm for Vietnamese Book Cover Images

Tham Nguyen Thi[1,2(✉)], Kieu-Hoa Vo[1,2(✉)], Minh-Tam Nguyen[1,2(✉)],
Truong-Thinh Nguyen[1,2(✉)], Gia-Phu P. Tran[1,2(✉)], Chi-Thanh Dang[1,2(✉)],
Co-Thai Quach[1,2(✉)], and Trong-Hop Do[1,2(✉)]

[1] University of Information Technology, Ho Chi Minh City, Vietnam
{18521384,18520767,20520748,20520783,20520694,
20520761,20520756}@gm.uit.edu.vn, hopdt@uit.edu.vn
[2] Vietnam National University, Ho Chi Minh City, Vietnam

Abstract. This study presents a novel deep learning based two-stage algorithm for extracting information from Vietnamese book cover images. First, the Vi-BCI dataset, which is the first dataset for Vietnamese book cover information extraction, was built. This dataset contains 7,875 images of Vietnamese book covers annotated with the text location on the cover, the text content, the text label Title, Author, and Publisher of the books. Next, a novel two-stages deep learning based algorithm is proposed and implemented on the Vi-BCI dataset to extract text information presented on book cover images. The experimental results show that the proposed algorithm archives an outstanding performance of information extraction with the accuracy of 95.50%.

1 Introduction

OCR (Optical Character Recognition) is a well-researched area in Computer Vision. The main goal of OCR is extracting text from images and documents, such as invoices, business cards, letters, and books,... Instead of typing documents manually, which takes up lots of time, the OCR application only needs a phone to scan images of documents. After that, they will be digitized and stored quickly with high accuracy. Parallelly, digitized documents will help save storage space and make it easier to manage. In libraries, bookstores, or personal bookshelves, to organize books and save their information, it is necessary to record the information on book cover about the title, author, and publisher. However, when there are a huge number of books in the library or bookstore, the work of recording and saving the information of books takes a lot of time and effort, while OCR can be used to save time and do it more efficiently.

© The Author(s), under exclusive license to Springer Nature Switzerland AG 2023
L. Braubach et al. (Eds.): IDC 2022, SCI 1089, pp. 160–167, 2023.
https://doi.org/10.1007/978-3-031-29104-3_18

In this study, we build the Vi-BCI dataset of 7,875 Vietnamese book cover images that is the first dataset for the information extracting problem. The Vi-BCI dataset can be downloaded from github [3]. Each book cover is labeled with coordinates and text. At the same time, we implement a pipeline to test the dataset. The input is the book cover image, the output is the text on that cover. The pipeline includes 2 sub-task as follows: text detection and text recognition. Text detection builds a model to locate text on book cover and text recognition builds a model to recognize the detected text.

2 Related Work

The datasets used to solve OCR problems can be divided into two categories: structured text and unstructured text. Unstructured text is sparse text with no proper row structure, complex background, random placement in the image, and no standard font. Some typical examples for that are English COCO-Text [5] at ICDAR2017 competition, Street View Text (SVT). The challenge of unstructured text image datasets to be solved is to accurately detect and recognize text content. On the other hand, the content of each book cover, including background image, font, and letter arrangement is uniquely presented. However, all book covers have some information in common: title, author, and publishing house, which are structured aspects of those data. We build the Vi-BCI dataset aiming to solve the problem of extracting information fields: book title, author, and publisher from Vietnamese book cover images. At the same time, we also suggest baselines for text detection and text recognition on our dataset. The datasets used to solve OCR problems can be divided into two categories: structured text and unstructured text. Unstructured text is sparse text with no proper row structure, complex background, random placement in the image, and no standard font. Some typical examples for that are English COCO-Text [5] at ICDAR2017 competition, Street View Text(SVT). The challenge of unstructured text image datasets to be solved is to accurately detect and recognize text content. On the other hand, the content of each book cover, including background image, font, and letter arrangement is uniquely presented. However, all book covers have some information in common: title, author, and publishing house, which are structured aspects of those data. We build the Vi-BCI dataset aiming to solve the problem of extracting information fields: book title, author, and publisher from Vietnamese book cover images. At the same time, we also suggest baselines for text detection and text recognition on our dataset.

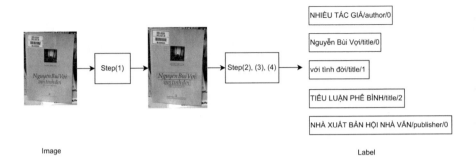

Label

Fig. 1. Labeling a book cover image in 4 steps.

3 Dataset Creation

3.1 Annotation Guidelines

To ensure data quality, the annotators joining in the annotation process have to follow instructions. The annotation tool used is PPOCRLabel which optimizes the data annotation process as it has support for rectangular and polygon-bounding boxes. An annotator with a complete image has to perform 4 steps: (1) label text objects in a book cover image with bounding boxes, (2) write the text seen in the image, (3) label the Title, Author, Publisher, Other class for the text seen, (4) label the linking number for the text seen, show in Fig. 1.

In step 1, the method to define the bounding box for text should be flexible, but still ensure the basic standard as following:

- The bounding boxes cover all the text objects but still need to minimize the redundancy that does not belong to that text object.
- The bounding boxes are annotated at the line level.

In step 2, the text content for the bounding boxes needs to be annotated correctly, including uppercase and lowercase letters. Then, the annotators determine the labels Title, Author and Publisher for the bounding boxes in step 3. In a case bounding boxes containing text content that does not belong to these three labels, it is annotated Other. For example, some book covers will have citations printed on the outside of the book cover. Despite being annotated in steps 1 and 2, if they do not belong to any of the labels: Title, Author, and Publisher, then Other label will be applied. In the last step, the text on the cover of the book is annotated to the line level, so the information fields are often divided into many bounding boxes. Therefore, annotators need numbering to link the contents of each field.

(a)   (b)   (c)

Fig. 2. Correct and incorrect labeled data examples. (a) Correct example and (b), (c) incorrcet examples.

3.2   Annotation Process

Our annotation process performed 5 rounds, each round annotating 1,600 images equally divided among 7 annotators. During the annotating process, if any annotator finds confusing points that are not covered in the guideline, the annotator is responsible for notifying and discussing that case with the team. The guideline is always updated after each discussion. In order to ensure the quality of the dataset, the low-quality images such as blur, and cropped content... will be detected and removed. At the end of each annotating round, the datasets of each annotator are double-checked by another annotator. Annotation errors detected at retest will be corrected and noted to notify the annotator of that dataset. If re-checking and many mistakes are found, the annotators are trained in the guideline to do better for the next round. Figure 3 shows us the progress of building this dataset.

3.3   Dataset Statistics

Our annotation process performed 5 rounds, each round annotating 1,600 images equally divided among 7 annotators. After 5 rounds of annotating, the Vi-BCI dataset has 7,875 annotated images. In the 7,875 images, there are 67,038 bounding boxes annotated. The bounding boxes have coordinates in the format[1] and the text seen, classification labels, and linking numbers separated by slashes. We divide this large dataset into three sets including train set, test set, and validation set with the ratio 8:1:1.

---

[1] [[left,top],[right,top],[right,bottom],[left, bottom]].

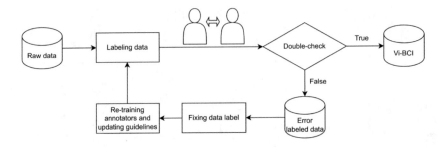

Fig. 3. The process of building the annotated dataset.

4   Baselines

We carried out a pipeline to extract the information from the book cover. First, the input book cover image will go through the text detection model to identify the location on book cover, then the detected text will be cropped into text-line images and passed through the recognition model to predict the text content, Fig. 4 illustrates our pipeline. Building this pipeline, we do two tasks of text detection and text recognition, which are detailed in Sects. 4.1 and 4.2.

4.1   Text Detection

In this task, the input of the model is an image of a book cover. After processing, the model outputs the bounding boxes containing the text. The model has drawn the boxes to the line level, shown in stage 1 in the Fig. 4.

We focus on implementing OCR algorithms provided by PaddleOCR. PaddleOCR which contains detection models to make OCR highly accurate offers a series of high-quality pre-trained models. A segmentation-based Single-shot Arbitrarily-Shaped Text detector (SAST) [4] which integrates both the high-level object knowledge and low-level pixel information in a single shot and detects scene text of arbitrary shapes with high accuracy and efficiency. SAST employs a context attended multi-task learning framework based on a Fully Convolutional Network (FCN) [2] to learn various geometric properties for the reconstruction of polygonal representation of text regions, including text center line (TCL), text border offset (TBO), text center offset (TCO), and text vertex offset (TVO). SAST consists of three parts, including a stem network, multi-task branches, and a post-processing part. The stem network is based on ResNet-50 with FPN [1] and CABs to produce context-enhanced representation. The TCL, TCO, TVO, and TBO maps are redicted for each text region as a multi-task problem.

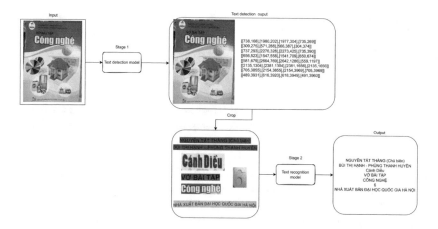

Fig. 4. Our pipeline which used to extract information from book cover.

4.2   Text Recognition

The input of the text recognition model is the bounding boxes drawn by the text detection model. The output of the model is the text content of these boxes. The recognized characters can be English or Vietnamese words and are case-sensitive, as illustrated in stage 2 in the Fig. 4. In this task, the input data is the text region images cut from the book cover images and their corresponding content label. VietOCR and PaddlOCR are used to build the Vietnamese text recognition model. In PaddleOCR, we use a text recognizer of the CRNN architecture. CRNN [2] uses both CNNs and RNNs to form an architecture along with CTC loss which is proven to be very accurate and fast. This network consists of three layers, CNNs followed by RNNs and then the Transcription layer. It adopts the Connectionist Temporal Classification(CTC) loss to avoid the inconsistency between prediction and label. ResNet-34 is used as the network backbone with pretrained weight on ImageNet.

For the second text recognizer we use, VietOCR, which is a very popular pre-trained text recognition library for Vietnamese with 10 million images. The OCR model of VietOCR integrates feature extraction and sequence modeling. In the feature extraction layer, VietOCR uses the VGG-16 model that is easier to finetune the pooling size parameter to extract features from input images. Then, feature maps that are outputs of producing from the feature extraction layer will be fed to the feature sequence layer. Transformer model is used as an alternative to RNN model as in CRNN architecture, with the aim of solving weaknesses such as training time is too long and sometimes inefficient for long-range dependencies of the RNN model.

5   Experiments

5.1   Experiment Setting

For the task of text detection, we train the SAST model on our dataset of 1875 images. ResNet50_vd_ssld is used as the network backbone with pretrained weight on ImageNet. To compare the effectiveness of SAST with existing OCR models, we are testing thorough experiments on our test set with popular OCR models, i.e., CRAFT and EasyOCR.

For the task of text recognition, text images cropped from book cover images are 1875 images annotated into 67,038 images. The pre-trained VietOCR model was used to re-train with our dataset. In addition, we also use the Paddle recognition model - latin_PP-OCRv3_rec that is a multilingual optical character recognition model for Latin fonts combined with VietOCR's Vietnamese character dictionary. The parameters used in the training of the three models above are detailed in Table 1.

Table 1. Describe the model parameters

Model	Parameters
SAST - text detector	$epoch\_num = 50$, $learning\_rate = 0.0001$, $batch\_size = 8$, $score\_thresh = 0.5$, $image\_shape = [512,512]$
CRNN - text recognizer	$CRNN$ - text recognizer $epoch\_num = 100$, $learning\_rate = 0.001$, $batch\_size = 8$, $image\_shape = [3, 48, 320]$
TransformerOCR- text recognizer	$iters = 50000$, $max\_lr = 0.0003$, $batch\_size = 32$, $image\_height = 32$, $image\_max\_width = 512$, $image\_min\_width = 32$

5.2   Experiment Results

As mentioned above, we use available models of EasyOCR and CRAFT to detect coordinates of text positions on book covers. Results of the text detection models task with three scores (Precision, Recall, H-mean) are presented in Table 2. The CRAFT model is trained with images in nature that are diverse in typefaces and fonts. That explains why the result of accuracy is quite good, 77.29% on the H-means score. When using pre-trained PaddleOCR on our data, the result is obviously outstanding by 80.73% on the H-mean score. However, some challenges exist in the abstract and complicated design of the background and text on the book cover. For the text recognition task, we use both the CRNN of the PaddleOCR text recognizer and the VietOCR text recognizer on our dataset. The results of text recognition models with accuracy on sequence level and character level are shown in Table 3. The VietOCR text recognizer gives the outstanding result of 95.50% accuracy on character level. This hopeful result can be explained by the pre-trained VietOCR being pre-trained on Vietnamese dataset and using Transformer architecture which can solve the disadvantages of the traditional sequential model such as long training time, weak term memory, week self-attention.

Table 2. Experimental results of the text detection task

Model	Precision	Recall	H-mean
EasyOCR	68.66%	43.95%	53.60%
CRAFT	85.79%	70.31%	77.29%
Paddle Detecion	**94.82%**	**70.28%**	**80.73%**

Table 3. Experimental results of the text recognition task

Model	Acc_full_seq	Acc_per_char
Paddle Recognition	64.87%	87.00%
VietOCR	**87.69%**	**95.50%**

6 Conclusion and Future Work

Our study has the following main contributions. First, the study creates the Vi-BCI dataset, which is the first dataset for the problem of extracting information on the cover of a Vietnamese book. Second, the study suggests a pipeline including the task of text detection and the task of text recognition on Vi-BCI dataset. For each task in the pipeline, we experiment with different models and get quite the expected results. Specifically, the highest model efficiency of text detection task and text recognition task achieved are 80.73% and 95.50% respectively. We will further improve the amount of data with the expected future goal of 10,000 photos. Besides, we continue to research and deploy a model to extract book cover contents according to specific fields.

Acknowledgment. This research was supported by The VNUHCM-University of Information Technology's Scientific Research Support Fund.

References

1. Lin, T.Y., Dollar, P., Girshick, R., He, K., Hariharan, B., Belongie, S.: Feature pyramid networks for object detection, pp. 936–944 (2017). https://doi.org/10.1109/CVPR.2017.106
2. Milletari, F., Navab, N., Ahmadi, S.A.: V-Net: fully convolutional neural networks for volumetric medical image segmentation, pp. 565–571 (2016). https://doi.org/10.1109/3DV.2016.79
3. Nguyen, T.T., et al.: Vi-BCI dataset (2022). https://github.com/kieuhoavo/BookCoverImage_Extraction.git
4. Wang, P., et al.: A single-shot arbitrarily-shaped text detector based on context attended multi-task learning, pp. 1277–1285 (2019). https://doi.org/10.1145/3343031.3350988
5. Veit, A., Matera, T., Neumann, L., Matas, J., Belongie, S.: COCO-text: dataset and benchmark for text detection and recognition in natural images (2016)

Enhancing 2D Hand Pose Detection and Tracking in Surgical Videos by Attention Mechanism

Quang-Dai Nguyen[1,2], Anh-Thuan Bui[1,2], Tri-Hai Nguyen[3],
and Trong-Hop Do[1,2(✉)]

[1] University of Information Technology, Ho Chi Minh City, Vietnam
{19521306,19521001}@gm.uit.edu.vn, hopdt@uit.edu.vn
[2] Vietnam National University, Ho Chi Minh City, Vietnam
[3] Department of Computer Science and Engineering,
Seoul National University of Science and Technology, Seoul, Korea
haint93@seoultech.ac.kr

Abstract. Hand pose estimation is a fundamental task for many human-robot interaction-related applications. In this work, we proposed novel hand pose estimation models, which leverage two types of attention mechanisms: self-attention and channel attention. By incorporating a simple yet efficient Squeeze-and-excitation (SE) block into Res152-CondPose, our best method, SERes152-CondPoseSE, successfully models interdependencies between channels. It outperforms the baseline, Res152-CondPose, by an absolute 9.56% in mean Average Precision and 17.78% in Multiple Object Tracking Accuracy.

1 Introduction

The emergence of machine learning applications in medicine has led to better and more widely available illness treatments, better and more effective patient care, and improved health care and disease prevention. Nowadays, computer vision has made significant advancements in the healthcare sector, resulting in a wide range of tasks such as tumor segmentation, the classification of cancer and diabetic retinopathy, clinical competence, and the identification and tracking of tools. In this study, we concentrate on articulated hand pose tracking in the operating room, which forms the core of many other beneficial tasks like temporal action identification or instructing surgical trainees. Currently, review from professionals is needed for these tasks, but thanks to the development of detecting and tracking hand poses, automating these processes is possible in the future.

Many studies have been done on articulated hand pose tracking. Our work is mainly based on Louis *et al.* [4] in which a novel dataset (Surgical Hands) and model (Res152-CondPose) for tracking and estimating multiple instances of articulated hands in the in-vivo surgical domain are introduced. The model makes predictions conditioned on the pose estimations of earlier frames. With the curiosity of investigating multi-head attention's performance, we replace the default attention mechanism in Res152-CondPose with Attention Augmented Inverted Bottleneck Block [5]. On the other hand, we also use Squeeze-and-excitation block [3] instead of the original BottleNeck block to enhance performance without significantly increasing model complexity and computational burden. In addition, we do a minor change in the rotation angle, increasing from [−40, 40] to [−180, 180] for better generalization. With these improvements, our model outperforms previous ones on the Surgical Hands dataset, on the mean Average Precision (mAP) metric for hand pose estimation and Multiple Object Tracking Accuracy (MOTA) metric for hand pose tracking.

2   Related Work

There are two approaches for human pose estimation: top-down and bottom-up. The top-down methods use single-person posture estimation to find human keypoints after obtaining human candidates via the detection module. The bottom-up approaches localize human keypoints throughout the entire image and then assemble these keypoints into each person. Due to better performance in practice, our methods follow a top-down paradigm. Existing works on 2D hand pose estimation are similar to human pose estimation. However, deep learning-based hand pose tracking methods are underresearched. Simon *et al.* [6] propose a technique that allows a network to detect occluded joints more accurately by training on multiple views of the same scene. A self-attention module was applied to the estimation network by Santavas *et al.* [5]. Louis *et al.* [4] published Surgical Hands, a novel video dataset for monitoring and estimating multiple instances of articulated hands in the surgical field. Although many existing datasets exist, temporal coherence between video frames is not fully supported. The only two datasets that claim to track are STB [7] and Surgical Hands [4]. However, STB's drawbacks include only having one hand during video sequences and allowing for no more than one detection each frame. Contrarily, Surgical Hands features a variety of lighting situations, quick movement, and scene appearances.

Detection and tracking in the medical field mostly concentrate on surgical instruments. In this area, there are other smaller tasks like surgical instrument recognition, localization, tracking, and skill ranking. For tracking surgeon's hands, there are quite a few studies; one of these is the study of Louis *et al.* [4]. They begin by detecting hands in the frame, estimate hands' joints based on bounding box detections, and then carry out tracking for those identifications. They use MOTA [2] to evaluate tracking performance.

3   Attention-Based 2D Hand Pose Detection and Tracking

3.1   Baseline

The baseline model (**Res152-CondPose**) [4] is based on ResNet-152 architecture with extra deconvolutional layers to construct heatmaps of key joints.
An image of hand crop I is passed through the model to attain a stack of 21
heatmaps H representing the locations of 21 joints in the human hand. The
model leverages the attention mechanism to exploit visual features from the
$conv\_1$ layer concatenated with prior heatmaps $H_{t-\delta}$. It helps relate previously
localized heatmap with current image features, ideally learning to weigh each
joint properly. The fusing module of the model takes the concatenation of the
newly produced heatmap and the output of the attention module. The model can
better estimate the human hand pose by utilizing information from the previous
steps.

$$H_t = M_{fus}(P(I_t), M_{att}(v_t, H_{t-\delta})) \tag{1}$$

3.2   Soft-Attention-Based Models

3.2.1   Res152-CondPoseAA

The attention module of the baseline only includes two convolution layers. Therefore, our first proposed model architecture uses Augmented Attention (AA). The
self-attention mechanism allows the model to attend to important regions of the
input. Each pixel of the image is ordered in such a way to give meaning to the
human eyes. So they are given positional embeddings so that the model can
tell where each pixel locates in the image. The Augmented Attention module is
called this way because it consists of Multihead Attention and the Convolution
layer that takes the input parallelly. The final output of the module is the concatenation of these two. This visual attention can help the model focus on the
features and spatial positions.

$$H_t = M_{fus}(P(I_t), M_{aug\_att}(v_t, H_{t-\delta})) \tag{2}$$
$$M_{aug\_att}(X) = concat[Conv(X), MHA(X)] \tag{3}$$

3.2.2   Res152-CondPoseAA+

We also replace the original attention mechanism of the baseline with the Attention Augmented Inverted Bottleneck Layer [5]. This layer is the wrap-around of
the Augmented Attention block presented earlier. It consists of Conv, Depthwise
Conv, and Augmented Attention. The proposed structure allows the model to
understand global constraints and correlations between keypoints, as shown in
Fig. 1.

$$H_t = M_{fus}(P(I_t), M_{aug\_att+}(v_t, H_{t-\delta})) \tag{4}$$
$$M_{aug\_att+}(X) = concat[Conv(X), M_{aug\_att}(X)] \tag{5}$$

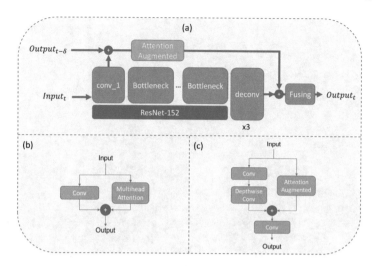

Fig. 1. The proposed architecture using the self-attention mechanism. (a) The self-attention mechanism is used as a side branch in place of a simple attention module in [4], called Res152-CondPoseAA. (b) The architecture of the Augmented Attention block. The original multi-head attention output is combined with the convolution layer's output. (c) The architecture of the Attention Augmented Inverted Bottleneck Layer includes an additional branch along with the Augmented Attention block in (b). The combined output is passed through the last convolutional layer.

3.3   Channel Attention-Based Models

3.3.1   SERes152-CondPose

Besides efforts to improve the baseline using the soft-attention mechanism, we also attach channel attention to the baseline architecture. We replaced the original BottleNeck blocks with Squeeze-and-excitation (SE) blocks from SENet [3], which was the winner of the ImageNet Classification Contest in 2017. Compared to the ResNet block unit, this block contains two additional fully-connected layers. It improves channel interdependencies without almost any computational cost. Each channel is assigned a newly recalibrated weight at the end of the block. Channels with a higher level of useful information will get bigger weighted scores.

SE layer performs two operations: (1) Squeezing: Input $X \in \mathbb{R}^{H \times W \times C}$ is passed through an 2D adaptive average pooling layer to produce output $Y \in \mathbb{R}^{1 \times 1 \times C}$. Each scalar, called a channel descriptor, represents the global spatial information of each channel. The collection of these local descriptors gives us expressive statistics for the whole image; and (2) Excitation: After being squeezed, the input flows through a bottleneck formed by two fully-connected layers with ReLU in between. This behaves as a gating mechanism that attempts to model the interdependencies of channels (Fig. 2).

$$H_t = M_{fus}(P_{SE}(I_t), M_{att}(v_t, H_{t-\delta})) \qquad (6)$$

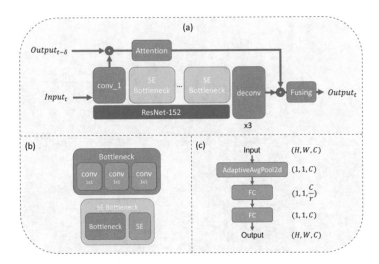

Fig. 2. The proposed architecture uses a channel attention mechanism. (a) The original bottleneck block of the ResNet is replaced with the SE bottleneck block, which is an attentive version of the Resnet bottleneck. (b) The architecture of the SE bottleneck block is enhanced by the additional SE layer in each bottleneck block. The components in (c) are the design of the SE layer.

3.3.2   SERes152-CondPoseSE

Having obtained favorable results in the channel attention approach, we proposed another model version that exploits channel interdependencies on the attention branch, called SERes152-CondPoseSE. The original attention is enhanced with an additional SE layer at the end. The output of the attention module in the baseline is $(W, H, 21)$, so by applying the SE layer to the output, we aim to produce better attention information from current visual features and the previous heatmap.

4   Performance Evaluation

4.1   Dataset

We use Mixed Hands Dataset [6], which consists of data from 2 sets: CMU Manual Hands and CMU Synthetic Hands, for pre-training our models. The first, CMU Manual Hands, contains the manually-annotated data in the MPII Human Pose [1] training set and images from the New Zealand Sign Language (NZSL) Exercises of the Victoria University of Wellington. It has a total of 2,758 hand annotations with various hand poses in an in-the-wild environment. The latter, CMU Synthetic Hands, is an initial set of synthetically rendered images of hands and contains 14,261 hand annotations.

We finetune our models on Surgical Hands [4], the first video dataset for estimation and tracking multiple hand poses in the surgical domain. First, 28 videos with visible surgeons' hands and patient cavities were selected from YouTube and other web sources. Then, we extracted 76 clips from these videos and sampled them at eight frames per second, resulting in 2,838 annotated frames with 8,178 unique hand annotations. On average, there are 2.88 hands in each frame. The number of hands in the dataset has a median of 3 and a maximum of 7.

4.2   Experimental Setup

Due to the limited training resources, we cannot report k-fold cross-validation (k = 28) as in [4]. So instead, we divide the dataset into train, validation, and test splits. The splitting procedure: (1) We use 28 pre-trained weights provided in [4] to evaluate 28 folds; (2) 4 best and worst-performing validation folds are selected; (3) We take two folds from each group (best and worst) to create the validation and test sets.

Training hyperparameters are followed as in [4], with a batch size of 12, learning of 8e−5, Adam optimizer, and 10 epochs. Heatmap priors come from the ground truths and the previous predictions. If only using true priors, the model may be biased towards perfect input data. In contrast, training the model using only the predicted heatmap causes those heatmaps to be ignored because it never provides any useful information. We set the probability of using predicted heatmaps to 0 and increased it by 1/3 per epoch. Therefore, the model can leverage ground truths priors first and then gradually switch to using the heatmap predictions. This teaches the model not to rely on perfect priors to generate predictions, which is impossible during the evaluation phase.

We presume the direction of the hand is largely dependent on the direction of the joint connection between the wrist and the metacarpophalangeal joint of the finger. Therefore, the hand direction is computed by calculating the angle between that bone and vector (1, 0). As there are many occlusions, we successfully calculate the direction of the hands in only 48% frames. Under the default setup: a random image rotation between −40 and +40° C and a 50% chance of horizontal flipping, the hand direction distribution is strongly imbalanced. This skewness may hurt the generalization ability of the model. As a result, we test our models' performance on a new rotation angle (from −180 to 180) along with the defaults in [4]. Figure 3 demonstrates our estimation strategy.

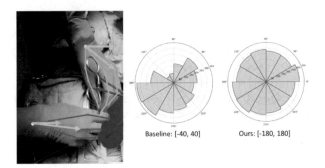

Baseline: [-40, 40]   Ours: [-180, 180]

Fig. 3. The illustration of the hand angle estimation approach.

4.3 Numerical Results

4.3.1 Detection Performance

We use mAP to measure the performance of our models versus the baseline. The metric is calculated using the Probability of Correct Keypoints (PCK) metric, which is the probability of correctly localizing keypoints within a threshold distance, σ:

$$PCK_\sigma^i = \frac{1}{D} \sum_{s=1}^{D} 1(\|y_p - \hat{y}_p\| \leq \sigma) \tag{7}$$

where the key joints indexes $i = [1, \ldots, 21]$ and s represent the hand sample index, and D is the total number of samples. PCK of each predicted heatmap (hand localization) in one frame versus all ground truth heatmaps in the same frame is calculated. The pose prediction with the highest PCK with one ground truth pose (among other predictions) is assigned to that pose. We reported Average Precision (AP) for each joint and mAP on the test set.

Most of our model experiments achieve better results than the baseline [4] at the total mAP, as shown in Table 1. This illustrates the effect of adding the attention mechanism to the original model (Res152-CondPose).

Effect of Attention Mechanism: Two models using self-attention as a replacement for the attention module of the baseline perform poorly under the same additional rotation angle in preprocessing phase. The third proposed model (SERes152-CondPose) yields better total mAP performance by 1.43%.

Effect of Adding more Rotation Angle: By successfully estimating angles of hands across the dataset, our proposed different rotation angle added during training time gives the average mAP improvement of 7.62% in our first three models. The best model we experiment with under [−180, 180] rotation angle is SERes152-CondPoseSE, an enhanced version of the second-best model (SERes152-CondPose). Compared with the baseline, our best model has a 9.56% improvement in total mAP over its 41.25%.

Table 1. mAP performance comparison.

Model	Rot. Ang	Wrist	Thumb	Index	Middle	Ring	Pinky	Total
Res152-CondPose (Baseline)	40	70.80	40.87	42.16	46.70	39.04	30.11	41.25
Res152-CondPoseAA		53.15	38.51	32.87	35.86	30.90	26.01	33.80
Res152-CondPoseAA+		61.44	33.90	33.85	38.40	28.80	32.89	34.89
SERes152-CondPose		68.41	50.65	42.82	41.83	39.73	31.94	42.68
Res152-CondPoseAA	180	73.74	49.30	41.60	42.38	45.63	28.69	43.05
Res152-CondPoseAA+		**75.90**	47.43	39.23	45.60	41.19	28.17	42.02
SERes152-CondPose		75.01	**55.99**	48.06	**50.67**	43.79	40.83	49.16
SERes152-CondPoseSE		69.49	52.68	**50.94**	45.93	**53.12**	**46.73**	**50.81**

4.3.2   Tracking Performance

Detection performance only considers the hand localization ability in a single frame. We use the MOTA metric to evaluate the consistency of localized keypoints across the video. The MOTA formula is shown as:

$$MOTA = 1 - \frac{\sum_t (FN_t + FP_t + IDSW_t)}{\sum_t G_t} \qquad (8)$$

where false negatives (FN) are joints present in ground truth but not getting detected, false positives (FP) are joints not present in the ground truth but got detected, identity switches ($IDSW$) are ground truth hands that are falsely related to some other hands due to false tracking, and G is the total of the ground truth joints.

To assign tracking ids to predicted hand poses across frames, we experiment with two matching strategies: L2 distance and IoU overlap. The average L2 distance is measured from the regressed keypoints between the previous and current frames' predictions, while IoU overlap is calculated between hand bounding boxes at time t and $t-1$. A hand detection at the current frame that shares a significant IoU score with another hand at the previous frame may indicate that they are the same. Table 2 shows the MOTA performance of the baseline and proposed models.

Comparison between Matching Strategies: Tracking using IoU overlap gives better overall results in total MOTA than L2 distance. For each key joint, MOTA varies among proposed methods and tops under training experiments with IoU matching strategy and proper hand angles distribution (Rotation angles: [−180, 180]).

Table 2. MOTA performance comparison. *Italic* denotes the best in each matching strategies group, while **bold** denotes the best of all experiments.

Model	Rot. Ang.	Mat. stra	Wrist	Thumb	Index	Middle	Ring	Pinky	Total
Res152-CondPose (Baseline)	40	IoU	*33.14*	*23.19*	*32.26*	*35.08*	*20.87*	9.16	*24.54*
		L2	32.56	22.77	31.77	34.74	20.78	9.16	24.26
Res152-CondPoseAA+		IoU	*30.11*	*31.32*	*25.08*	*28.87*	*22.25*	*16.46*	*25.05*
		L2	28.98	30.14	24.21	28.46	21.97	16.09	24.40
Res152-CondPoseAA		IoU	19.08	25.67	*27.11*	*31.85*	20.3	*26.18*	*25.88*
		L2	*28.98*	*30.14*	24.21	28.46	*21.97*	16.09	24.40
SERes152-CondPose		IoU	31.79	36.05	31.93	35.50	30.66	20.11	30.89
		L2	31.79	36.05	31.93	35.50	30.66	20.11	30.89
Res152-CondPoseAA	180	IoU	**59.32**	*40.34*	*36.31*	*35.26*	*31.28*	15.87	*33.12*
		L2	**59.32**	40.04	36.10	35.13	30.91	15.87	32.93
Res152-CondPoseAA+		IoU	47.40	*34.13*	31.79	40.84	32.66	22.05	33.01
		L2	47.40	33.66	31.44	*40.97*	32.45	22.05	32.84
SERes152-CondPose		IoU	49.43	**42.02**	37.77	*45.08*	34.82	22.71	*37.10*
		L2	*50.00*	41.79	*37.86*	45.04	34.61	22.58	37.02
SERes152-CondPoseSE		IoU	40.12	*41.51*	44.28	40.82	**45.30**	**40.21**	**42.32**
		L2	40.12	41.19	**44.30**	40.82	**45.30**	**40.21**	42.26

Comparison with the Baseline: Our top-performing methods obtain a significant increase in total MOTA metric compared with the baseline model. Our latest model experiment (SERes152-CondPoseSE) outperforms the baseline by 17.72% MOTA, indicating that successfully modeling interdependencies between channels in the main and the attention branches could lead to higher tracking accuracy.

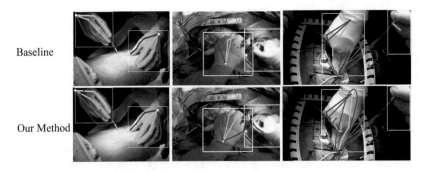

Fig. 4. Comparison between the baseline and our top-performing model (SERes152-CondPoseSE).

Figure 4 visualizes the comparison between the baseline model and our top-performing model (SERes152-CondPoseSE)[1]. Our method successfully estimates well-occluded joints and does not confuse between joints.

5 Conclusion

In this paper, we proposed four novel models based on the Res152-CondPose architecture: Res152-CondPoseAA, Res152-CondPoseAA+, SERes152-Cond Pose, and SERes152-CondPoseSE. We showed that incorporating the attention mechanism could significantly enhance the existing model. Our best model, SERes152-CondPoseSE, improved 9.56% in mAP and 17.78% in MOTA metrics compared to the baseline. We also addressed the importance of setting up proper pre-processing techniques, which strongly depend on the dataset characteristics. A comparison of the proposed model with other modern models can be conducted in future work.

References

1. Andriluka, M., Pishchulin, L., Gehler, P., Schiele, B.: 2D human pose estimation: new benchmark and state of the art analysis. In: 2014 IEEE Conference on Computer Vision and Pattern Recognition, pp. 3686–3693. IEEE (2014). https://doi.org/10.1109/cvpr.2014.471
2. Bernardin, K., Stiefelhagen, R.: Evaluating multiple object tracking performance: the CLEAR MOT metrics. EURASIP J. Image Video Process. **2008**(1), 1–10 (2008). https://doi.org/10.1155/2008/246309
3. Hu, J., Shen, L., Albanie, S., Sun, G., Wu, E.: Squeeze-and-excitation networks (2020). https://doi.org/10.1109/tpami.2019.2913372
4. Louis, N., et al.: Temporally guided articulated hand pose tracking in surgical videos (2021). https://doi.org/10.2139/ssrn.4019293
5. Santavas, N., Kansizoglou, I., Bampis, L., Karakasis, E., Gasteratos, A.: Attention! A lightweight 2D hand pose estimation approach (2021). https://doi.org/10.1109/jsen.2020.3018172
6. Simon, T., Joo, H., Matthews, I., Sheikh, Y.: Hand keypoint detection in single images using multiview bootstrapping. In: 2017 IEEE Conference on Computer Vision and Pattern Recognition (CVPR). IEEE (2017). https://doi.org/10.1109/cvpr.2017.494
7. Zhang, J., Jiao, J., Chen, M., Qu, L., Xu, X., Yang, Q.: 3D hand pose tracking and estimation using stereo matching (2016)

---

[1] Tracking performance comparison can be found at: https://youtu.be/k8ioKqLlSms.

An Autonomy-Based Collaborative Trust Model for User-IoT Systems Interaction

Alessandro Sapienza[✉] and Rino Falcone

Institute of Cognitive Sciences and Technologies,
National Research Council of Italy (ISTC-CNR), 00185 Rome, Italy
{alessandro.sapienza,rino.falcone}@istc.cnr.it

Abstract. IoT technologies continue to spread in our society. Nevertheless, there are still different critical point to be faced and solved. Among the many, given the ever-growing potential for action, a major issue concerns the autonomy of these devices. In this work, we propose a possible solution on how IoT devices may act in order to get more autonomy, in relation to the specific user they interface with. After introducing a theoretical model, we considered a possible implementation in a simulation context, showing how the proposed approach works.

1 Introduction

Recent research has focused on introducing new methodologies that incorporate autonomy within smart objects. Indeed, next generation IoT must be autonomous as well as cooperative in order to efficiently coordinate supporting actions towards meeting common high-level goals [1]. As Shneiderman reports [2], expansion from an algorithm-focused view to a human-centered perspective can shape the future of technology so as to better serve human needs. More in detail, Miranda and colleagues [3] talk about Internet of People (IoP), highlighting that, since technology will increasingly be integrated into the our daily lives, users should be at the center of the system. They stress the fact that IoT systems must be able to adapt to the user, taking people's context into account and avoiding user intervention as much as possible. Similarly, Ashraf [4] talks about autonomy in the Internet of things, pointing out that in this context it is necessary to minimize user intervention. Nonetheless, research into the user acceptance of IoT services is still scarce and the relevance of technological autonomy in determining its acceptance has not been fully recognized [5].

Indeed, there may be several approaches that favor the increase of the autonomy of the devices. In this work, we focus our attention on a specific solution: the device must adapt its degree of autonomy according to the needs that it can detect in the collaborative approach. This requires various potentialities for our smart devices: i) to be able to manage its own different degrees of autonomy in the collaboration with the user; ii) be able to evaluate and adapt, in each interaction, the right degree of autonomy to carry out collaborative behavior; iii) knowing how to evaluate its one's ability to act in the various degrees

L. Braubach et al. (Eds.): IDC 2022, SCI 1089, pp. 178–187, 2023.
https://doi.org/10.1007/978-3-031-29104-3_20

of autonomy. In other words, all those scenarios in which it lacks the user's collaboration, which would be necessary to complete the task, could turn into opportunities to implement the potential of the devices. Therefore, the device can evaluate whether to increase its autonomy to successfully complete a task that would otherwise fail. This is anything but an uncommon situation in the field of human-artificial system interaction. As a remarkable example, we could think of the interaction and monitoring of the elderly [6,7]. In these cases, it becomes essential to have a non-intrusive system, which is however able to estimate the user's capabilities and possibly compensate for any shortcomings.

However, to identify these situations during interaction and to quantify the chances of success, the IoT devices needs to produce a mental representation of the user.

With this work, we want to propose a general IoT system able (1) to adapt to the specific user characteristics in order to identify the proper level of autonomy and (2) to push the user to accept this technology more. After introducing a theoretical model, we considered a possible implementation in a simulation context, aiming to show how the proposed approach works. Our idea is that of providing a general abstract model, which can then be applied into many different IoT context.

2   State of the Art

Although some solutions have been proposed in the past, very few contributions have given a key role to autonomy in the interaction with human users.

As a first example, in the HRI domain, the authors of Optimo [8] proposed a system capable of perceiving how much the user trusts the device and then adapting its behaviors dynamically, to actively seek greater trust and greater efficiency within future collaborations.

In the IoT domain, we introduced a model for user's acceptance [9], in which devices evaluated how much the user trusts them, in order to perform tasks with an adequate level of autonomy.

Recently, the authors of [1] introduced the concept of autonomy in large-scale IoT ecosystems, by making use of cognitive adaptive approaches.

Furthermore, Sifakis [10] states that IoT is a great opportunity to reinvigorate computing by focusing on autonomous system design. His idea is to compensate the lack of human intervention by introducing adaptive control.

Starting from this concept, our work focuses on the estimation of the degree of user collaboration to adjust at run-time the autonomy in the execution of tasks.

3   Simulation Model

In this work, we are interested in identifying the complex relations between a user u and an IoT system S consisting of n devices $\{s_1, s_2 \ldots, s_n\}$. On the one hand, the user u grants an initial level of autonomy to the devices and makes use of them to perform some tasks. On the other hand, the devices want satisfy at best the user's requests. Nevertheless, in order to do so, they need to increase the autonomy the user grants them, since higher level of autonomy correspond to more room for action. Specifically, in this scenario we introduce the possibility that the devices analyze the degree of user collaboration to determine if and when it is possible to go beyond the autonomy granted [11,12].

3.1 The User

In the simulation, we considered a single user u dealing with a number of IoT devices. The user makes use of such devices to pursue its own goals, granting them an initial autonomy level, which limits their actions. The user is described in terms of:

1. *predisposition*: how much autonomy it is willing to grant to the devices at the beginning;
2. *attention*: the degree of attention to the requests for intervention of the devices;
3. *competence*: how competent it is in carrying out its part of the task.

Together, *attention* and *competence* define the degree of *collaboration* of the user (see Eq. 1).

$$collaboration = mean(attention, competence) \qquad (1)$$

The *predisposition* determines the initial level of autonomy granted to the devices, taking into account the fact that different users may have different attitudes towards this type of technology. For instance, the authors of [13] found that household members with high technical skills are more willing to adopt smart home services and products.

Starting from this initial value, the autonomy granted may change over time in response to certain situations and may vary from device to device. In addition, while dealing with the device d_i, u will update the trust evaluation of d_i on the specific task level.

3.2 The Devices

There can be a variable number of devices in the world. All of them possess two purposes. The first one is to pursue the user's task to satisfy its need (even if he has not explicitly requested them). The second one consists in trying to increase the autonomy (compatibly with its own abilities) the user grants them, so that they can perform increasingly smart and complex functions for the user. Remarkably, this second goal is functional to the first one.

Then, on each turn, a given device d_i will identify at which level it can execute the task, according to 1) the user's trust, i.e. its estimation of how much the user trusts it, which value should at least be equal to a given threshold σ and 2) to the granted autonomy. Then it will try to perform that task.

At the end of the task, u checks d_i's performance, which can be positive or negative. The actual performance of the devices depends on the *error probability* of the considered task level. Indeed, we suppose this probability to be linearly related to the level, as tasks with a greater autonomy usually imply a greater complexity, and so it should more difficult to get carry them out. Resuming, a device is characterized by:

1. Its *trustworthiness* estimation on the various levels, to evaluate its own skills in the different cases;
2. The estimation of the *autonomy* granted by the user;
3. The estimation of the user's *attention* and *competence*;
4. $\Delta autonomy$ that, as we will explain later, represents the extent to which autonomy should be increased to perform a task when the user's cooperation is lacking;
5. *error probability* on each level: this is an intrinsic characteristic of the device, neither it nor the user can directly access it, but it can be estimated through interaction.

Concerning the estimation of user's *attention* and *competence*, we suppose that this is shared knowledge among the devices, so that they can collaboratively work to estimate such dimensions. Therefore, when a device interacts with the user, it updates the common estimations of these two variables. As we will see better in Sect. 3.3, based on the type of task, it is possible to estimate *attention* and/or *competence*. These values are updated by weighted mean as in Eqs. 2 and 3. In the case of *attention*, the performance (*performance1*) concerns whether the user perceived the indication of the device while, in the case of competence, it (*performance2*) accounts for the user's success or failure in doing its part of the task. Concerning α and β, they represent the weights of the old and the new value.

$$newAttention = \frac{attention * \alpha + performance1 * \beta}{\alpha + \beta} \tag{2}$$

$$newCompetence = \frac{competence * \alpha + performance2 * \beta}{\alpha + \beta} \tag{3}$$

3.3 The Tasks

We have considered the presence of three abstract types of tasks:

1. *Individual tasks*, which the device performs in complete autonomy.
2. *Mild collaborative tasks*, in which the user and device play their part independently. In this case, devices can exploit the user's action to deduce *competence*.
3. *Close collaborative tasks*, for which the device requires user intervention. In this case, it is possible to deduce user's *attention* and *competence* both.

Moreover, the same task can be carried out at different level of autonomy. We refer to the classification introduced in [9], which divides tasks into 5 levels according to autonomy:

1. *Level 0*: to operate according to its basic function;
2. *Level 1*: to communicate with other agents inside and outside the environment, but just in a passive way;
3. *Level 2*: to autonomously carry out tasks, but without cooperating with other devices;
4. *Level 3*: to autonomously carry out tasks cooperating with other devices. The cooperation is actually a critical element, as it involves problems like the partners' choice, as well as recognition of merit and guilt;
5. *Level 4*: the possibility of going beyond the user's requests, proposing solutions that it may not have even imagined (over-help [11]).

3.4 Trust and Autonomy

Trust is a key element in every aspect of social cooperation. Its importance has also been clearly recognized in Human-Machine interaction [14] and, more in details, also in the IoT domain [15].

In this paper, we refer to the socio-cognitive model of trust [16]. Trust is taken into account both by the user, who estimates the reliability of the devices, and by the devices, who need to know how trustworthy they are (and are perceived by the user) to choose at what level of autonomy to operate.

Concerning competence, this dimension is implemented through the *error probability*, i.e. the probability that a device will not be able to successfully complete the requested task. Of course, this probability also depends on the level of autonomy at which the task takes place, as the task becomes more and more complex.

In the light of such premise, we have modeled the autonomy as a scalar variable defined in $[0, 1]$, in which 0 allows the device to work just at Level 0, while 1 grants them the maximum autonomy (Level 4). The thresholds 0.25, 0.5 and 0.75 allow devices to work respectively at Level 1, 2, and 3.

Instead, the trustworthiness of a device is modeled as a 5 elements vector, precisely to evaluate its skill at every level of autonomy.

Of course, working at a given level of autonomy gives devices the opportunity to show their competence at that level. When a task ends, the trustworthiness is updated through a weighted mean, according to Eq. 4. Here, *performance3* is the binary outcome of the task (success/failure), while α and β represent the weights of the old and the new value.

$$newTrustworthiness = \frac{trustworthiness * \alpha + performance3 * \beta}{\alpha + \beta} \quad (4)$$

Autonomy update follows a different procedure. Indeed, the autonomy should not increase if the device does not push beyond the level it normally performs. However, if it independently increases such level autonomy without a real reason, this could provoke a negative reaction from the user [17]. Conversely, the devices will detect those situations in which going further would represent an advantage for the user while not doing so would result in a loss and they will limit their increase of autonomy only to such situations. This is in accordance with their primary goal: to best satisfy the user's requests.

Therefore, in case of close collaborative tasks, if the user does not collaborate or fails its part of the task, the device evaluates whether to go further. At first, the device estimates $\Delta autonomy$, i.e. how much more autonomy it would be necessary to complete the task. Since this increase in the autonomy is meant to bridge the collaboration gap, we assume that $\Delta autonomy$ depends on the level of collaboration of the user (see Eq. 5).

$$\Delta autonomy = 1 - autonomy - collaboration \qquad (5)$$

In Eq. 5, $autonomy$ is the current level of autonomy granted to the devices, while $collaboration$ depends on the user's actions.

Thus, the device estimates if it is able to perform a task of this level, namely if it is sufficient trustworthy to do it. If so, it tries executing the task. Otherwise, it avoid doing it. Therefore, similarly to what happens to trustworthiness, the autonomy of the device is updated following Eq. 6. In this case, $performance4$ tells us whether, in the specific case in which the device decided to increase its autonomy, it has managed to successfully complete the task or not.

$$newAutonomy = \frac{autonomy * \alpha + performance4 * \beta}{\alpha + \beta} \qquad (6)$$

4 Simulations and Results

In this section, we present the results of the agent-based simulation experiments, implemented on the NetLogo platform [18]. The experiments were conducted using the following setting:

- $predisposition = 0$;
- $attention = [0, 10, 20, 30, 40, 50, 60, 70, 80, 90, 100]$;
- $competence = [0, 10, 20, 30, 40, 50, 60, 70, 80, 90, 100]$;
- number of device: 10;
- $\sigma = 0.5$;
- $\alpha = 0.9$ and $\beta = 0.1$;

The $predisposition$ is set to 0. By doing so, we place ourselves in the most restrictive scenario. As far as it concerns $attention$ and $competence$, we decided to investigate several values throughout the definition range, as these are the variables that interest us most. The threshold σ has been set to 0.5, to ensure that devices provide an average performance greater than 50% when they decide to execute a task. Then, α and β were set to give more importance to past experience and to avoid excessive fluctuations in the variable values during simulations.

Starting from this setting, we have considered a baseline scenario without error, in order to compare it with a second scenario characterized by error increasing with the level of autonomy. At the end of the simulation, we analyzed the average perception of the user about the devices, in terms of autonomy and trustworthiness. Specifically, we took into account the average values after 1000 rounds. This time window is reasonably long to let the user make experience with the system and the devices. In addition, what we considered are the results averaged over 100 simulations, in such a way as to eliminate the variability introduced by the random effects on the individual runs.

In the first scenario, we investigated an ideal case, that is in the absence of error. In this case, the performance of the devices is always positive. Figure 1a shows the average autonomy granted to devices, depending on attention and competence. Such result suggests that for attention levels of 60% and below devices manage to obtain the maximal autonomy. Yet, increasing further users' attention, the autonomy decreases, even to 0%. This happens because, in the presence of a perfectly attentive and competent user, this approach does not leave the devices room for additional interventions.

In a similar way, even the perception of trustworthiness is affected by the user's profile. Figure 1b reports the average value for Level 4. Even in this case, starting from a value of 70% for the attention, the user's evaluation of the device starts decreasing, until reaching the value 0.5, which correspond to the σ threshold.

Certainly, this first case analyzes a hypothetical and unrealistic scenario. However, it allows us to isolate the effect of the adjustable autonomy, in relation to the different users' profile. Thus, in the second experiment we investigated what happens in the presence of error. Specifically the error probability was set to 0, 5, 10, 15, and 20 respectively for the five levels of autonomy.

Figure 1c shows the evolution of autonomy in this case. In the presence of error, the lower performance of the devices acts as a filter. This, in turn, prevents the maximum value of autonomy from being reached by the devices. With attention values below 60%, the average autonomy varies between 0.87 and 0.98. This implies that some of the devices cannot execute Level 4 tasks, but they are limited to Level 3 tasks. Increasing the value of attention, we find ourselves in the presence of the same phenomenon seen before: autonomy decreases to 0.

In order to be able to make a comparison, we report also in this case the average evaluation of trustworthiness for Level 4 (Fig. 1d), which is around 0.6. This stands in contrast to what we would have expected, as with an error probability of 20%, the average trustworthiness should be 0.8. However, by combining this data with the autonomy analysis, it can be observed that some devices are not able to perform the Level 4 tasks, thus having a trustworthiness of 0.5 or below. Thus, the presence of error, combined with the constraints on autonomy, can have a strongly negative impact on the perception of trustworthiness.

Then, even in this case, for high value of attention and competence the trustworthiness evaluation lower down to 0.5.

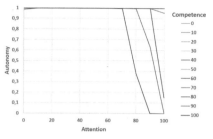

(a) Average autonomy granted by the user in the absence of error

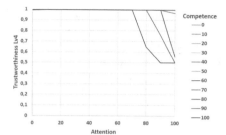

(b) Average trustworthiness evaluation of the user in the absence of error

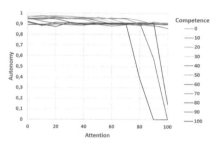

(c) Average autonomy granted by the user in the presence of error

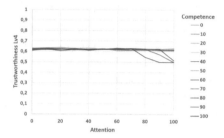

(d) Average trustworthiness evaluation of the user in the absence of error

Fig. 1. The average values of autonomy (left) and trustworthiness evaluation (right) in the absence (up) or presence (down) of error

5   Discussion and Conclusions

In this work, we investigated one of the possible approaches to increase user's acceptance of IoT system, thus resulting in a greater autonomy for IoT devices.

Indeed, these devices have enormous potential to offer to end users. However, the user may not necessarily be willing to accept all the functionalities offered from the beginning. In light of such consideration, it becomes essential to understand how the device should behave in order to stimulate the user's trust and gradually ensure that the latter is willing to grant more autonomy. Of course, many approaches are possible to solve this problem [19,20]. In this article, we have proposed a methodology based on user collaboration, in which devices analyze user characteristics and exploit situations of non-collaboration to show their capabilities and obtain greater autonomy.

Summarizing the results of this analysis, we can say that *attention* and *competence* affect both the autonomy granted to the devices and the user's perception of the devices' trustworthiness. More in detail, as *attention* and *competence* increase, autonomy decreases. This happens because the proposed approach is based on exploiting situations in which the user fails to perform its part of the collaborative tasks (either due to lack of attention or lack of competence). Therefore, in the presence of collaborative users, there are fewer opportunities to take advantage of. In extreme cases, with fully collaborative users, this approach does not provide benefits to the system.

As attention and competence increase, the perceived trustworthiness of devices decreases, falling down even to the neutral value 0.5. Indeed two opposing phenomena influence the perception of trustworthiness:

1. The more the devices act alone, the more they are able to show their actual skills at the highest levels. This phenomenon leads to favoring situations in which one acts alone.
2. However, frequently acting alone at high levels of autonomy, in the presence of incremental errors, will increase the error rate and the possibility to fail the task. This introduces a disadvantage in performing higher-level tasks.

To conclude, the results of this work, albeit more theoretical, aimed at proving the importance of the proposed approach, providing interesting insights for the evolution of IoT systems. Indeed, the experiments suggest that the approach used actually makes it possible to improve the levels of autonomy at which devices can operate. It is worth noting that the proposed framework is not intended as an alternative to the others, but rather it could provide support to other valid solutions. Certainly, among the limitations, it must be considered that this model does not provide benefits when the devices interface with fully attentive and competent users.

References

1. Michailidis, I.T., et al.: Embedding autonomy in large-scale IoT ecosystems using CAO and L4G-CAO. Discov. Internet Things **1**(1), 1–22 (2021). https://doi.org/10.1007/s43926-021-00003-w
2. Shneiderman, B.: Human-Centered AI. Oxford University Press, Oxford (2022)
3. Miranda, J., et al.: From the internet of things to the internet of people. IEEE Internet Comput. **19**(2), 40–47 (2015)
4. Ashraf, Q.M., Habaebi, M.H.: Introducing autonomy in internet of things. In: Recent Advances in Computer Science, pp. 215–221. WSEAS Publishing (2015)
5. Constantinides, E., Kahlert, M., de Vries, S.A.: The relevance of technological autonomy in the acceptance of IoT services in retail. In: 2nd International Conference on Internet of Things, Data and Cloud Computing, ICC 2017 (2017)
6. Valero, C.I., et al.: EBASI: IoT-based emotion and behaviour recognition system against elderly people social isolation. In: Camacho, D., Rosaci, D., Sarné, G.M.L., Versaci, M. (eds.) IDC 2021. SCI, vol. 1026, pp. 3–13. Springer, Cham (2022). https://doi.org/10.1007/978-3-030-96627-0_1
7. Mincolelli, G., Imbesi, S., Marchi, M.: Design for the active ageing and autonomy: the role of industrial design in the development of the "habitat" IOT project. In: Di Bucchianico, G., Kercher, P.F. (eds.) AHFE 2017. AISC, vol. 587, pp. 88–97. Springer, Cham (2018). https://doi.org/10.1007/978-3-319-60597-5_8
8. Xu, A., Dudek, G.: OPTIMo: online probabilistic trust inference model for asymmetric human-robot collaborations. In: 2015 10th ACM/IEEE International Conference on Human-Robot Interaction (HRI), pp. 221–228. IEEE (2015)

9. Falcone, R., Sapienza, A.: On the users' acceptance of IoT systems: a theoretical approach. Information **9**(3), 53 (2018)
10. Sifakis, J.: System design in the era of IoT—meeting the autonomy challenge. arXiv preprint arXiv:1806.09846 (2018)
11. Falcone, R., Castelfranchi, C.: The human in the loop of a delegated agent: the theory of adjustable social autonomy. IEEE Trans. Syst. Man Cybern.-Part A: Syst. Hum. **31**(5), 406–418 (2001)
12. Falcone, R., Castelfranchi, C.: Levels of delegation and levels of adoption as the basis for adjustable autonomy. In: Lamma, E., Mello, P. (eds.) AI*IA 1999. LNCS (LNAI), vol. 1792, pp. 273–284. Springer, Heidelberg (2000). https://doi.org/10.1007/3-540-46238-4_24
13. Kowalski, J., Biele, C., Krzysztofek, K.: Smart home technology as a creator of a super-empowered user. In: Karwowski, W., Ahram, T. (eds.) IHSI 2019. AISC, vol. 903, pp. 175–180. Springer, Cham (2019). https://doi.org/10.1007/978-3-030-11051-2_27
14. Sapienza, A., Cantucci, F., Falcone, R.: Modeling interaction in human-machine systems: a trust and trustworthiness approach. Automation **3**(2), 242–257 (2022)
15. Fortino, G., Fotia, L., Messina, F., Rosaci, D., Sarné, G.M.: Grouping IoT devices by Trust and Meritocracy. In: 2021 International Conference on Cyber-Physical Social Intelligence (ICCSI), pp. 1–5. IEEE (2021)
16. Castelfranchi, C., Falcone, R.: Trust Theory: A Socio-Cognitive and Computational Model, vol. 8. Wiley, Hoboken (2010)
17. Przegalinska, A.: Wearable Technologies in Organizations: Privacy, Efficiency and Autonomy in Work. Palgrave Macmillan, Cham (2019)
18. Wilensky, U.: NetLogo. Center for connected learning and computer-based modeling, Northwestern University, Evanston (1999)
19. Kuhn, C., Lucke, D.: Supporting the digital transformation: a low-threshold approach for manufacturing related higher education and employee training. Proc. CIRP **104**, 647–652 (2021)
20. Janiesch, C., Fischer, M., Winkelmann, A., Nentwich, V.: Specifying autonomy in the internet of things: the autonomy model and notation. Inf. Syst. e-Bus. Manage. **17**(1), 159–194 (2019)

Efficient Approaches to Categorize Unstructured Documents into Sustainable Categories by Using Machine Learning

Vikrant$^{(\boxtimes)}$, Sheikh Sharfuddin Mim, and Doina Logofatu$^{(\boxtimes)}$

Frankfurt University of Applied Sciences, Frankfurt am Main, Germany
vikrant@stud.fra-uas.d, logofatu@fb2.fra-uas.de

Abstract. Each business must submit an annual sustainability report to assess its influence on the economy, environment, society, and human rights. In the sustainability report, fuel, water, heating, waste disposal, materials, and electricity use are calculated annually. To do so, assess and record all invoices for the aforementioned categories. Classifying and processing a large quantity of unstructured documents to extract information is needed. Manually or using machine learning methods. This paper aims to investigate different approaches using machine learning algorithms to allocate each submitted invoice to a sustainable category and extract the required information.

Keywords: Sustainability · Document Classification · Machine Learning · Naive Bayes · K-Means · Bayesian Gaussian mixture model · Elbow Method

1 Introduction

Sustainability is a consummately effective 'boundary term'. In policy-oriented research, the term describes what public policies should accomplish [8]. Social, economic, and environmental performance are assessed in sustainability reporting. Public bodies generate quantitative and qualitative reports to address stakeholder concerns is called sustainable reporting. International reporting requirements are followed. Popular ones include GRI. UN employs GRI. EU-commission mandates that from 2023, all companies with 250 or more employees must produce a sustainability report [2]. GRI SaaS is in demand because of these reports. GRI must follow formulae and norms, but outcomes are only feasible with proper invoice inspection. Human mistake may affect the reliability of created reports. This questioned the process and led to the necessity to construct an algorithm utilizing ML to identify a company's documents, images, and invoices to a certain sustainability class, such as electricity, water, fuel, waste, materials, and mobility. This paper compares ML techniques to classify unstructured materials into sustainable categories for a sustainability report.

© The Author(s), under exclusive license to Springer Nature Switzerland AG 2023
L. Braubach et al. (Eds.): IDC 2022, SCI 1089, pp. 188–194, 2023.
https://doi.org/10.1007/978-3-031-29104-3_21

2   Related Work

A survey covers document categorization using keyword extraction [5]. Bayes-theorem-based Bayesian classification is discussed. Initially, a bag of words was extracted and converted to a list, then words were annotated using posterior probability. Word frequency classifies documents. Explaining decision tree. In this method, document classification rules are established in a tree. Document class determined through tree traversal.

Zehra and Eser clustered Yahoo and Karypis Labs data in 2007 [3]. They sped up K-means using centroids. Instead of document similarity, they employed Euclidean-distance and cosine metrics. Consine similarity enhanced centroid-based techniques. Centroid-based classification is faster than KNN.

3   Text and Document Classification

Optical character recognition (OCR) makes scanned, printed, or hand-written text editable [7]. An OCR machine can read text like a person. The brain understands text using lexical and linguistic abilities. Structured texts and documents are needed for **Feature Extraction**. First, text categorization apps process features. Many approaches exist to clean data: Noise Removal, Stemming, Lemmatization, and Tokenization [6]. **Classification Techniques -** "Naive Bayes", the most common text classification technique based on Bayes theory [6]. NB classifier computes a posteriori probability from labeled documents. Unlabeled documents are allocated to the class with the highest posteriori probability. "K-Means" groups centroid-based data into k clusters with equal points. k-means clustering allocates each data point to one of the k-sets with the least Euclidean distance. **Evaluation** - Evaluation metrics and performance measures and highlight ways in which the performance of classifiers can be compared. These metrics (recall, precision, accuracy, micro-average, macro average etc.) are based on a "confusion matrix" that includes true positives (TP), false positives (FP), false negatives (FN), and true negatives (TN) [6]. **Accuracy** is defined as the proportion of correct forecasts among all predictions (Eq. 1 [6]). Accuracy is only useful when datasets are symmetric (false negative and positive counts are equivalent) and the costs of both are comparable. **Recall** (Eq. 2 [6]) is defined as the fraction of known positives that are correctly predicted i.e., the true positive rate. **Precision** (Eq. 3), i.e., positive predictive value is the ratio of accurately anticipated positives to all positives [6]. Choose precision if you want to be more confident in your true positives. **F1-score** (aka**F-score**/**F-measure**) (Eq. 4) is preferable if precision and recall are balanced; if false positives and false negatives cost differently, F1 is your savior if you have an unbalanced class distribution.

$$accuracy = \frac{(TP + TN)}{(TP + FP + FN + TN)} \tag{1}$$

$$recall = \frac{\sum_{l=1}^{L} TP_l}{\sum_{l=1}^{L} TP_l + FN_l} \tag{2}$$

$$precision = \frac{\sum_{l=1}^{L} TP_l}{\sum_{l=1}^{L} TP_l + FP_l} \tag{3}$$

$$F1score = \frac{2 * (Recall * Precision)}{(Recall + Precision)} \tag{4}$$

4   Implementation Details

4.1   Dataset Preparation

Previous sections included OCR text extraction, data cleaning, text preparation, and text-to-number representation. While the previous techniques were utilized on a focused dataset, the dataset was first prepared using a rule-based classification algorithm that looks for a given set of declared keywords in document text and tags them with the related document category (Fig. 1). The document isn't labeled if no criteria match. Django Python and aws's Postgres are used. Postgres tables hold AWS S3 data. This paper won't explain the application further. The present approach creates ML training and test datasets but is a poor classifier.

Document Types as per Sustainability							
Mobility	Waste	Electricity	Material	District Heati	Commuting	Water	Fuel
traveling	müll	strom	Sand	fernwärme	pendler	trinkwasser	Benzin
bahn	abfall	elektrizitat	Kies	wärme	zug	leitungswasse	Diesel
flug	entsorgung	energie	Ton	joule	auto	wasserabrech	LPG
airline	entsorgungsu	spannung	Stein	kwh	bus	wasserrechnu	CNG
airlines	abfallmenge	kwh	Zement	heizen	mitarbeiter	zählerstand	E10
boading	abfalltonne	kilowattstunc	Beton	heizkosten	angestellte	frischwasser	Super
zugticket	abfallentsorg	joule	Salz	fernwärme	öffentliche	frischwasserg	Super 95
zug	abfallwirtsch	eon	Eisen	heating	ÖNV	wasserzähler	Aral
hotel	restmüll	vattenfall	Stahl			abwassergebü	Agip
spa	papiermüll	kilowattstunc	Aluminium			abwasser	Shell
unterkunft	biomüll	naturstrom	Kulper				ESSO
plane	abfallentsorger		Blei				tanken
train			Zink				tankstelle
accommodation			Lithium				Shell
hostel			Brom				OMV
trip			seltene erden				
departure			Öl				
arrival			Öle				
deutsche bahn			Gummi				
fare			Polymer				
fahrpreis			Polymere				
			Kohle				

Fig. 1. Document Types and corresponding Keywords.

4.2   Data Cleaning and Preparing

Rules-based algorithms categorize documents. Documents with one tag are extracted to generate a new Django query set. It has training and test arrays. Training sets include 823 and 211 documents. Most documents were on electricity and fuel. Data cleaning involves deleting stop words, capitalization, numbers, patterns, extra spaces, and symbols. German papers were cleaned up using spacy's German module. In order to create a list of stop words, nltk's library 'stopwords' was utilized and 686 more were added. 'Pyspellchecker' corrected spelling.

4.3   Optimal Number of Clusters

The faulty rule-based system initially grouped documents into 7 clusters. So, we wonder if there are actually 7 data clusters. Hierarchical Clustering and Elbow method were utilized. **Hierarchical Clustering** explains how data points are clustered using a "proximate distance matrix". Ward method, which is based on an agglomerative hierarchical clustering process, blends neighboring locations based on a *function* [4]. In this approach, the function is residual. To use ward, calculate each document's "distance matrix". The TF-IDF matrix is used to calculate document-corpus cosine similarity. **Elbow Method** selects the ideal number of clusters by fitting the model with a range of k values. If the plot resembles an arm [1], the curve's elbow indicates that the model fits best with that number of clusters. In our situation, 4 clusters fit well, proving the Ward method.

4.4   K-Means Unsupervised Algorithm

K-Means creates k non-overlapping clusters. K-Means clusters data points based on their squared distance to the centroid, reducing variation. TF-IDF training uses K-Means from SK-Learn Python. After modeling training and test values, scatter plots were constructed (Fig. 2(a, b)). Figure 2(a) displays 4 projected clusters, Fig. 2(b) shows fitting prediction around test data centroids. Let's plot predicted top features using k-means (first 20 words) according to their importance scores (Fig. 3). Four clusters produce good results because the words correlate well and most top words aren't common.

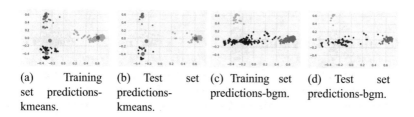

(a)   Training set predictions-kmeans.

(b)   Test set predictions-kmeans.

(c) Training set predictions-bgm.

(d)   Test set predictions-bgm.

Fig. 2. K-Means (a, b) and Bayesian Gaussian Mixture (c, d) scatter plots.

4.5   Bayesian Gaussian Mixture Algorithm

It's an unsupervised learning approach for multi-class label data, including classification, outlier detection, etc. To determine cluster location, use means and precision, the inverse of variance. When a Gaussian model is given a mean and a precision, it tries to plot the data points across the mean with the supplied precision [9]. The Gaussian model is improved using Naive Bayes. Gamma distribution, or prior value, helps Gaussian model find clusters. The SK-Learn

Python library's Bayesian Gaussian Mixture algorithm fit the TF-IDF training set. Scatter plots were created after fitting the model and predicting the training and test set values (Fig. 2(c, d)). Figure 2(c) displays 4 projected clusters, and Fig. 2(d) illustrates that test data predictions fit. Notably, the initial 2 groups are predicted differently by k-means (Fig. 2(a, b)). Let's display the top features predicted by Bayesian Gaussian Mixture (first 20 words) according to their importance scores (Fig. 4). Four clusters give superior outcomes due to better correlation between words and most of the top words aren't common. Next, we show how we evaluate several methods to see whether model is a "potential classifier".

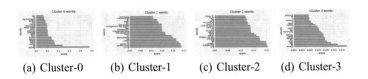

(a) Cluster-0   (b) Cluster-1   (c) Cluster-2   (d) Cluster-3

Fig. 3. K-Means - Top features.

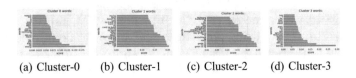

(a) Cluster-0   (b) Cluster-1   (c) Cluster-2   (d) Cluster-3

Fig. 4. Bayesian Gaussian Mixture - Top Features.

5   Results and Evaluation

Although the initial training data was created from 8 distinct categories of documents, electricity (244) and fuel (607) records dominated. These sorts of documents have more TF-IDF features, but others were still important. To enhance

Cluster	train		test	
	Kmeans	BayesianGaus	Kmeans	BayesianGaussian
0	116	239	28	69
1	118	228	27	59
2	247	109	65	26
3	342	247	91	57

Fig. 5. Output-grouping 4 cluster models.

Test	K-Means-Revised		Bayesian Gaussian Mixture-Revised	
Scores/Experiments	Training phase	Testing and Validation phase	Training phase	Testing and Validation phase
Accuracy	0.9720534629404617	0.9478672985781991	0.8432563791008505	0.838862559241706
F1-score Weighted	0.9719017989579976	0.9441567820786133	0.842514784196826	0.835694783940841
F1-score Macro	0.9669367201109602	0.9266864881177557	0.8658690390051049	0.860660116295362
F1-score Micro	0.9720534629404617	0.9478672985781991	0.8432563791008506	0.838862559241706

Fig. 6. Evaluation for 4 cluster models.

algorithms, fit the training set into more clusters. Rewriting the model with 4 clusters improves performance using the Ward and Elbow methods. Before comparing accuracy and F1-scores, let's analyze how many documents were assigned to each algorithm cluster to determine how comparable the models are.

In Fig. 5, it can be noticed that K-Means cluster (0, 2) predicts a similar number of documents as Bayesian Gaussian cluster (2, 0) for both training and test sets predictions, which can also be verified from Fig. 2(a, b) and Fig. 2(c, d), where 2 clusters are perfectly aligned to each other while the other two are not. Improved assessment utilizing cross-validation and logistic regression. All methods used so far will be cross-validated against the Logistic regression model to measure accuracy and F1-score. In Fig. 6, K-Means has an accuracy of 97.2% for training and 94.72% for test & validation, and a weighted F1-score of 0.97 for training and 0.94 for test set, which is better than Bayesian model's 0.84 F1-score and K-Means predicts proper clusters. We classified 4 of 8 documents. The training data was badly structured, with only four sorts of documents. K-Means should replace rule-based classification despite drawbacks.

6   Conclusion and Future Work

The paper covers text classification algorithms and helps choose a classifier to predict document sustainability. K-Means and Bayesian Gaussian Mixture missed 7 clusters. Excellent cross-validation accuracy and F1-score, however 4 of 7 clusters overlapped, resulting in deceptive class predictions. Ward (hierarchical) and Elbow methods revealed 4 clusters, increased accuracy to 97% and F1-score to 0.97. Revised algorithm anticipated test data. This approach can be enhanced in the future so it can create seven classes. Documents from 2 classes were collected. This method may be enhanced by supplying balanced data for all sustainability categories and by combining several classification techniques to increase 'precision'. This might establish a sustainable SaaS. While the study was originally done to categorize invoices, images, and other data into sustainability categories, the algorithm may predict additional sub-categories; for example, if a document related to fuel, what kind of fuel (petroleum, benzene, etc.) was used. A sustainability reporting SaaS with "collected to analyze" data may help organizations notice patterns and prevent excessive resource usage.

References

1. Dutta Baruah, I: Mastering clustering with a segmentation problem (2021). https://www.kdnuggets.com/2021/08/mastering-clustering-segmentation-problem.html. Accessed 6 Jan 2022
2. Amfori.org: Sustainability Disclosure Becoming the Norm (2021). https://www.amfori.org/news/sustainability-disclosure-becoming-norm. Accessed 7 July 2022
3. Cataltepe, Z., Aygun, E.: An improvement of centroid-based classification algorithm for text classification. In: 2007 IEEE 23rd International Conference on Data Engineering Workshop, pp. 952–956 (2007)

4. Wikipedia.org: Ward's method (2020). https://en.wikipedia.org/wiki/Ward%27s_method. Accessed 5 Jul 2022

5. Yoganand, C.S., Praveen, N., Saranya, N., Ganesh Karthikeyan, V.: Survey on document classification based on keyword and key phrase extraction using various algorithms. Int. J. Eng. Res. Technol. (IJERT), **3**(2) (2014)

6. Kowsari, K., Jafari Meimandi, K., Heidarysafa, M., Mendu, S., Barnes, L., Brown, D.: Text classification algorithms: a survey. Information **10**(4), 150 (2019). https://doi.org/10.3390/info10040150

7. Patel, C., Patel, A., Patel, D.: Optical character recognition by open source OCR tool tesseract: a case study. Int. J. Comput. Appl. **55**(10) (2012)

8. Scoones, I.: Sustainability. Dev. Pract. **17**(4–5), 589–596 (2007). https://doi.org/10.1080/09614520701469609. November 2010

9. du Toit, J.: Bayesian Gaussian mixture models (without the math) using Infer.NET (2020). https://towardsdatascience.com/bayesian-gaussian-mixture-models-without-the-math-using-infer-net-7767bb7494a0. Accessed 8 July 2022

Prediction Models Applied to Lung Cancer Using Data Mining

Rita Sousa⬥, Regina Sousa⬥, Hugo Peixoto$^{(\boxtimes)}$⬥, and José Machado⬥

ALGORITMI/LASI, University of Minho, Braga, Portugal
a88333@alunos.uminho.pt, regina.sousa@algoritmi.uminho.pt,
{hpeixoto,jmac}@di.uminho.pt

Abstract. Lung cancer is the most common cause of cancer death in men and the second leading cause of cancer death in women worldwide. Even though early detection of cancer can aid in the complete cure of the disease, the demand for techniques to detect the occurrence of cancer nodules at an early stage is increasing. Its cure rate and prediction are primarily dependent on early disease detection and diagnosis. Knowledge discovery and data mining have numerous applications in the business and scientific domains that provide useful information in healthcare systems. Therefore, the present work aimed to compare several prediction models as well as the features to be used, with the help of Weka and RapidMiner tools. Both classification and association rules techniques were implemented. The results obtained were quite satisfactory, with emphasis on the Naive Bayes model, which obtained an accuracy of 95.03% for cross-validation 10 folds and 94.59% for percentage split 66%.

Keywords: Lung Cancer · Data Mining · Classification · Association Rules

1 Introduction

The term cancer refers to a group of diseases characterized by the development of abnormal cells anywhere in the body, which divide and grow uncontrollably [1]. Lung cancer is one of the leading cancers for both genders all over the world. It is the most common cause of cancer death and reaches 19.4% of the total [2]. Mortality and morbidity due to tobacco use is very high, about 90% of cases of lung cancer are related to exposure to tobacco smoke due to cigarettes that contains over 70 cancer-causing chemicals. This disease mostly affects people between the ages of 55 and 65 and often takes many years to develop [3]. Being a disease that is commonly misdiagnosed and a disease that impacts do many people, an early and correct diagnosis can save a lot of lives. This can be achieved with the development of predictive models of lung cancer that can help in decision

This work has been supported by FCT-Fundação para a Ciência e Tecnologia within the R&D Units Project Scope: UIDB/00319/2020.

making software [4]. This models are developed using Data Mining approaches, that is based on the identification of anomalies, patterns, and correlations, which are difficult to find and detect with traditional statistical methods, in large data sets to predict outcomes. This work aims to compare several prediction models as well as the features to be used, with the help of Weka and Rapid Miner.

2   Related Work

V. Krishnaiah developed a paper named Diagnosis of Lung Cancer Prediction System Using Data Mining Classification Techniques [3], whose objective was to summarize various review and technical articles on diagnosis of lung cancer. This work compared the models are Naïve Bayes, Decision Trees (J48/C4.5), OneR and Neural Network and reached the conclusion that Naïve Bayes the most effective model at predicting patients with lung cancer disease, followed by Association Rules.

In 2015, Haofan Yang wrote a paper named Data mining in lung cancer pathology staging diagnosis: Correlation between clinical and pathology information [2]. The goal was to demonstrate the feasibility of applying the clinical information to replace the pathology report especially in diagnosing the lung cancer pathology staging. For this purpose, the Apriori algorithm was used to extract association rules and the significance of each generated rule was examined using support, confidence, and lift. The evaluation results demonstrated that the proposed *framework* can provide insight into solutions for support diagnosis of lung cancer pathological staging.

3   Materials and Methods

The dataset used was "Lung Cancer" [5] which contains data of 309 patients where 270 were diagnosed with lung cancer. To reach the main goal, a set of tools and techniques was used. Cross Industry Standard Process for Data Mining (CRISP-DM) is a structured approach to data mining and its steps were followed during the development of this work [7]. Through the use of learning or classification algorithms based on neural networks and statistics, one is able to explore a set of data, extract or help to highlight patterns and assist in the discovery of knowledge, using the Data Mining process [6]. Both Data Mining tools Weka and Rapid Miner were used in the process:

Weka is a Machine Learning workbench initially intended to aid in the application of machine learning technology to real-world data sets, specifically data sets from agricultural sector. It contains tools for data preparation, classification, regression, clustering, association rules mining and visualization [8,9].

Rapid Miner that is a commercial tool for data analysis using machine learning that can be considered an alternative to Weka. The function of this tool is to speed up the process of creating predictive analyses and make it easier to apply them in practical business scenarios. In this tool, association rules will be performed in order to understand which symptoms manifest together [10].

4   Data Mining Implementation

Business Understanding - The main goal of this work is to develop a prediction model of lung cancer in order to assist health care professionals in making decisions to prevent diagnostic errors. As peripheral approach a set of association rules, in order to determine attribute relations was also performed.

Data Understanding - The "Lung Cancer" dataset has a cc0 license, is in the public domain and also has public visibility. This dataset has 5091 downloads and 34388 views and its source is "Online Cancer Prediction System". This dataset is composed by 16 Nominal attributes:

- Gender (M-Male, F-Female)
- Age (Integers - 21 to 87)
- Smoking, Yellow Fingers, Anxiety, Peer Pressure, Chronic Disease, Fatigue, Allergy, Wheezing, Alcohol Consuming, Coughing, Shortness of Breath, Swalloing Difficulty, Chest Pain (yes-2, no-1)
- Lung Cancer (yes, no) - label attribute.

Data Preparation

- **Weka** - Through the analysis of the attributes, it was possible to notice that there were no missing values or duplicated data. Moreover, Interquartile Range filter, showed that there were no values statistically considered outliers. Weka Attribute Selection filter, constructed a subset with the attributes age, anxiety, peer pressure, chronic disease, fatigue, allergy, wheezing, alcohol consuming, coughing, swallowing difficulty and chest pain. Since the lung cancer class had 270 patients with lung cancer and only 39 without lung cancer, it was necessary to balance the class. To do this the Smote method was used, which consists of generating synthetic data of the minority class from neighbors. Using the filter "SMOTE", it was possible to increase the number of patients that weren't diagnosed with lung cancer to 273 [11].
- **Rapid Miner** - There are values that are considered outliers in the age attribute. In order to eliminate the instances whose age corresponded to an outlier, the operator called "Delete Outlier (Distance)" was used in Rapid Miner, followed by a filter to only let through instances that are not outliers.

Modeling

- **Weka** - To achieve the prediction model a global Data Mining Model (DMM) was constructed. This is a Classification with two set of scenarios, one with all attributes and another with the attributes that resulted from applying the "Attribute Selection" filter (allergy, wheezing, alcohol consuming, coughing, swallowing difficulty and lung cancer). Data mining techniques applied were: OneR, JRip, J48, Naive Bayes and Part. In addiction, there is only one data approach which is with smote, two sample methods, cross-validation and percentage split, and one target that corresponds to the attribute lung cancer [12,13].

– **Rapid Miner** - In order to access attribute relationship, Association Rules
was implemented. In order to find the most frequent item sets operator "FP
Growth" was used. This operator efficiently computes all frequently occurring
item sets in an ExampleSet, using the FP-tree data structure. Before adding
this operator, it was necessary to transform the data that was imported as
integer to binominal, since the Rapid Miner association rule operators require
that the attributes of this type, by adding the "Numerical to Binominal"
operator. In this operator it was necessary to create a subset consisting of
the attributes whose type was wanted to be transformed. Then, the operator
"Select Attributes" was added in order to reduce the number of attributes,
remaining all attributes except age, gender and lung cancer. To investigate
these relationships we can use the "Create Association Rules" operator. This
operator uses the data from the pattern frequency matrix and looks for any
patterns that occur often enough to be considered rules.

Evaluation and Discussion

– **Weka** - To evaluate the classification algorithms, performance metrics were
calculated, such as accuracy, sensitivity and precision. These metrics are calculated through confusion matrices that provide the amount of true positives
(TP), true negatives (TN), false positives (FP) and false negatives (FN) [14].
Table 1 shows the performance metrics achieved for the algorithms. However,
the OneR algorithm had the worst percentages, which can be explained by the
fact that this algorithm creates rules taking only one attribute into account.
The best algorithm was, by a wide margin, Naïve Bayes, although this is
considered a less accurate algorithm compared to other algorithms since it
assumes that the attributes are independent of each other and that there is
no correlation between them [15].

Table 1. Performance metrics for the dataset consisting of all attributes

DMT	SD	Accuracy (%)	Sensitivity (%)	Precision (%)
OneR	Cross-Validation 10 folds	77,90	77,42	77,42
	Percentage Split 66%	77,84	71,20	71,20
JRip	Cross-Validation 10 folds	93,37	95,40	95,40
	Percentage Split 66%	93,51	96,59	96,59
J48	Cross-Validation 10 folds	94,29	94,49	94,49
	Percentage Split 66%	93,51	96,59	96,59
Naive Bayes	Cross-Validation 10 folds	95,21	96,25	96,25
	Percentage Split 66%	94,59	96,67	96,67
Part	Cross-Validation 10 folds	94,11	93,50	93,50
	Percentage Split 66%	94,59	94,68	94,68

The results drawn from the performance metrics for the subset, represented
in Table 2, are the similar to those drawn from the dataset with all attributes,
meaning that the percentages of accuracy, sensitivity and precision were high
for all techniques, with the percentages for the OneR technique being the

Table 2. Performance metrics for the subset

DMT	SD	Accuracy (%)	Sensitivity (%)	Precision (%)
OneR	Cross-Validation 10 folds	77,90	77,42	77,42
	Percentage Split 66%	77,84	71,20	71,20
JRip	Cross-Validation 10 folds	94,29	95,15	95,15
	Percentage Split 66%	93,51	96,59	96,59
J48	Cross-Validation 10 folds	94,11	94,79	94,79
	Percentage Split 66%	93,51	94,56	94,59
Naive Bayes	Cross-Validation 10 folds	95,03	95,56	94,57
	Percentage Split 66%	94,59	96,67	96,67
Part	Cross-Validation 10 folds	93,74	91,93	91,93
	Percentage Split 66%	94,59	95,65	95,65

lowest and the ones for the Naïve Bayes the highest. More over, there is not much difference in results between the different sample data. Comparing the two scenarios created, it is possible to verify that there is not much difference in the values of the performance metrics. This shows that, although the "attribute selection" filter selects the attributes with the highest predictive ability, the resulting attributes are also relevant for lung cancer prediction.

– **Rapid Miner** - Since the number of association rules generated is high due to the high number of attributes, only the rules whose confidence is higher than 85% were selected. For example, the first rule of association says that if a patient's symptoms are fatigue, coughing, and wheezing, the patient in question also has shortness of breath. This rule has a support of 27.8% which means that these symptoms arise simultaneously in 27.8% of the transactions, and a confidence of 86.5% which means that in 86.5% of the transactions, patients who have fatigue, coughing and wheezing as symptoms also have shortness of breath.

5   Conclusions and Future Work

Nowadays, cancer has become devastating and is a threat to our lives. Thus, experts have introduced many useful methods to diagnose the disease at earlier stages. The high percentages of accuracy, sensitivity and precision in both sample data show that all the algorithms used are optimal to apply to the "Lung Cancer" dataset. Still, one can see that these percentages have lower values for the OneR algorithm and higher values for the Naive Bayes algorithm. The results of the association rules are ordered in increasing order of the percentage of confidence, and therefore the premise with the highest confidence is the one in the last row of the table. However, it can be seen that this is not the premise with the most support. Thus, an association could be made between rules number thirteen and fourteen because they have the most support and confidence, respectively, within the rules present in the table.

In the future, it is essential to gather more data and create new subsets to further investigate the relevance of each attribute in lung cancer.

References

1. Cancer Online: What is lung cancer?. https://www.cancro-online.pt/cancro-do-pulmao/informacao-basica/o-que-e-o-cancro-do-pulmao/. Accessed 5 Jan 2022
2. Yang, H.: Data mining in lung cancer pathologic staging diagnosis: Correlation between clinical and pathology information (2015). Accessed 29 Dec 2021
3. Krishnaiah, V.: Diagnosis of lung cancer prediction system using data mining classification techniques (2013). Accessed 2 Jan 2022
4. Reis, R., Peixoto, H., Machado, J., Abelha, A.: Machine learning in nutritional follow-up research (2017). https://www.degruyter.com/document/doi/10.1515/comp-2017-0008/html. Accessed 29 Mar 2022
5. Bhat, M.A.: Lung Cancer (2021). https://www.kaggle.com/datasets/mysarahmadbhat/lung-cancer. Accessed 18 Dec 2021
6. DevMedia: Data Mining: concepts and use cases in healthcare. https://www.devmedia.com.br/data-mining-conceitos-e-casos-de-uso-na-area-da-saude/5945. Accessed 21 Dec 2021
7. Horácio, J.: Data driven mindset - O modelo de mineração CRISP-DM. https://jorgeaudy.com/2021/01/29/data-driven-mindset-o-modelo-de-mineracao-crisp-dm/. Accessed 22 Dec 2021
8. Damasceno, M.: Introduction to Data Mining using Weka. http://connepi.ifal.edu.br/ocs/anais/conteudo/anais/files/conferences/1/schedConfs//papers/258/public/258-4653-1-PB.pdf. Accessed 30 Dec 2021
9. Garner, S.R.: Weka: The waikato environment for knowledge analysis. In: Proceedings of the New Zealand Computer Science Research Students Conference, vol. 1995, pp. 57–64 (1995). Accessed 29 Mar 2022
10. iMasters: Data Mining: Association Rules. https://imasters.com.br/back-end/data-mining-na-pratica-regras-de-associacao. Accessed 30 Dec 2021
11. Santana, R.: Dealing with unbalanced classes - machine learning (2020). https://minerandodados.com.br/lidando-com-classes-desbalanceadas-machine-learning/. Accessed 10 Jan 2022
12. Fonceca, F., Peixoto, H., Mirande, F., Machado, J., Abelha, A.: Step towards prediction of perineal tear (2017). https://repositorium.sdum.uminho.pt/bitstream/1822/51692/1/3.pdf. Accessed 10 Jan 2022
13. Neto, C., Peixoto, H., Abelha, V., Abelha, A., Machado, J.: Knowledge discovery from surgical waiting lists (2017). https://www.sciencedirect.com/science/article/pii/S1877050917323438. Accessed 29 Mar 2022
14. iMasters: Machine Learning: Metrics for Classification Models (2019). https://imasters.com.br/desenvolvimento/machine-learning-metricas-para-modelos-de-classificacao. Accessed 11 Jan 2022
15. Rodrigues, M., Peixoto, H., Machado, J., Abelha, A.: Understanding stroke in dialysis and chronic kidney disease (2017). https://www.sciencedirect.com/science/article/pii/S1877050917317052. Accessed 29 Mar 2022

Social Systems

Sentiment Analysis of Social Network Posts for Detecting Potentially Destructive Impacts

Diana Gaifulina, Alexander Branitskiy, Dmitry Levshun, Elena Doynikova[✉],
and Igor Kotenko

SPC RAS, 39, 14th Liniya, St. Petersburg, Russia
{gaifulina,branitskiy,levshun,doynikova,ivkote}@comsec.spb.ru

Abstract. The paper considers the posts' sentiment in the social network communities in the context of correlation with possible destructive impacts. Under the destructive impacts the authors understand the impacts that can "provoke aggressive actions and aggressive behavior in relation to others or yourself". The paper describes an approach to the determination of posts' sentiment. The authors propose an additional feature for the detection of destructive impacts of social network communities based on posts' sentiment. The experiments were conducted to test the sentiment analysis models, to analyse the proposed feature based on posts' sentiment, and test the classifier for the detection of the potentially destructive impacts. The analysis of the correlation of the proposed feature with the communities that have potentially destructive impacts on anxiety is conducted. The analysis of the obtained results is provided. During the experiments, the authors found out that consideration of the posts' sentiment allows increasing accuracy of the classifier for anxiety destructive impacts on 12.24%.

1 Introduction

Nowadays people spend a lot of time in the information space communicating within various information platforms. The authors suppose that information can influence people's feelings and personalities. Especially it is relevant for young people. In the research, the authors make attempts to prove this hypothesis. The authors selected for the analysis the Vkontakte social network [1] and the results of the psychological tests passed by the students as the representatives of the young generation. The common idea is to find correlations between the information provided within the social network communities and the users' personalities. For this goal the authors applied the following approach in [4,8]:

1. Get the results of the psychological test that represent the dynamics of the users' personality for the representatives of the young generation. This part of the research was conducted in conjunction with the expert psychologists

[5]. They proposed using the Ammon's test [2] that allows detecting the constructive, destructive, and deficient manifestations of the following personality Ego-functions: aggression, anxiety, external Ego-delimitation, internal Ego-delimitation, narcissism, and sexuality.

2. Collect information about the social network communities in which users who passed the test participate. The authors labeled the communities considering the results of Ammon's test for their participants. Currently, the authors considered only the destructive manifestations of the anxiety Ego-function. The community was marked as "low destructive anxiety", "medium destructive anxiety" or "high destructive anxiety" depending on the test results of the majority of its participants.

3. Train the classifier by the "low destructive anxiety", "medium destructive anxiety" and "high destructive anxiety" for the social network communities using the posts of the communities as input data to detect if there is any correlation among the information provided within the social network communities and the dynamics of the users' personality.

The experiments conducted in [4] for the 244 communities showed the following results: the best accuracy of 55,72% for the community class forecasting was obtained using soft voting ensembles. The authors used as input data only the text of the posts.

In this paper, the authors attempt to enhance the results of the classifier accuracy using the sentiment of posts in the social network communities.

The sentiment of the text reflects the emotional attitude of the author to any object expressed in this text. As a rule, the main goal of sentiment analysis is to find opinions in the text and identify their properties. The main approaches usually classify texts into three categories: positive, negative, and neutral.

There are three main types of approaches to sentiment analysis [15].

- *Rule-based approaches* use expert classification rules that define emotional keywords and sentiment dictionaries. Examples of Russian-language sentiment dictionaries are RuSentiLex [11] or KartaSlovSent [10].
- *Machine learning-based approaches* automatically extract features from text and subsequently use machine learning classification algorithms. To extract features from the text, such approaches can use both text-to-vector space mapping models, such as Bag of Words, and word embedding models, such as BERT [7], FastText [9], or ELMo [12].
- *Hybrid approaches* combine both rule-based and machine learning-based approaches.

Sentiment analysis is used in various areas of text analysis, such as product reviews and media news. One of the most popular areas is the analysis of user-generated content in social networks as important and publicly available sources of public opinion. The authors supposed that the sentiment of the text can represent its potential destructive influence on the reader as well.

The contribution of this paper is as follows:

- the approach for detecting potentially destructive impacts considering a sentiment-based feature;
- the original sentiment-based feature for detection of the potentially destructive impacts;
- enhancement of the classifier of the potentially destructive impacts on the users feeling and personality using the proposed feature (namely, increasing the results of the destructive impacts' classifier accuracy on 12.24%);
- the results of feature analysis and experiments for detection of the potentially destructive impacts on the example of the destructive anxiety.

The novelty of this paper is as follows: (1) the proposed sentiment-based feature for detection of the potentially destructive impacts; (2) enhancement of the classifier of the potentially destructive impacts on the users feeling and personalities using the proposed feature.

The paper is organized as follows. Section 2 describes the methodology of the research. Section 3 includes existing datasets for the experiments, and existing methods for sentiment determination. Section 4 provides our approach to the sentiment analysis, the selected dataset, method, the proposed feature, and the results of the experiments. Section 5 contains the description of the enhanced classifier considering the proposed feature, the results of the experiments, and a discussion of the work done. Section 6 contains the main conclusions and the future work directions.

2   Research Methodology and Input Data

The results of the community class forecasting obtained in [4] are not satisfactory. Thus, the authors aim to enhance them. The authors supposed that the sentiment of posts in the social network communities can correlate with possible destructive impacts generated by the community. To approve this hypothesis in this paper the following research methodology is used: (1) analyse datasets and models for sentiment analysis and select the dataset for training; (2) specify and calculate the sentiment-based feature for the same communities in the social network that were previously used in [4]; (3) train the classifier for sentiment analysis of posts in the social network community using the selected dataset; (4) reconstruct the classifier for the social network communities using the proposed sentiment-based feature; (5) conduct the experiments using an additional sentiment-based feature.

3   Datasets and Models for Sentiment Analysis

Analysis of the Datasets and Models for Sentiment Analysis and Selecting the Dataset for Training. Popular datasets containing Russian-language texts from social networks are RuTweetCorp [14], RuSentiment [13], and Russian Language Toxic Comments (RLTC) [3]. RuTweetCorp is based on posts in Russian from the social network Twitter and contains more than 200,000 lines of text, automatically labeled by positive and negative sentiment. RuSentiment is a set of posts from the social network VKontakte (VK), popular among the Russian audience. The dataset contains about 27,000 lines of text, manually labeled into five categories: negative, positive, neutral, speech, and skip. The speech act category includes greetings, thank you messages, and congratulatory messages, which may or may not express the actual feelings. The skip category is suitable for obscure cases, noisy posts, and non-authored content (poetry, lyrics, jokes, etc.). RLTC is a dataset of 14,000 comments from 2ch.hk and pikabu.ru marked as negative and positive.

In this study, the authors use the posts of communities (groups) of the VKontakte social network as initial data. The used dataset is described in more detail in Sect. 3. To determine the sentiment of posts, the authors suggest using a classifier pretrained on the RuSentiment dataset, because it is obtained from a similar social network.

The authors analyze RuSentiment pre-trained models for sentiment analysis from two libraries: DeepPavlov [6] and Dostoevsky [13]. Classifiers in these libraries are based on deep neural networks (DNN), such as convolutional neural networks (CNN) or bidirectional gated recurrent units (BiGRU), and use word embedding methods.

Table 1 shows a comparison of the models in terms of classification accuracy and processing time. As a validation sample, the authors use a set of 4,000 rows from the RuSentiment dataset. Based on the results obtained, the authors choose the Dostoevsky model, since it provides the highest performance with sufficient accuracy.

Table 1. Comparison of sentiment analysis models

Library	Model	Accuracy	Time (sec)
DeepPavlov	BERT[a]	0.842	568.18
	Conversational RuBERT[b]	0.891	605.38
	FastText + CNN[c]	0.92	41.65
	ELMo + CNN[d]	0.839	1049.35
	FastText + BiGRU[e]	0.793	31.96
Dostoevsky	**FastText + DNN[f]**	**0.847**	**0.807**

[a]https://github.com/deepmipt/DeepPavlov/blob/0.17.2/
deeppavlov/configs/classifiers/rusentiment_bert.json
[b]https://github.com/deepmipt/DeepPavlov/blob/0.17.2/
deeppavlov/configs/classifiers/rusentiment_convers_bert.json
[c]https://github.com/deepmipt/DeepPavlov/blob/0.17.2/
deeppavlov/configs/classifiers/rusentiment_cnn.json
[e]https://github.com/deepmipt/DeepPavlov/blob/0.17.2/
deeppavlov/configs/classifiers/rusentimen_bigru_superconv.json
[f]https://github.com/bureaucratic-labs/dostoevsky

4   Sentiment-Based Feature for Destructiveness and Experiments

Specify and calculate the sentiment-based feature. To determine the potentially destructive impacts of communities in social networks based on machine learning, the authors can use the features that characterize the community. One such important feature that the authors offer is the post's text sentiment metric.

Publication (*post*) in a social network is a collection of original content of the owner (*orig*) and quoted content of another owner (*repost*): *post* = (*orig, repost*). Let's denote the sentiment score of the publication as $S(post)$. This score can take a value from a limited list of categories or from a given sentiment range. The authors use the following sentiment categories: positive $(+1)$, negative (-1), and neutral (0). Consequently, $S(post) \in \{-1, 0, +1\}$.

The owner (user or community) U publishes a set of posts $P^U = \{post_1, post_2, ..., post_n\}$, where n is the number of posts. For a given set of publications, the corresponding set of sentiment scores can be defined as $S(P^U) = \{S(post_1), S(post_2), ..., S(post_n)\}$, where $S(post_k) \in \{-1, 0, +1\}, k \in [1, n]$.

The integral sentiment metric of the owner's publications SP^U is specified as:

$$SP^U = \frac{1}{n} \sum_{k=1}^{n} S(post_k). \tag{1}$$

The value of this metric is in the interval $[-1, +1]$, where SP $= -1$ denotes an absolute negative sentiment, and SP $= +1$ denotes an absolute positive sentiment. SP values close to 0 indicate a neutral sentiment.

In this study, the authors experiment only on the text content of the post. The authors also allow the following simplification: the authors do not consider posts containing both original text content and repost text content. Determining the sentiment of such a publication is complicated by the need for additional study of the relationship between the sentiment of the two components. In other words, the authors analyze the publication *post* = (*orig, repost*), where *orig* = $\varnothing$ or *repost* = $\varnothing$.

Train the classifier for sentiment analysis of posts in the social network community using the selected dataset. In the experiment, the authors conducted a sentiment classification of 162,607 posts from the communities of the VK social network. These communities are selected as the most popular (with a large number of subscribers) among users who have passed Ammon's test. The experimental dataset contains information about 244 communities, namely, their unique id, name, screen name, status, photo, and activity categories. An activity category is a category assigned by a community creator or administrator, such as "entertainment", "music", or "art". It is also known how many users that passed the Ammon's test are members of the community. For each community post, the following information is stored: owner id, post id, post status, post text, and repost text.

The selected Dostoevsky model was trained on RuSentiment and classifies the text sentiment into five classes: negative, positive, neutral, speech, and skip.

The authors combine the categories of neutral and speech into one category - neutral. Figure 1 shows the result of sentiment analysis of post texts. One can note that most of the posts in the sample have a neutral sentiment.

Figure 2 shows the relationship between the popularity of a post and its sentiment. Popularity is reflected in the ratio of views and likes of social network users. The number of reactions to the post increases depending on the number of views. In addition, one can note that posts with positive (blue solid line) or negative (orange dashed line) sentiment receive more likes than posts with a neutral sentiment (green dotted line). The authors can conclude that the sentiment of a post influences its perception by users.

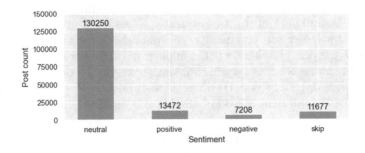

Fig. 1. Sentiment distribution of posts in the experimental sample

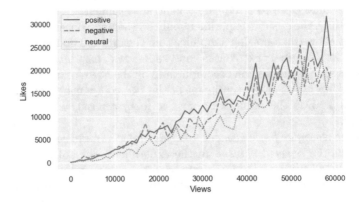

Fig. 2. The relationship between the number of views, likes and sentiment of a post

To analyze the integral sentiment metric, the authors selected 200 posts from 212 communities. The authors excluded communities with fewer posts. The authors also do not include posts in the "skip" category in this selection, because they do not participate in the calculation of the integral sentiment metric. The sentiment metric of community posts is determined by the Eq. (1). Figure 3 shows the distribution of the values of this metric for the selected communities. The Y-axis reflects the number of communities of a certain tonality on the X-axis. The line indicates the trend of values. Figure 4 shows the percentage relationship between the sentiment of community posts and their category. The authors defined the positive sentiment threshold as $SP > 0.1$ (blue color), and the negative sentiment threshold as $SP < -0.1$ (orange color). The neutral values (green color) of the metric are in the interval $SP \in (-0.1, -0.1)$. The white values in bars show the actual number of communities in a given category.

One can see that more popular communities have neutral sentiment posts (distribution around 0). At the same time, the number of communities with positive sentiment posts (distribution above 0) is greater than communities with negative sentiment posts (distribution below 0).

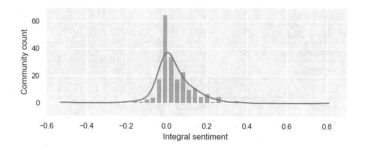

Fig. 3. Distribution of values for the integral sentiment metric of community posts

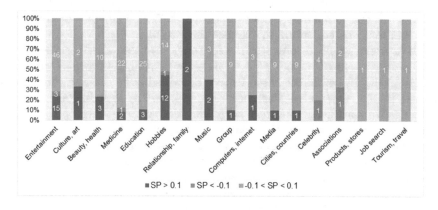

Fig. 4. The sentiment of posts by community activity categories

5   Classifier for the Detection of the Potentially Destructive Impact Considering the Sentiment-Based Feature and Discussion

Reconstruct the classifier for the social network communities. In [4] the 244 labeled communities were used to train the classifier based on the posts of the communities. The 10-block cross-validation was used: 90% of the dataset was used as the training sample and 10% - as the testing sample. The authors tested the bag of words, a weighted bag of words, a continuous bag of words, skip-gram and fastText classifiers, as well as max-wins, weighted voting, and soft voting ensembles. The obtained results are provided in Table 2.

Table 2. Results of the classification of groups by the destructive anxiety without sentiment-based feature [4]

Classifiers	Ego-function: destructive anxiety			
	Accuracy	Precision	Recall	F1-score
Bag of words	55.72%	56.22%	55.86%	56.04%
Weighted bag of words	55.11%	55.66%	55.28%	55.47%
Continuous bag of words	50.74%	50.80%	50.81%	50.80%
Skip-gram	55.68%	54.86%	55.65%	55.25%
FastText	50.37%	38.82%	49.83%	43.64%
Max-wins	54.61%	56.81%	54.59%	55.68%
Weighted voting ensemble	54.61%	56.81%	54.59%	55.68%
Soft voting ensemble	54.50%	59.74%	54.38%	56.93%

It can be seen from Table 2 that the soft voting ensembles showed the best accuracy for the community class forecasting: 55,72%. The authors propose to use a sentiment-based feature to enhance these results.

The authors specified three features for the posts-based classifier: the feature based on the result of the posts-based classifier used in the previous version of the approach, the sentiment-based feature described in Sect. 2 and the number of posts in the community. The authors conducted the experiments using the following five classifiers: k-nearest neighbors, decision tree, perceptron, linear support vector machine, and Gaussian naive Bayes classifier.

Conduct the experiments using an additional sentiment-based feature. In this research, the authors conducted experiments for the class "destructive anxiety". The 244 labeled communities were used to train the classifier based on the posts of the communities. 80% of the dataset (195 communities) was used as the training sample and 20% (49 communities) - as the testing sample.

The results of the experiments demonstrated that the k-nearest neighbors classifier based on the three introduced features allows increasing accuracy compared with the post-based FastText classifier used in the previous version on 12.24% resulting in final accuracy of 62.21% for the classification of the communities according to the possible destructive anxiety impacts, decision tree – on 6.12%. The perceptron, linear support vector machine, and Gaussian naive Bayes classifier do not increase the accuracy.

The authors calculated the Pearson correlation coefficient for the forecasted class label and post-based FastText feature (0.094), the sentiment-based feature

(0.069), and the number of posts in the community (0.02). As soon as the number of posts in the community doesn't influence the classifier accuracy, it can be excluded from the consideration.

It should be noticed that the conducted experiments have some limitations. Thus, in the approach specified in Sect. 1, the authors label the communities considering the results of Ammon's test for the community participants. Not all the participants passed the test and thus the consideration of test results for all participants can change the class labels for the communities. Besides, Ammon's test should be used in dynamics. Thus, not the test results of users in statics should be considered but the changes in their results in time. And, finally, the test results of the majority of participants do not necessarily determine the possible destructive impacts generated by the community. The authors plan to research these questions in future works and validate the labels for the communities with the help of experts.

Besides, in future work the authors plan to conduct experiments on the other Ego-functions (aggression, external Ego-delimitation, internal Ego-delimitation, narcissism, and sexuality), validate the results of sentiment-based posts classification by the experts, and extend the list of features for the communities classification by the possible destructive impacts. Detection of the communities that can spread the destructive impacts is required to generate constructive stimuli to prevent their negative influence on the personal identity. It should be noticed that all automated results should be validated by the experts before making the final conclusions.

6   Conclusion

In the paper, the authors analyzed the possibility to enhance the results of classification of the communities in the social networks considering their possible destructive impacts on the participants' personalities. The authors focused on the representatives of the young generation as one of the most vulnerable groups. Our previous classifier took as input the text of posts in the community but the obtained results were not satisfactory. The authors supposed that posts' sentiments can affect people's feelings and personalities as well. To check this hypothesis the authors used 244 communities labeled considering the results of a psychological test that represent the dynamics of the users' personality. The authors proposed the new sensitivity-based feature and used it together with the output of the post-based FastText classifier, and the number of posts in the communities. The authors used these features within the k-nearest neighbors, decision tree, perceptron, linear support vector machine, and Gaussian naive Bayes classifier. The authors found out that consideration of the post's sentiment allows increasing the results of the previous destructive impacts post-based FastText classifier accuracy on 12.24%. In future work the authors plan to conduct additional experiments, validate the results of sentiment-based posts classification and communities classification by the experts, and extend the list of features for the communities classification by the possible destructive impacts.

Acknowledgements. This research is being supported by the grant of RSF #21-71-20078 in SPC RAS.

References

1. Social network vkontakte. https://vk.com. Accessed 15 Mar 2022
2. Ammon, G., Finke, G., Wolfrum, G.: Ich-Struktur-Test nach Ammon (ISTA). Swets, Zeitlinger, Frankfurt (1998)
3. Belchikov, A.: Russian language toxic comments (2019). https://www.kaggle.com/blackmoon/russianlanguage-toxic-comments. Accessed 15 Mar 2022
4. Branitskiy, A., Doynikova, E., Kotenko, I.: Technique for classifying the social network profiles according to the psychological scales based on machine learning methods. J. Phys: Conf. Ser. **1864**, 012121 (2020)
5. Branitskiy, A., et al.: Determination of young generation's sensitivity to the destructive stimuli based on the information in social networks. J. Internet Serv. Inf. Secur. (JISIS) **9**(3), 1–20 (2019)
6. Burtsev, M.S., et al.: Deeppavlov: open-source library for dialogue systems. In: ACL, vol. 4, pp. 122–127 (2018)
7. Devlin, J., Chang, M.-W., Lee, K., Toutanova, K.: Bert: pre-training of deep bidirectional transformers for language understanding. arXiv preprint arXiv:1810.04805 (2018)
8. Doynikova, E., Branitskiy, A., Kotenko, I.: Detection and monitoring of the destructive impacts in the social networks using machine learning methods. In: Pissaloux, E., Papadopoulos, G.A., Achilleos, A., Velázquez, R. (eds.) ICT for Health. Accessibility and Wellbeing, pp. 60–65. Springer, Cham (2021). https://doi.org/10.1007/978-3-030-94209-0_6
9. Joulin, A. , Grave, E., Bojanowski, P., Mikolov, T.: Bag of tricks for efficient text classification. arXiv preprint arXiv:1607.01759 (2016)
10. Kulagin, D.: Publicly available sentiment dictionary for the Russian language kartaslovsent. In: Computational Linguistics and Intellectual Technologies: Papers from the Annual International Conference "Dia-logue", vol. 16, pp. 19–23 (2021)
11. Loukachevitch, N., Levchik, A.: Creating a general Russian sentiment lexicon. In: Proceedings of the Tenth International Conference on Language Resources and Evaluation (LREC 2016), pp. 1171–1176 (2016)
12. Peters, M.E., et al.: Deep contextualized word representations. In: Proceedings of the 2018 Conference of the North American Chapter of the Association for Computational Linguistics: Human Language Technologies, vol. 1 (Long Papers), pp. 2227–2237, New Orleans, Louisiana, June 2018. Association for Computational Linguistics (2018)
13. Rogers, A., Romanov, A., Rumshisky, A., Volkova, S., Gronas, M., Gribov, A.: RuSentiment: an enriched sentiment analysis dataset for social media in Russian. In: Proceedings of the 27th international conference on computational linguistics, pp. 755–763 (2018)
14. Rubtsova, Y.: Reducing the deterioration of sentiment analysis results due to the time impact. Information **9**(8), 184 (2018)
15. Smetanin, S.: The applications of sentiment analysis for Russian language texts: current challenges and future perspectives. IEEE Access **8**, 110693–110719 (2020)

SABRE: Cross-Domain Crowdsourcing Platform for Recommendation Services

Luong Vuong Nguyen and Jason J. Jung[(⊠)]

Department of Computer Science and Engineering,
Chung-Ang University, Seoul, Korea
j2jung@gmail.com

Abstract. Existing recommendation services place emphasis on personalization to achieve promising accuracy of recommendations. This study aims to exploit the user cognition similarity across multiple domains. The purpose is to leverage this information to enhance the user-based collaborative filtering algorithm for cross-domain recommendation services. The main idea of this is i) to collect feedback from users across multiple domains to represent user cognition; ii) to establish a user cognition-based collaborative filtering (UCCF) model for the multi-domain recommendation; iii) generating recommendations in the target domain. The experimental results demonstrate that the prediction performance of the proposed model outperforms in comparison with all baseline methods.

1 Introduction

Recommendation systems (RS) have applied to many web applications nowadays, such as sharing videos on Youtube, e-commerce on Amazon, and social networking Facebook because they can help to avoid the information overload problem [2]. Collaborative filtering (CF) has been shown to be the most promising technique in the existing RS [1, 2, 5, 16]. The core idea behind CF is to generate top suitable products for a target user based on the preferences of other users who have similar preferences to the target user.

However, in the real-world scenarios, most applications meet the data-sparse problem since only a few users give ratings or review items [2]. This makes the accuracy in generating recommendations of collaborative filtering algorithms reduced. Almost CF-based recommendation systems suffer from this data sparsity problem to several levels (the cold-start problem), especially for new products or users. This issue may result in over-fitting when training a CF-based model, which affects recommendation accuracy dramatically. Cross-domain recommendation system(CDRS) has emerged to address the data sparsity problem by utilizing relatively richer information, e.g., user/item information, thumbs-up, tags, reviews, and observed ratings, from the richer (source) domain to improve recommendation accuracy in the sparser (target) domain [3]. For example, because a common user in several domains is likely to have similar likes, the RS can propose books to users based on their movie reviews.

L. Braubach et al. (Eds.): IDC 2022, SCI 1089, pp. 213–223, 2023.
https://doi.org/10.1007/978-3-031-29104-3_24

To deal with these problems above of the cross-domain recommendation, one of the simple solutions is to extend the CF algorithms. Specifically is to change the issue in CF into multi-task learning handled by propagating similarities between multiple domains [6,7]. Surprisingly, the setting of multi-domain frequently offers similarities to online user behaviors that cannot be disregarded. Let's consider an example, in the movie domain of the SABRE platform, the system will suggest similar/related products from the book and fashion domain when users rate/like any item in the movie domain. When users click on any similar/related products, we implicit that is the similar items in the auxiliary domains (book and fashion) of users in the source domain (movie). The purpose of this is to collect cognitive similarity data of users across user activities in different domains. To do that, we deploy the crowdsourcing platform [11], called SABRE, where users can share and leave feedback (explicit and implicit) about the similar/related items from multiple different domains. Then, we can represent the user cognition based on these cognitive similarity data in order to improve the accuracy of the clustering method of the similar users from multiple domains. Regarding to this purpose, we propose the User Cognition-based Collaborative Filtering model (UCCF) for a top-N cross-domain recommendation task. In general, the UCCF is the basis of the adaptive K-Nearest-Neighbor framework. It means to learn a set of nearest neighbors who are sufficiently similar to the active user in terms of cognitive similarity data over multiple domains. The main contributions of this research are described as follows

- Deploying the online crowdsourcing platform (SABRE[1]) to collect cognitive similarity data of items from users in multiple different domains.
- Proposing the novel User Cognition-based Collaborative Filtering (UCCF) model with the basis of the adaptive KNN algorithm.

The remaining structure of this manuscript is as follows. The next section introduces the related work on the CF cross-domain recommendation systems and existing studies on improving the performance of the user-based collaborative filtering approach. Section 3 presents the cross-domain crowdsourcing platform for recommendation services based on user cognition. The experimental results and evaluation are details described in Sect. 4. In Sect. 5, we conclude the current research results and give the future work.

2 Related Work

The earliest works on Cross-domain Collaborative Filtering (CDCF) were mentioned in [3,10,12], which proposed several approaches for aggregating rating vectors of users from different domains. The CDCF methods were categorized into two types.

---

[1] http://sabre.cau.ac.kr.

- The first type assumes shared users or items in different domains (neighborhood-based models) [3,8,9]
- The second type does not require shared users or items in different domains (latent factor models) [10,12]

Specifically, in [3], the authors mention the neighborhood-based CDCF (N-CDCF). The main purpose of this approach is to estimate the similarity between users or between items. This approach can be subdivided into the user-based nearest neighbor (NCFU) and the item-based nearest neighbor (NCFI). Following this categorization, the algorithms for N-CDCF are also separated into two types: the user-based neighborhood CDCF model (N-CDCF-U) and the item-based neighborhood CDCF model (N-CDCF-I). Our research focuses on improving the user-based method and we only provide a details review of N-CDCF-U model in the rest of this section.

The problem formulation of N-CDCF-U in [3] is stated as follows. Given m domains, the set $\mathbf{D} = \{D_1, D_2, \ldots, D_m\}$ is represent all domains. The user set of $\mathbf{D}$ is represented as $U = \{u_1, u_2, \ldots, u_n\}$, where n is number of users. The items set $I_k = \{i_k^1, i_k^2, \ldots, i_k^{n(k)}\}$ belonging to the domain D_k $(0 \leq k \leq m)$, where $n(k)$ denotes the size of item set in D_k. Following the user-based CDCF algorithm, we first estimate the similarity of the users u and v $(Sim(u,v))$ who have co-rated the same set of items. $Sim(u,v)$ can be calculated by the Pearson correlation which is shown in the equation as follows.

$$Sim(u,v) = \frac{\sum_{i \in i_{u,v}} (r_{u,i} - \bar{r}_u)(r_{v,i} - \bar{r}_v)}{\sqrt{\sum i \in i_{u,v} (r_{u,i} - \bar{r}_u)^2} \sqrt{\sum i \in i_{u,v} (r_{v,i} - \bar{r}_v)^2}} \tag{1}$$

where co-rated item of user u and v is $i_{u,v} = i_u \cap i_v$ $(i_u = \cup_{d \in D_u^{id}}$ and $i_v = \cup_{d \in D_v^{id}})$ denotes the items of domains $\mathbf{D}$. $r_{u,i}$ and $r_{v,i}$ are the ratings on item i given by users u and v, respectively while r_u and r_v are the average ratings of users u and v for all the items rated, respectively. In second step, the predicted rating of an item p for user u can be calculated by the following equation

$$\hat{r}_{u,p} = \bar{r}_u + \frac{\sum_{v \in U_{u,p}^k} Sim(u,v) * (r_{v,p} - \bar{r}_v)}{\sum_{v \in u U_{u,p}^k} |Sim(u,v)|} \tag{2}$$

where $U_{u,p}^k$ is the set of top-k users (k neighbors) that are most similar to user u who rated item p.

Besides the above model, the traditional matrix factorization (MF) model is also employed to handle the CDCF problems. In [13], the Funk-SVD model has been proposed as the most commonly used MF model for a single-domain

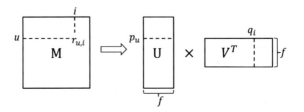

Fig. 1. Illustrate the decomposition of the Funk-SVD model

collaborative filtering recommendation system. The idea of this model is to map users and items into a joint latent factor space of dimensional f. The decomposition of Funk-SVD model is described in Fig. 1, where each item i is associated with a latent vector $q_i \in R^f$. The predicted rating of user u on item i is denoted as $\hat{r}_{u,i}$ and formulate as

$$\hat{r}_{u,i} = q_i^T p_u \tag{3}$$

The Funk-SVD model minimizes the regularized squared error on the set of known ratings to learn the latent vectors and shown in the following equation

$$\min_{q*,p*} \sum_{(u,i) \in K} \left(r_{ui} - q_i^T p_u\right)^2 + \alpha \left(|q_i|^2 + |p_u|^2\right) \tag{4}$$

where, K is the set of the (u, i) pairs for each known $r_{u,i}$. α is the constant to handle the extent of regularization that avoid over-fitting. This constant is defined by cross-validation [14]. To minimize optimization problems, the stochastic gradient descent was used which repeats through all ratings in the training dataset. In each training step, the associated prediction error of the predicted rating $r_{u,i}$ is computed by the following equation

$$e_{u,i} = r_{u,i} - q_i^T p_u \tag{5}$$

Then, the parameters are modified by magnitude proportional to θ (i.e., the learning rate) in the opposite direction of the gradient

$$q_i \leftarrow q_i + \theta(e_{u,i} p_u - \alpha q_i) \qquad p_u \leftarrow p_u + \theta(e_{u,i} q_i - \alpha p_u) \tag{6}$$

We can easily answer the CDCF issues using the standard MF approach. We may combine all of the elements from multiple domains to create an enhanced rating matrix by horizontally concatenating all matrices. As a result, we may use the MF model to detect the latent user factors and latent item factors. These latent elements are utilized to forecast. The MF model on CDCF issues is referred to as MF-CDCF in this study. The N-CDCF and MF-CDCF methods are derived directly from the single-domain neighborhood and MF-based CF methods, respectively. The single-domain approach, on the other hand, implies item homogeneity. Because objects in various domains might be highly diverse, N-CDCF and MF-CDCF fail to account for this. As a result, performance is not always adequate.

In the other research, they applied the Cross-Domain Triadic Factorization (CDTF) model [8], which takes into account the whole relationship between users, products, and domains in order to efficiently exploit user preferences on goods across domains. They employ a tensor of order three to describe the user-item-domain interaction and a tensor factorization approach to factorize people, items, and domains into latent feature vectors. The element-wise product of user, item, and domain latent factors is used to generate a rating of user for an item in a specific domain. The temporal complexity of tensor factorization, on the other hand, is exponential, being $O(k^m)$, where k is the number of factors and m is the number of domains.

In addition, the work in [9], presents a Transfer by Collective Factorization (TCF) model. The goal is to reduce the data sparsity problem in numerical ratings by transferring knowledge (like and dislike) from auxiliary domains. TCF assumes that user-item-specific latent feature matrices are exactly similar, and uses the 5-star (rating) numerical target data as well as binary like/dislike auxiliary data. Different from CBT model [10] and RMGM model [12], the TCF aims to share latent features and uses two inner matrices to analyze the data-dependent information. This purpose requires the users/items in the target matrix (rating) and the auxiliary binary matrix (like/dislike) to be aligned. Clearly, this approach only deals with the scenario of one auxiliary domain.

3   SABRE: Cross-Domain Crowdsourcing Platform

3.1   Overview

We develop the SABRE with Java 8 under Spring Framework. The architecture of SABRE is based on the MVC model which contains web service and background service. We run the Tomcat 11 on the web service side and use MySQL on the background service. All data collected from users will be stored in the MySQL database and then extract into CSV files in order to provide for our experiments. The architecture of the SABRE platform was shown in Fig. 2.

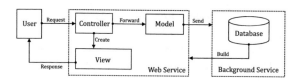

Fig. 2. Architecture of SABRE Crowdsourcing Platform

Besides, the purpose of the SABRE platform is to collect user cognition effectively. Hence, we aim to design a user-friendly interface so that normal users can use SABRE will different skill levels. To do that, we observe the essential guide for the user-interface design mentioned in [4]. In particular, when using SABRE, users no need to remember the predefined sequence, just memo URLs.

In addition, the platform provides alerts and roll-back methods for users in case they have a mistake. In case users must have stopped by a sudden situation (e.g., solving other problems), they will continue their work easily. Because SABRE has been designed very simple, and it is our main strategy when we design SABRE.

3.2   User Cognition-Based Collaborative Filtering Model

Given d different domains with a user set $U = \{u_1, .., u_n\}$, the matrix $X_d \in R^{m_d \times n}$ contain the interactions of all users U_d on items V_d in domain d-th where $d \in \{1, .., D\}$ and m_d is the size of V_d. In our hypothesis, the matrix X_d is binary because we note that the action like clicks, and purchases are easier to collect from users than the rating value. This means when the user has any interaction with an item, the element value of X_d is 1 and 0 if otherwise (no interaction).

Consider the series of matrices $\{X_1, .., X_D\}$ that indicate the user-item of multiple domains, the goal of our research is to generate a personalized rank list of the conformity items for users in each domain. To formulate the problem, we first introduce the model for a recommendation based on general similarity in a single domain. In this model, the preference values are determine with the function $f(X, \Theta)$. In which, Θ is a parameter of the model and X is the user-item interaction matrix. Our model relies on the adaptive K-Nearest-Neighbor(KNN) algorithm, one of the very popular collaborative filtering which is simple but has good performance in generating top-N recommendations [15].

In this user-based KNN problem, we set parameter Θ as matrix $\mathbf{W} \in R^{n \times n}$, and the function $f(.)$ is an aggregation of interaction values of the nereast neighbors k. Thus, in mathematically, the predicted scorce $\widehat{x}_{i,j}$ is calculated by

$$\widehat{x}_{ij} = \mathbf{x}_i^T \mathbf{w}_j \tag{7}$$

where $\mathbf{x}_i \in R^n$ is the interaction indicator of item v_i and $\mathbf{w}_j \in R^n$ is the similarity coefficients of user u_j. In the next optimization step, since the parameter $\mathbf{W}$ is considered a similarity between users and optimized by the following function

$$\widetilde{\mathbf{W}} = \arg\min_{\mathbf{W}} \mathcal{L}(\mathbf{W}) + \Omega(\mathbf{W}) \tag{8}$$

where the loss function $\mathcal{L}$ is defined as least squares and the regularization penalty Ω is used to enforce parameters with specific structures.

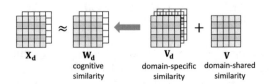

$$X_d \quad \approx \quad \underset{\substack{\text{cognitive}\\\text{similarity}}}{W_d} \quad \longleftarrow \quad \underset{\substack{\text{domain-specific}\\\text{similarity}}}{V_d} \quad + \quad \underset{\substack{\text{domain-shared}\\\text{similarity}}}{V}$$

Fig. 3. Ilustration of the UCCF Model

Considering the scenario of CDRS that existing the common users across multiple domains with the purpose to exploit the similarity between these common users based on their cognition with similar items. The key problem of this idea is to define the group of the nearest neighbors of the target user in terms of cognitive similarity between these users over different domains. Hence, we can represent the cognition of the target user by the set of similar neighbors of this target user. Hence, we can represent the cognition of the target user by the set of similar neighbors of this target user. Normally, in the single domain, the cognitive similarities are shared to represent the personalized neighbors for each user independently (domain-specific cognitive similarity). However, in the multiple domains context, we aim to combine cognitive similarities that are shared between different domains, called domain-shared cognitive similarity, with the domain-specific cognitive similarity. Thus, the proposed model is described in Fig. 3 and formulated as follows

$$\mathbf{W_d} = \mathbf{V} + \mathbf{V_d}, \quad d \in \{1, \ldots, D\} \tag{9}$$

where parameter matrix $\mathbf{V}, \mathbf{V_d} \in R^{n \times n}$ denotes the personalized neighbors with similar cognition over multiple domains of the target user and the representative neighbors of the target user in the specific domain, respectively. The similarity is then explored by regularizing the structure of the set $\mathbf{V_d}$ (set of domain-specific cognitive similarity) with group lasso. This show that the domain-specific can represent the users better and lead to the results being denser for the corresponding rows in $\mathbf{V_d}$. The group lasso $\Omega_{g-lasso}$ of $\mathbf{V_d}$ is defined as follows.

$$\Omega_{g-lasso}(\mathbf{V}_d) = \delta_d |\mathbf{V}_d|_{2,1} = \sum_{i=1}^{n} \delta_d |\mathbf{v}_d^i|_2 \tag{10}$$

where $|\mathbf{v}_d^i|$ denotes the l_2-norm value for each row in parameter set $\mathbf{V_d}$, and δ_d is the contributions of group lasso in each specific domain. In order to keep the generality, we suppose that $\delta_d = \delta$ with all d. Besides, in the overall user set, personalized neighbors of the target users are sparse. Hence, the constraint, lasso, is used to reduce noise coefficients in $\mathbf{V}$ and also formulated as follows

$$\Omega_{lasso}(\mathbf{V}) = \gamma |\mathbf{V}|_1 = \sum_{i=1}^{n}\sum_{j=1}^{n} \gamma |v_{ij}| \tag{11}$$

Finally, the cognitive similarity can be unified into a loss function with the least square loss defined as follows equation

$$\mathcal{L}(\mathbf{W}_d) = \frac{1}{2}|\mathbf{X}_d - \mathbf{X}_d\mathbf{W}_d|_F^2 \tag{12}$$

Following the Eq. 8, the UCCF model is proposed as follows

$$\min_{\Theta} \sum_{d=1}^{D} \alpha_d \left(\mathcal{L}(\mathbf{W}_d) + \Omega_{g-lasso}(\mathbf{V}_d)\right) + \Omega_{lasso}(\mathbf{V}) \quad \forall d, \quad d = 1, \cdots, D \tag{13}$$

In which, the set Θ is used to present succinctly the parameter set $\{\mathbf{W_d}, \mathbf{V_d}, \mathbf{V}\}$ and α_d balance the effect of different multiple domains. The entire model is described in Fig. 3.

4   Experiments

4.1   Datasets

To evaluate the effectiveness of the proposed method, we used the dataset collected from SABRE. We first import the collected data from Kaggle and IMDB to the SABRE platform as initial datasets. The initial data consist of user, item, and rating of Movie[2], Book[3], and Fashion[4] domains. The detailed initial dataset is shown in Table 1.

Table 1. Description of Datasets

Domains	#Items	#Users	Sparsity
Movie	14,235	1,337	0.9887
Book	32,577	4,332	0.9975
Fashion	44,121	3,221	0.9994

The SABRE platform collects cognitive similarity data of the users by allowing the users to give their implicit feedback (e.g. click on products, click on similar products) from the suggestion functions. The example of user interaction in SABRE is described as Fig. 4. We then remove all data of users who give under 10 feedback to ensure a good evaluation. The final dataset that collected from SABRE consisted of approximately 7,000 interactions of 3,210 users.

4.2   Evaluation

Our proposed model performance is analyzed by comparing the top recommendations with the true behaviors of users (click on similar products). The two metrics, mean average precision (MAP) and Normalized Discounted Cumulative Gain (NDCG), were used to evaluate our experiments. In which, the setting of NDCG was set as $N = 5$. The higher results of the metrics demonstrate the better recommendations. Regarding the baseline methods, we compare our proposed model (UCCF) with existing methods that are mrSLIM, mrBPR, PopRank, NCDCF-U. For constructing the training and testing set, we separate the

---

[2] http://recsys.cau.ac.kr:8084/ourmoviesimilarity.
[3] http://www.kaggle.com/lukaanicin/book-covers-dataset.
[4] http://www.kaggle.com/paramaggarwal/fashion-product-images-dataset.

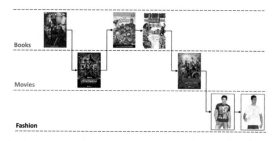

Fig. 4. An example of interactions of user across multiple domains in the SABRE platform

feedback dataset into 2 parts with 80% for training and 20% for testing. Then we randomly take samples and repeated this step 10 times. The weights combination of three domains in UCCF was set as a respectively set $\{0.2, 0.3, 0.5\}$. The optimal parameters for UCCF model are $\delta = 0.5$ and $\gamma = 0.01$. As shown in Table 2, the proposed model UCCF is consistently better than other baselines. Specifically, because PopRank is the method that just predicts products based on their popularity, this leads to all other baselines being better performance than it. This demonstrates the significance of personalization in recommendation systems because, in all baseline methods, only PopRank is not a personalized algorithm. With the experimental results on three domains movie, book, and fashion. We recognize that generating movie recommendations seems an easy task because, except PopRank, all methods have good performance in the movie domain. This means that using knowledge in the movie domain as the auxiliary domain in order to transfer information to other domains (e.g. book and fashion) would be better beneficial. Besides, the NCDCF-U (with $k = 100$ and $k = 50$ nearest neighbors) cannot obtain satisfactory results in the book and fashion domains n the absence of a well-constructed mechanism for multi-task learning. We construct mrSLIM by setting the elastic net weights to $\delta = 5$ and $\alpha = 0.1$ to get best performance for the comparison. Furthermore, UCCF outperforms the cutting-edge ranking algorithm mrBPR, with a learning rate

Table 2. The prediction performance of the proposed model (UCCF) and mrSLIM, mrBPR, PopRank, NCDCF-U on three domains. The best results are in the bold text.

	Movie		Book		Fashion	
	MAP	NDCG	MAP	NDCG	MAP	NDCG
mrSLIM	0.3114	0.1006	0.3961	0.1553	0.6215	0.4341
mrBPR	0.3433	0.1171	0.4280	0.1654	0.6306	0.3477
PopRank	0.2470	0.0719	0.2979	0.1022	0.4884	0.2349
NCDCF-U	0.2926	0.0913	0.3747	0.1425	0.5982	0.3259
UCCF	**0.3846**	**0.1341**	**0.4445**	**0.1803**	**0.6431**	**0.3552**

0.01 and weights of $l2$-norm of $1e-3$. In cap, the proposed method UCCF is outperform in comparison with all baselines methods even in the case of best setting to baselines.

5 Conclusion

This manuscript proposes a novel UCCF model for the cross-domain recommendation, which integrates the user cognition effect with personalization to construct a collective similarity parameter. Regarding this purpose, the online crowdsourcing platform was deployed to collect cognitive similarity data of items from users on multiple domains. By exploiting the similarity between users based on their cognitive similarity data across different domains, the cognitive similarity data of the active users can be predicted by the optimized neighbors. Experiments on cross-domain datasets show our method outperforms baselines for top-N recommendations.

Acknowledgment. This research was supported by the MSIT (Ministry of Science and ICT), Korea, under the National Program for Excellence in SW (20170001000061001) supervised by the IITP (Institute of Information & communications Technology Planning & Evaluation) in 2022.

References

1. Nguyen, L.V., Jung, J.J., Hwang, M.: OurPlaces: cross-cultural crowdsourcing platform for location recommendation services. ISPRS Int. J. Geo-Inf. **9**(12), 711 (2020)
2. Ricci, F., Rokach, L., Shapira, B.: Introduction to recommender systems handbook. In: Ricci, F., Rokach, L., Shapira, B., Kantor, P.B. (eds.) Recommender Systems Handbook, pp. 1–35. Springer, Boston, MA (2011). https://doi.org/10.1007/978-0-387-85820-3_1
3. Berkovsky, S., Kuflik, T., Ricci, F.: Cross-domain mediation in collaborative filtering. In: Conati, C., McCoy, K., Paliouras, G. (eds.) UM 2007. LNCS (LNAI), vol. 4511, pp. 355–359. Springer, Heidelberg (2007). https://doi.org/10.1007/978-3-540-73078-1_44
4. Galitz, W.O.: The Essential Guide to User Interface Design: An Introduction to GUI Design Principles and Techniques. Wiley, Hoboken (2007)
5. Nguyen, L.V., Hong, M.S., Jung, J.J., Sohn, B.S.: Cognitive similarity-based collaborative filtering recommendation system. Appl. Sci. **10**(12), 4183 (2020)
6. Krohn-Grimberghe, A., Drumond, L., Freudenthaler,C., Schmidt-Thieme, L.: Multi-relational matrix factorization using Bayesian personalized ranking for social network data. In: Proceedings of the Fifth ACM International Conference on Web Search and Data Mining, pp. 173–182 (2012)
7. Singh, AP., Gordon, GJ.: Relational learning via collective matrix factorization. In: Proceedings of the 14th ACM SIGKDD International Conference on Knowledge Discovery and Data Mining, pp. 650–658 (2008)
8. Hu, L., Cao, J., Xu, G., Cao, L., Gu, Z., Zhu, C.: Personalized recommendation via cross-domain triadic factorization. In: Proceedings of the 22nd International Conference on World Wide Web, pp. 595–606 (2013)

9. Pan, W., Liu, NN., Xiang, EW., Yang, Q.: Transfer learning to predict missing ratings via heterogeneous user feedbacks. In: Twenty-Second International Joint Conference on Artificial Intelligence (2011)
10. Li, B., Yang, Q., Xue, X.: Can movies and books collaborate? cross-domain collaborative filtering for sparsity reduction. In: Twenty-First International Joint Conference on Artificial Intelligence (2009)
11. Nguyen, L.V., Jung, J.J.: Crowdsourcing platform for collecting cognitive feedbacks from users: a case study on movie recommender system. In: Pham, H. (ed.) Reliability and Statistical Computing. SSRE, pp. 139–150. Springer, Cham (2020). https://doi.org/10.1007/978-3-030-43412-0_9
12. Li, B., Yang, Q., Xue, X.: Transfer learning for collaborative filtering via a rating-matrix generative model. In: Proceedings of the 26th Annual International Conference on Machine Learning, pp. 617–624 (2009)
13. Koren, Y., Bell, R., Volinsky, C.: Matrix factorization techniques for recommender systems. Computer **42**(8), 30–7 (2009)
14. Kohavi R. A study of cross-validation and bootstrap for accuracy estimation and model selection. In: InIjcai, vol. 14, no. 2, pp. 1137-1145 (1995)
15. Ning, X., Karypis, G.: Slim: sparse linear methods for top-n recommender systems. In: 2011 IEEE 11th International Conference on Data Mining, pp. 497–506 (2011)
16. Vuong Nguyen, L., Nguyen, T.H., Jung, J.J., Camacho, D.: Extending collaborative filtering recommendation using word embedding: a hybrid approach. Concurrency Comput. Pract. Experience (2021)

Balance vs. Contingency: Adaption Measures for Organizational Multi-agent Systems

Michael Köhler-Bußmeier[(✉)] and Jan Sudeikat

Department for Informatics, University of Applied Science Hamburg,
Hamburg, Germany
{michael.koehler-bussmeier,jan.sudeikat}@haw-hamburg.de

Abstract. Designing adaptive multi-agent Systems (MAS) is a challenging development effort. A key point of adaptive systems is that they provide *alternative* options for acting and designers have to weight the number and elaboration of these alternatives. Here, we concentrate on organisation-oriented MAS and show that organisational models provide suitable means for identifying key measures of this adaptivity. We define such measures, namely the *balance* and *contingency* of organizations based on our specification formalism for MAS-organisations, called SONAR.

1 Adaptivity in Agent Architectures

Usually, it seems a natural idea to implement each building block in a system's architecture exactly once. But, for complex adaptive systems (CAS), the famous BACH group (Burks, Axelrod, Cohen, and Holland) advocate for having a *(gene) pool of alternatives* for each decision at hand [1]. The choice of one of the alternatives depends on relative weights and short-term learning is implemented via adapting these weights; while long-term learning is established via means of transforming the pool itself, e.g. by genetic programming, i.e. via cross-over and mutation.

In our research we study the design of adaptive multi-agent systems (MAS) using the approach of organisation-oriented mas (Org-MAS) architectures [2] where one considers large-scale and agent-independent aspects of mas, like roles, norms, positions, interaction protocols etc. We developed a model for these Org-MAS, called SONAR [3]. The formal model of SONAR [4] is based on Petri nets [5], which facilitates the definition of the semantics as well as the analysis of concrete architectures.

The adaptivity of a system allows for *robustness* or *flexibility* [6]. A rich body of research is available for adapting systems, e.g. see [7]. In this respect, the SONAR framework provides a formal context for describing how organizations of agents can adapt. While it describes organizational structures as design elements which can be manipulated at run-time, it is agnostic towards the type

L. Braubach et al. (Eds.): IDC 2022, SCI 1089, pp. 224–233, 2023.
https://doi.org/10.1007/978-3-031-29104-3_25

of monitoring and adaption techniques (e.g. discussed in [7,8]). Here, we are particularly addressing *compositional* adaptations [9], where functional components, i.e. agents, are replaced at run-time. Examples are (micro-)Service-based Systems [10] and Distributed Execution Frameworks [9]. We particularly consider cyber-physical systems (CPS), e.g. *virtual power plants* [11] and energy communities [12], as an application domain for highly flexible organizations (cf. [13]). Allowing these systems to be adjusted at run-time, e.g. to form energy trading coalitions, requires regulating the interplay of producers, consumers, and energy storage, e.g. in order to adjust supply dynamically when new consumers arise etc. One design question is which adaption schemes are appropriate given a certain configuration of reaction times, size of batteries, failure rates of components, etc.

In this contribution we will focus on the analysis aspect of designing adaptive systems. Our SONAR-model allows to identify these decision points of the system (and also the pool of alternatives) easily. Here, our main analysis question is whether the size of all the gene pools that handle these decision points is well-designed, i.e. it is somehow *balanced* and reflects the *relevance w.r.t. the environmental dynamics* properly. We will give the demands a precise formalisation a show how they could be measured using key values derived from our SONAR-model.

The contribution has the following structure. Section 2 presents our Org-MAS model SONAR and describes how it is used to enable concrete MAS. Section 3 recalls the basic definition of the formalism and shows how it is used to specify interaction, teamwork etc. Section 4 defines the identification of decision situations and how to evaluate the balance of the system. Then, Sect. 5 integrates statistical information about the environmental behaviour to quantify how well the flexibility of a SONAR-model is adjusted. The paper closes with a conclusion.

2   SONAR: Organisations as Run-Time Models for MAS

In the following we give a short introduction into the SONAR formalism as well as its use for implementing MAS. For the overall analysis of such systems we advocate the organisational approach [2]. Here one defines the general structure of a system – including roles, positions, protocols etc., which establishes a specification for agent implementation. In previous research we developed the Org-MAS model SONAR [3] which is based on a rigorous formalism [4]. The SONAR-model defines which *tasks* may arise in the systems; it defines the *positions* in the system; it imposes constraint on the negotiation process; it specifies the allowed transformation operations; etc.

SONAR-models are used as run-time to configure a general Org-MAS engine [14]. The SONAR-engine instantiates each position as an agent: the organisation position agent (OPA). The network of OPAs defines the so-called *formal organisation.* The OPA roughly defines the *type* of agents expected at the given position. The implementation of this position is done by coupling another agent, the organisational member agent (OMA), with the OPA (Fig. 1).

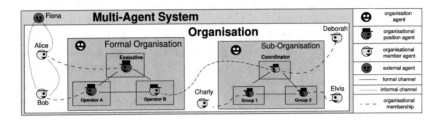

Fig. 1. A SONAR Organisation MAS: The Interplay of Formal and Informal Organisation

Figure 1 shows an example system with the OMAs Alice, Bob, Charlie etc., which are paired with their OPA. The SONAR-model specifies the *formal organisation.* In combination with the member agents we obtain the whole organisational MAS. Note, that the complete MAS may contain additional agents, e.g. the agent Fiona does not belong to the Org-MAS. In this model Alice and Bob may communicate "inoffically" or they may communicate "officially" as being members of the positions *Executive* and *Operator A.* This difference may be relevant, since in general Alice will have access to organisational resources due to her membership, i.e. due to the coupling to the OPA *Executive.*

The SONAR-engine defines a general *execution loop* (cf. [15]), where a *task* triggers the *formation* of a team, the team develops a team *plan* via negotiation (see Sect. 3). As a special feature the execution of the team-plan may involve *transformation* statements, which modify the SONAR-model at run-time, i.e. this enables the *cooperative, self-organised adaption* of the organisation:

Organisation $\rightarrow$ Team $\rightarrow$ Plan $\rightarrow$ Transformation $\rightarrow$ new Organisation $\rightarrow$ ...

The formalism of SONAR [4] defines a notion of well-formedness to guarantee that each task may be handled by at least one team etc. Additionally, transformations in well-formed organisations preserve the well-formedness.

3   The SONAR-Formalism

The SONAR organisational model [16] is based on Petri nets [5] and the execution loop is defined on top of the operational semantics of Petri nets. The interaction of agents are defined by a *distributed workflow nets* (DWFN) $D = (P, T, F, r :$

$T \to \mathcal{R}$), which is a multi-party version of the well-known workflow nets [17] where the interacting parties are called *roles*. Let $\mathcal{R}$ be a universe of roles. Each transition of a distributed workflow net is mapped by r to a role with the meaning that a transition t can executed only by an agent that implements the role $r(t)$. If we restrict a DWFN D to transitions belonging to a given set of roles R we obtain the so-called *role-component* $D[R]$, which defines the *service* provided by D w.r.t. the role set R. We assume R to be a subset of the roles used in D, i.e. $R \subseteq R(D)$. For singletons $D[\{r\}]$ we write $D[r]$.

Each partition $R_1, \ldots, R_k$ on the set of roles in D also decomposes D into its role-components: $D = D[R_1] \| \ldots \| D[R_k]$ where $\|$ denotes composition of net components by fusion of input/output places. We define a behavioural equivalence: $D[R] \simeq D'[R']$ denotes the fact that a component $D[R]$ cannot be distinguished from another component $D'[R']$ with the same interface, i.e. they are bisimilar with respect to the input/output behaviour at the message interface.

Assume that $\mathcal{A}$ is the set of position agents within the organisation. Let $\mathcal{D}$ be a set of DWF nets. In SONAR, the organisation is a Petri net, where each place is of the form $p = \mathsf{task}^a_{D[R]}$, which describes a *task* for the agent a to establish the service $D[R]$. The tasks are either generated in the environment or are sub-tasks, generated by the organisation itself.

$$\mathcal{P} := \{\mathsf{task}^a_{D[R]} \mid D \in \mathcal{D} \land R \subseteq R(D), a \in \mathcal{A}\}$$

The mapping $\alpha : \mathcal{P} \to \mathcal{A}$ returns the owner of a task: $\alpha(\mathsf{task}^a_{D[R]}) := a$.
    Each task is handled recursively via *team-operators*:

1. Delegate ($\mathcal{T}_{deleg}$): The task to implement DWFN $D[r]$ is delegated from agent a to b. Only the delegation operation delegates the ownership of a task.
2. Split ($\mathcal{T}_{split}$): The DWFN $D[r_1, \ldots, r_n]$ is split into the components $D[r_1], \ldots, D[r_n]$.
   As $D[r_1, \ldots, r_n] \simeq (D[r_1] \| \cdots \| D[r_n])$, this operation does not alter the behaviour.
3. Refinement ($\mathcal{T}_{refine}$): The DWFN $D[r]$ is replaced by $D'[r_1, \ldots, r_n]$, which has to be a refinement, i.e. $D[r] \simeq D'[r_1, \ldots, r_n]$ must hold.
4. Assign execution ($\mathcal{T}_{assign}$): The DWFN $D[r]$ is executed by the agent that is responsible for the task.

Each team-operator t defines a *constraint* $\psi(t)$ for the execution of DWFN D, e.g. given in temporal logic [18]. We define $\mathcal{T} := \mathcal{T}_{deleg} \cup \mathcal{T}_{split} \cup \mathcal{T}_{refine} \cup \mathcal{T}_{assign}$ as:

$$
\begin{aligned}
\mathcal{T}_{deleg} := \big\{ &\mathsf{d}^\psi(\{\mathsf{task}^a_{D[r]}\}, \{\mathsf{task}^b_{D[r]}\}) \\
&\big| \; D \in \mathcal{D} \land r \in R(D) \land a, b \in \mathcal{A} \land a \neq b \land \psi \in \Psi \big\} \\
\mathcal{T}_{split} := \big\{ &\mathsf{s}^\psi(\{\mathsf{task}^a_{D[r_1, \ldots, r_n]}\}, \{\mathsf{task}^a_{D[r_1]}, \ldots, \mathsf{task}^a_{D[r_n]}\}) \\
&\big| \; D \in \mathcal{D} \land \{r_1, \ldots, r_n\} \subseteq R(D) \land n > 1 \land a \in \mathcal{A} \land \psi \in \Psi \big\} \\
\mathcal{T}_{refine} := \big\{ &\mathsf{r}^\psi(\{\mathsf{task}^a_{D[r]}\}, \{\mathsf{task}^a_{D'[R]}\}) \\
&\big| \; D, D' \in \mathcal{D} \land D \neq D' \land D[r] \simeq D'[R] \land a \in \mathcal{A} \land \psi \in \Psi \big\} \\
\mathcal{T}_{assign} := \big\{ &\mathsf{e}^\psi(\{\mathsf{task}^a_{D[r]}\}, \emptyset) \mid D, D' \in \mathcal{D} \land a \in \mathcal{A} \land \psi \in \Psi \big\}
\end{aligned}
$$

The flow relation $\mathcal{F}$ for an operator $t = \mathrm{op}^\psi(X, Y) \in \mathcal{T}$ where $\mathrm{op} \in \{\mathsf{d}, \mathsf{s}, \mathsf{r}, \mathsf{e}\}$ is given by $^\bullet t := X$ and $t^\bullet := Y$. Note, that by construction we have $|^\bullet t| = 1$ for each transition t. Then, each operator set $T \subseteq \mathcal{T}$ induces an SONAR-*Organisation Net* $N = (P, T, F)$ given as $P = {}^\bullet T \cup T^\bullet$ und $F = \mathcal{F} \cap (P \cup T)^2$.

Definition 1. A SONAR-*Organisation* is $Org = (N, \mathcal{O}, \mathcal{R}, \mathcal{D})$ where the set of OPAs $\mathcal{O}$ is the set of agents owning the operators: $\mathcal{O} = \alpha(T)$.

The places with empty pre-set are those tasks that the organisation is responsible for, i.e. tasks that are generated externally: $P_0(Org) := {}^\circ P := \{p_0 \in P \mid {}^\bullet p_0 = \emptyset\}$

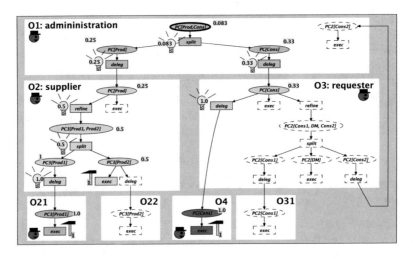

Fig. 2. A SONAR-Organisation Model (highlighted: a SONAR-Team)

Figure 2 shows an organisation net where the OPAs $O1$, O_2 etc. are highlighted as shaded regions. The organisation has exactly one task place $p = \mathsf{task}^{O_1}_{PC[Prod, Cons]}$, i.e. the task p is assigned to O_1 (the administration) and it is specified that triggers a *producer-consumer* DWFN $D = PC$ with the roles $R(PC) = \{Prod, Cons\}$. (The DWFN D is omitted here.) The DWFN PC has two refinements PC_2 and PC_3, which refine one role each: $PC[Cons] \simeq PC_2[Cons_1, DM, Cons_2]$ and $PC[Prod] \simeq PC_3[Prod_1, Prod_2]$. The producing OPAs are O_2, O_{21} and O_{22} on the left, while the consuming OPAs O_3, O_4 and O_{31} are shown on the right.

Some nodes are shown as unfilled elements. This will be explained in the context of team formation. The numerical inscriptions denote probabilities for team formation and will be explained later on.

Note, that the model does not have cyclic delegations, since this is impossible for well-formed nets. However, whenever we abstract the model to the level of positions we have cycles, since O_1 delegates some role part to O_3 who again delegates some sub-role back to the initiator O_1. The OPAs have a different spectrum of options. The OPA O_1 has only one operator to handle the initial task. For the generated sub-tasks O_2 has $(2+1+1+2)/4 = 1,5$ options in average, while O_3 has only $(3+3+1+1+1)/5 = 0,4$ options.

From a Petri net point of view the whole process of forming a team corresponds to a process that starts with a token on one of the initial task places and empties the net, since this means that all sub-tasks have finally be assigned for execution to agents. Formally, a team is defined as a maximal *partial-order run* of the organisation net [19]. Partial-order runs are Petri nets again, so a SONAR-team is also a Petri net (details can be found in [16]). The set of all these teams is denoted *Teams(Org)*.

The organisation in Fig. 2 the highlighted nodes define such a partial-order run, i.e. a team G. The three transition labelled with a hammer at the agent O_{21}, O_2 and O_4 show the assignment operators, i.e. these agents really implement the roles in the team-DWFN D. The transition labelled with light bulbs denote "inner" agents. Since the constraints of inner agents have to be fulfilled in the execution of D, too, these agents fulfill a coordinating purpose. The team G shown here will execute a team-DWFN $D(G)$, that is generated as the composition of the services of final transitions: $D(G) = \big(PC_3[Prod_1] \parallel PC_3[Prod_2] \parallel PC[Cons]\big)$.

4   Analysis of SONAR-Models Assuming a Static Environment

Our model has a natural representation of decision points: the task places. In general, each task has several operators in their post-set, i.e. there are several options how to react on tasks. This set of options will be used as the formal basis to define the *pool of options*. Which of these alternative options is chosen in a concrete situation is handled by the selection logic of OMAs. This logic is deliberately *not* part of the organisation model, to keep individual (micro) and macro aspects of the system apart. Instead we use the inherent non-determinism of Petri nets to express the options. But, we use run-time statistics how often the different operators are used. These are used as an approximation of $P[\text{ACT}_p = op]$, i.e. the probability that for a given task place p we chose on of the operators op from the post-set of p. We will extend this to a probability distribution for the set of teams *Teams(Org)*.

Assume a fixed team $G \in Teams(Org)$. We like to measure how well the options are balanced. The *option value* Opt_p of a task p is simply the number of available operators in the post-set of p, i.e. $Opt_p := |p^\bullet|$. For each team G we define the *contingency* as the expected value:

$$Contingency(G) := E[Opt_p] = \frac{1}{|P_G|} \cdot \sum_{p \in P(G)} Opt_p \tag{1}$$

We assume that in general a higher contingency is desirable. Of course, this key value is only a first approximation, since the impact of an decision is higher the earlier it is placed in the team net.[1]

We define the *balance* of G as the inverse of the expected deviation:

$$Balance(G) := (Var[Opt_p])^{-1/2} \qquad (2)$$

Due to the inverse, a smaller deviation of the option value results in a higher balance value. We assume that in general a higher balance value is desirable.

For the team G shown in Fig. 2 we obtain the contingency as:
$Contingency(G) = \frac{1}{10} \cdot \left(1 + (1+1) + (2+3) + 1 + (1+2) + (1+1)\right) = 1,4.$
The variance $Var[Opt_p]) \approx 0,44$ results in a balance of $Balance(G) \approx 1,51$.

In the following we define *contingency* and *balance* for the whole organisation net. These values are defined as a weighted sum over all teams $G \in Teams(Org)$.

It is possible to extend the probabilities of choices between operators, i.e. $P[\text{ACT}_p = op]$ to a distribution over the team set $Teams(Org)$ whenever we also the distribution how the environment generates tasks. From the run-time data we approximate the *environment model* $P[\text{TASKS} = p_0]$, i.e. the probability that from the set of all initial tasks $P_0(Org)$ the environment choses a given intial task $p_0 = \text{task}^a_{D[R]}$.

As we already said, we approximate the random variable $(\text{ACT}_p)_{p \in P}$ from the execution of a SONAR-MAS For each sub-task p $P[\text{ACT}_p = op]$ denotes the probability that the team formation process is extended by the operator $op = \text{op}^\psi(X, Y)$.

We can extend this random variable to a probability $P[\text{TEAM} = G]$ for each $G \in Teams(Org)$. This probability is defined recursively over the team structure, which is well-defined since teams are always acyclic Petri nets.

1. The final transitions have the probability $pr(t) = 1$.
2. For each place p there is exactly one (p, t)-arc, since the process has resolved the conflict between the alternative operators. We multiply the path probability $pr(t)$ with the selection probability $P[\text{ACT}_p = t]$, i.e. $pr(p) := P[\text{ACT}_p = t] \cdot pr(t)$. (Note, that transitions have exactly one place in the pre-set.)
3. For a transition t all paths starting from the places in the post-set evolve independently, i.e. the probabilities multiply: $pr(t) := \prod_{p \in t^\bullet} pr(p)$.

This leads to the recursive definition of pr:

$$
\begin{aligned}
pr(t) &:= 1, && \text{if } t^\bullet = \emptyset \\
pr(t) &:= \prod_{p \in t^\bullet} pr(p) && \text{if } t^\bullet \neq \emptyset \\
pr(p) &:= P[\text{ACT}_p = t] \cdot pr(t) && \text{operator } t \text{ in the post-set of } p \text{ (unique in } G)
\end{aligned}
\qquad (3)
$$

---

[1] If we want to express this aspect, we have to replace the absolute option value Opt_p in (1) by a value weighted with the distance of the task p from G's initial task p_0, i.e. $\gamma^{dist(p_0,p)} \cdot Opt_p$ where $\gamma \in [0; 1]$ is a decay factor.

Then $P[\text{TEAM} = G]$, which is the probability that we select the team G for a given initial place p_0, is:

$$P[\text{TEAM} = G] \quad := \quad P[\text{TASKS} = p_0] \cdot pr(p_0), \qquad \text{where } {}^\circ G = \{p_0\} \quad (4)$$

The numerical labels in Fig. 2 illustrate this recursive calculation of $P[\text{TEAM} = G]$. Here, we assume a uniform selection distribution, i.e. $P[\text{ACT}_p = t] = 1/|p^\bullet|$.

We define the *contingency of an initial task* $p_0 \in P_0$ as the weighted sum over all teams G which are generated for the initial task p_0 (and analogously for $Balance(p_0)$):

$$Contingency(p_0) = \sum_{G \in Teams(Org):{}^\circ G = \{p_0\}} P[\text{TEAM} = G] \cdot Contingency(G) \quad (5)$$

Definition 2. We define the *organisational contingency* as the weighted sum over initial tasks p_0 (Analogously for $Balance(Org)$.):

$$Contingency(Org) = \sum_{p_0 \in P_0} Contingency(p_0) \quad (6)$$

In general, a higher balance is desirable. In principle, a higher contingency is desirable, too. But, usually the management of options, including learning processes, introduce complexity that rises with the number of options, so high values are not always desirable.

5 Analysis of SONAR-Models Assuming a Dynamic Environment

In the last section we favored a balanced organisation under the assumption that the organisation is static. But, in general, the environment is dynamic and the SONAR-organisation adapts upon this change. It seems reasonable that those decision points with a high dynamics also need higher option values. In the following we define an indicator that expresses this alignment.

We define the environment as a stochastic process, i.e. the random variables are time dependent: $\text{TASKS} = \text{TASKS}_t$. In general, the concrete dynamics is unknown. So, we use previous observations $\text{TASKS}_{-1}, \text{TASKS}_{-2}, \ldots$ to estimate it. (Here, we use an infinite history for mathematical convenience. In practice, we set non-existing values to the oldest value.) We define the dynamics of $f(t)$ by its discrete second derivative $\frac{\Delta^2}{\Delta t^2} f = \frac{\Delta}{\Delta t}(\frac{\Delta}{\Delta t} f)$ where $\frac{\Delta}{\Delta t} f(t) := f(t) - f(t-1)$. Here we consider the generation of the an initial task p_0, i.e. $f_{p_0}(t) := P[\text{TASKS}_t = p_0]$.

Here, we are interested in the maximal absolute value of $\frac{\Delta^2}{\Delta t^2} P[\text{TASKS}_t = p_0]$ for the past times $t = 0, -1, -2, \ldots$ as this quantifies the dynamics of p_0. Since older values are less relevant we introduce a decay factor γ^t:

$$EnvDyn(p_0) := \max_{t \in \mathbb{N}} \left(\gamma^t \cdot \left| \frac{\Delta^2}{\Delta t^2} P[\text{TASKS}_{(-t)} = p_0] \right| \right) \quad (7)$$

Note, that "good" choice of $\gamma \in [0; 1]$ usually is application domain dependent.

Consider $EnvDyn(p_0)$ for all the initial tasks $P_0 = \{p_{0,1}, \ldots, p_{0,n}\}$. Let us assume that $p_{0,1}, \ldots, p_{0,n}$ is an ordered sequence, i.e. $EnvDyn(p_{0,i}) \leq EnvDyn(p_{0,j})$ for all $i \leq j$. Since we expect the option values $Contingency(p_0)$ to be aligned to this ordering, we can express the degree of alignment by the the number of *inversions*. An inversion is a pair (i, j) such that $EnvDyn(p_{0,i}) \leq EnvDyn(p_{0,j})$ but $Contingency(p_{0,i}) > Contingency(p_{0,j})$. If our alignment of options is perfect with respect to the environmental dynamics, the two sequences are identical and we have no inversions. Contrarily, if the sequence w.r.t. the option values is the complete reverse of the dynamics sequence, then we have the maximum of $\binom{n}{2}$ inversions.

Definition 3. Let $inv := inv(EnvDyn, Contingency)$ denote the number of inversions. The normalised *dynamic alignment* is defined as: $alignment(Org) = 1 - inv/\binom{n}{2}$.

So, $alignment(Org) = 1$ indicates that the organisation is well aligned w.r.t. the environmental dynamics.

6   Conclusion

In this contribution we studied adaptive systems, namely multi-agent systems. Here, we concentrated on organisation based MAS. Our aim was to define key values that indicate whether the design of the decision pools is reasonable. For our SONAR-model it was easy to identify these decision points in the architecture. Additionally, the option value was naturally represented in the Petri net model. In current work we also include the *diversity* of each pool.

On top of this we defined two values – namely, *contingency* and *balance* – which express how the number of option is distributed over the the organisation. The definition of these two values already include statistical information about the SONAR teamwork such that operators chosen more often have a greater impact. We extended our analysis also for a dynamic environment and defined the *alignment* of contingency to the environmental dynamics. The question whether these key measures are useful in specific application domains, e.g. for cyber-physical systems, is subject to ongoing research (cf. [13]).

References

1. Holland, J.H.: Hidden Order: How Adaptation Builds Complexity. Helix Books (1995)
2. Dignum, V., Padget, J.: Multiagent organizations. In: Weiss, G., (ed.) Multiagent Systems, 2nd ed. Intelligent Robotics, Autonomous Agents Series, pp. 51–98. MIT Press (2013)

3. Köhler-Bußmeier, M., Wester-Ebbinghaus, M.: SONAR*: a multi-agent infrastructure for active application architectures and inter-organisational information systems. In: Braubach, L., van der Hoek, W., Petta, P., Pokahr, A. (eds.) MATES 2009. LNCS (LNAI), vol. 5774, pp. 248–257. Springer, Heidelberg (2009). https://doi.org/10.1007/978-3-642-04143-3_27

4. Köhler-Bußmeier, M., Wester-Ebbinghaus, M., Moldt, D.: A formal model for organisational structures behind process-aware information systems. Trans. Petri Nets Other Models Concurrency **5460**, 98–114 (2009)

5. Reisig, W.: Petri Nets: An Introduction. Springer, Heidelberg (1985)

6. Schmeck, H., Müller-Schloer, C., Çakar, E., Mnif, M., Richter, U.: Adaptivity and self-organization in organic computing systems. ACM Trans. Auton. Adapt. Syst. **5**(3), 1–32 (2010)

7. Krupitzer, C., Roth, F.M., VanSyckel, S., Schiele, G., Becker, C.: A survey on engineering approaches for self-adaptive systems. Pervasive Mob. Comput. **17**, 184–206 (2015)

8. Salehie, M., Tahvildari, L.: Self-adaptive software: landscape and research challenges. ACM Trans. Auton. Adapt. Syst. **4**(2), 1–42 (2009)

9. Dean, P., Porter, B.: The design space of emergent scheduling for distributed execution frameworks. In: 2021 International Symposium on Software Engineering for Adaptive and Self-Managing Systems (SEAMS). IEEE (2021)

10. Filho, M., Pimentel, E., Pereira, W., Maia, P.H.M., Cortes, M.I.: Self-adaptive microservice-based systems - landscape and research opportunities. In: 2021 International Symposium on Software Engineering for Adaptive and Self-Managing Systems (SEAMS). IEEE (2021)

11. Sudeikat, J.O., Heitmann, O.: Towards modular assembling of virtual power plant control systems - the smart power Hamburg platform. In: Proceedings of the 27th Conference on Environmental Informatics. Shaker Verlag (2013)

12. Huang, Q., et al.: A review of transactive energy systems: concept and implementation. Energy Rep. **7**, 7804–7824 (2021)

13. Sudeikat, J., Köhler-Bußmeier, M.: On combining domain modeling and organizational modeling for developing adaptive cyber-physical systems. In: ICAART 2022 (2022)

14. Köhler-Bußmeier, M., Wester-Ebbinghaus, M., Moldt, D.: Generating executable multi-agent system prototypes from SONAR specifications. In: De Vos, M., Fornara, N., Pitt, J.V., Vouros, G. (eds.) COIN -2010. LNCS (LNAI), vol. 6541, pp. 21–38. Springer, Heidelberg (2011). https://doi.org/10.1007/978-3-642-21268-0_2

15. Köhler-Bußmeier, M., Wester-Ebbinghaus, M.: Model-driven middleware support for team-oriented process management. Trans. Petri Nets Other Models Concurrency **8**, 159–179 (2013)

16. Köhler, M.: A formal model of multi-agent organisations. Fund. Inform. **79**(3–4), 415–430 (2007)

17. Aalst, W.M.P.: Verification of workflow nets. In: Azéma, P., Balbo, G. (eds.) ICATPN 1997. LNCS, vol. 1248, pp. 407–426. Springer, Heidelberg (1997). https://doi.org/10.1007/3-540-63139-9_48

18. Clarke, E.M., Peled, D.A., Grumberg, O.: Model Checking. MIT Press, Cambridge (1999)

19. Esparza, J., Heljanko, K.: Unfoldings - A Partial-Order Approach to Model Checking. EATCS Monographs in Theoretical Computer Science. Springer, Heidelberg (2008). https://doi.org/10.1007/978-3-540-77426-6

Technique for Investigating Attacks on a Company's Reputation on a Social Media Platform

Maxim Kolomeets$^{(\boxtimes)}$, Andrey Chechulin , and Lidia Vitkova

St. Petersburg Federal Research Center of the Russian Acadeny of Science,
18th line of V.O., 39, St. Petersburg, Russia
{kolomeec,chechulin,vitkova}@comsec.spb.ru

Abstract. In this paper, we present the technique for investigating attacks on a company's reputation on a social media platform as a part of an arsenal of digital forensics investigators. The technique consists of several methods, including (1) identifying the attack based on sentiment analysis, (2) identifying the actors of the attack, (3) determining the attack's impact, and (4) determining core actors to identify the strategy of the attacker, including (4a) usage of bots, (4b) attempts to conflict initiation, (4c) competitor promotion, (4d) uncoordinated user attack. In the paper we also present the evaluation of this technique using the real investigation of use-case, where we investigate the attack on a retail company X, that occurs after the company changed its policy dedicated to COVID-19 QR codes for their visitors.

Keywords: Digital forensics · social media attacks · disinformation · social media bots · conflict initiation · competitor promotion

1 Introduction

Social networks have become a convenient tool for establishing feedback and communication between commercial companies, non-profit organizations, activists, civil society, governments, and other actors. For that reason, social platforms became very attractive to attackers, where they can try to manipulate opinion, spread misinformation, rumors, and conspiracy theories, create a fake reputation, fraud, and even suppress political competitors. Therefore, one of the fastest-growing attack vectors in information security are attacks on social media platforms and interference in social relations.

In this paper, we propose the technique for investigation of attacks on a companies' reputation, when an attacker creates a negative opinion about the company among visitors of social media platform. To do this, we use a number of selected computer science methods that already exist in order to answer a number of questions that form the proposed investigation pipeline: (1) is there an attack? (2) is there any damage? (3) what was the strategy of an attacker? The scientific novelty of the proposed method is in the union of investigation techniques into a single pipeline for the purposes of real-life attack investigations.

L. Braubach et al. (Eds.): IDC 2022, SCI 1089, pp. 234–243, 2023.
https://doi.org/10.1007/978-3-031-29104-3_26

The scientific contribution of this paper lies in the technique that allows an investigator to (1) determine the fact of an attack, (2) identify the actors of the attack, (3) determine the attack's impact, and (4) determine the core actors to identify strategy of the attacker, including (4a) usage of bots, (4b) attempts to conflict initiation, (4c) competitor promotion, (4d) uncoordinated user attack. We also present a real-world example – we used the proposed technique to investigate the case when the company's page on VKontakte social media was attacked by the anti-vaccination community.

The paper has the following structure. In the 2nd section, we present related research in the field of social media attacks investigation. In the 3rd section, we present the proposed technique. The 4th section contains the use-case evaluation. The 5th section finalizes the paper with discussion.

2   Related Research

From the point of view of information security, social networks threats can be considered as part of the CIA triad, including:

1. Confidentiality. Attacks are aimed at deceiving users through social engineering in order to obtain their personal data. This data can be used by an attacker for blackmail, bypassing security systems based on secret knowledge, competitive intelligence, etc. Separately, one can single out attacks on privacy, including a non-ethical collection of user data, which some researchers consider as part of confidentiality.
2. Integrity. Attacks aimed at deceiving users by changing social network metrics. Such attacks include various ways to cheat on reputation (can also damage the reputation of competitors), create bots that look like real people, interfere with internet voting/polls/contests, etc.
3. Availability. Attacks are aimed at disinformation, and social DoS. Such attacks include the spread of fake news to make information less available/trusted, or attack methods when the defense mechanisms of a social network block the victim (for example, because of the attacker's bot activity).

In a real-world scenario, such attacks can be multi-step and include several components of the CIA triad at once. For example, bots are often used to spread disinformation [5] and fake news [9,10] (availability). But in order to efficiently spread such information, bots need to distort the metrics of the social network to reinforce the user's trust (integrity).

At the same time, most of the research on attacks in social media is devoted to the creation of certain tools to solve one specific problem. While real-life analysis is usually a combination of such tools. There are many separate methods and technologies for analyzing attacks and impact [7] on social network users including analysis of sentiments [1,8,11], fake news [4,9,10] detection, bot detection [2,4,5] and so on.

In this paper, we present a methodology that combines such different tools to solve a specific problem - discrediting a company in a social network.

3   Proposed Methodology

We consider attacks whose main impact is damage to the company's reputation. The main consequence of such an attack is a bad impression about the company among users who have visited its page on the social network.

To analyze the attack, the analyst should answer 3 questions: (1) is there an attack? (2) is there any damage? (3) what was the strategy of an attacker? To answer these questions, we proposed the technique in form of an investigation pipeline. Decomposition of this pipeline consists of the following methods:

1. Attack identification - the 1st step is to fix the moment of the attack to determine the start time of the attack and narrow the time frame for analysis.
2. Attack actors identification - the 2nd step is to identify the accounts that are attacking among all accounts.
3. Impact assessment - the 3rd step is to assess the impact on the reputation of the company, which was done by attacking accounts.
4. Attack strategy identification - the last step is to determine which methods the attacker is using, among: (a) bots usage - attackers use many fake profiles; (b) conflict initiation attempts - the attacker uses a small number of accounts for conflict initiation; (c) competitor promotion - the attacker uses the attack to promote their goods or services; (d) uncoordinated user attack - many users attack the company on the call of someone, or on their own.

3.1   Attack Identification

To determine the presence of an attack, it is proposed to evaluate the dynamics of negative content on the company page. Such an analysis can be carried out using sentiment analysis of each message that was posted by visitors. For this can be used a number of models [8] such as GPT3 [11], Dostoevsky [1] and other. These models can assess the probability of belonging of message to "negative", "positive", "neutral", or other classes. As we suspect that shift in dynamics of negative content is unnatural and caused by attack, it is logical to select users who generate content with negative sentiments as a target for further analysis. It is proposed to use the *message toxicity metric* T_m:

$$T_m = \begin{cases} -S_n(m), & \text{if } S_n(m) > S_p(m), m \in M \\ S_p(m), & \text{if } S_n(m) \leqslant S_p(m), m \in M, \end{cases} \tag{1}$$

where m - a single message, $S_n(m)$ - negative assessment of message (probability that message is negative), where $S_p(m)$ - positive assessment of message (probability that message is positive), and M - set of messages.

As the next step, we propose to form a timeline from T_m values and calculate the average toxicity metric $\overline{T_m}$ over an N-hour window. Based on this timeline, the expert verifies that there was a shift in dynamic of negative content (that can be caused by attack and verified in the following steps), and can choose the time interval for further analysis. We also suggest giving the expert visual analytic tools (as shown in the experiments section, in Fig. 1), to simplify this process.

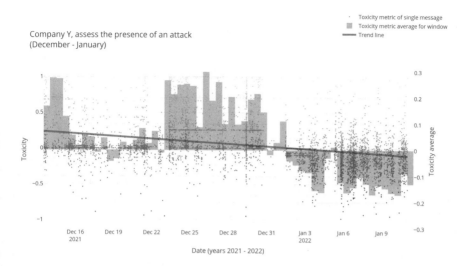

Fig. 1. Timeline of toxicity message T_m.

3.2   Attack Actors Identification

In order to identify potential attackers that post negative messages, it is proposed to determine the summary of *toxicity metric for user's account* T_u. Since positive messages have a toxicity metric $T_m > 0$, and negative ones $T_m < 0$, the toxicity metric for each account was calculated as the sum of the toxicity of the account's messages. A toxic account is defined as an account that has a more negative impact, and a non-toxic user is a user who has a more positive impact. Therefore:

$$T_u = \sum_{i=1}^{N_u} T_{m_i}$$
$$U_{toxic} = \{u | T_u < 0\}$$
$$U_{!toxic} = \{u | T_u \geqslant 0\}, \tag{2}$$

where u - user's account, N_u - number of messages for user, U_{toxic} - set of toxic accounts, $U_{!toxic}$ - set of non-toxic accounts.

3.3   Assess Impact

We assess impact with the following measures:

1. *Metric of accounts number* for toxic (N_{toxic}) and not toxic ($N_{!toxic}$) accounts:

$$N_{toxic} = |U_{toxic}|$$
$$N_{!toxic} = |U_{!toxic}| \tag{3}$$

2. *Metric of activity* for toxic (A_{toxic}) and not toxic ($A_{!toxic}$) accounts:

$$A_{toxic} = \sum_{u \in U_{toxic}} N_u$$
$$A_{!toxic} = \sum_{u \in U_{!toxic}} N_u \tag{4}$$

3. *Metric of impact* for toxic (I_{toxic}) and not toxic (I_{toxic}) accounts:

$$I_{toxic} = \sum_{u \in U_{toxic}} T_u$$
$$I_{!toxic} = \sum_{u \in U_{!toxic}} T_u \tag{5}$$

Using these metrics, one can determine how large the impact of an attack is.

3.4 Identify Core Actors

To identify an attack strategy, it is proposed to analyze the core of the attacking users. To do this, it is proposed to reduce the sample of toxic accounts U_{toxic} in order to eliminate the errors of the sentiment classifier. It is proposed to choose *core of toxic accounts* $\widehat{U_{toxic}}$ that: (a) left more than 3 comments (had more than one activity); (b) had high toxicity (account's impact < -1). Therefore:

$$\widehat{U_{toxic}} = \{u | u \in U_{toxic}, N_u > 3, T_u < -1\} \tag{6}$$

Based on $\widehat{U_{toxic}}$ it is proposed to identify an attack strategy.

3.5 Identify Bots Usage

In the study, a bot is understood as an account with an automatic or automated behavior model, as well as real users who perform the necessary actions for money (when an attacker on an exchange platform hires real users to perform malicious actions).

It is proposed to use a "bot/not a bot" AI classifier with a known value of balanced accuracy (as ex., Botometer [12], BotSentinel [13], etc. depending on analysed social network). After, taking into account the balanced accuracy of classifier, it is necessary to verify the results of classifier by testing the right-tailed hypothesis of the binomial distribution, where the number of success is the number of bots detected by the classifier. It is necessary to make sure that the found result is not false-positive inference. The result of this test is a p-value. With a p-value >0.05, it can be argued that the number of classifier errors lies in the expected error range. With a p-value <0.05, it can be argued that it was possible to detect a statistically significant number of bots, which is not a classifier error. Therefore:

$$H_0 : \frac{|U_{bots}|}{|U|} \leqslant 1 - Balanced\ Accuracy,$$
$$H_1 : \frac{|U_{bots}|}{|U|} > 1 - Balanced\ Accuracy, \tag{7}$$
$$p\text{-}value < 0.05 \Rightarrow reject\quad H_0,$$

where H_1 - attacker used bots, H_0 attacker does not use bots, U_{bots} - set of users accounts that were identified as bots by AI classifier, U - set of accounts, *Balanced Accuracy* - balanced accuracy of the bot detection classifier.

3.6 Identify Conflict Initiation Attempts

To identify an attempt to initiate a conflict, it is proposed to assess the communications of accounts with each other (which accounts responded to which accounts by mentioning in the messages).

All accounts who mentioned another account in their messages are selected for analysis. It is proposed to build the *communication graph* G_c. From graph G_c analyst extract clusters CL_{toxic}, that there is at least one toxic account ($u \in U_{toxic}$) connected to all another users in that cluster. If analyst extracts such cluster, it can be argued that toxic account initiate a conflict by responding to many accounts who do not seek to communicate. Therefore,

$$CL_{toxic} = \{cl|cl \subset G_c, center(cl) \in U_{toxic}, G_c = \{V, E\}\}, \qquad (8)$$

where CL_{toxic} - set of toxic cluster, cl - one cluster extracted by analyst, $center(cl)$ - returns the center of the cluster (user connected to all other users in the cluster), G_c - communication graph, V - an account, E - response of one account to another.

To analyze such a graph by an analyst, we suggest using visual analytics methods (as in the experiment, in Fig. 2). The hypothesis can be defined as:

$$H_0 : |CL_{toxic}| \leqslant c,$$
$$H_1 : |CL_{toxic}| > c,$$
$$|clusters(G_c)| > c \Rightarrow reject \quad H_0, \qquad (9)$$

where H_1 - there is conflict initiation, H_0 there is no conflict initiation, c - criteria that defined by analyst, $clusters()$ - function for clusters extraction.

3.7 Identify Competitor Promotion

To identify possible competitor promotions, we analyze users' messages. Since the operator cannot analyze the entire set of messages, we propose to reduce the sample. For this, a communication graph G_c is used. On this graph using the PageRank algorithm $PR_{toxic}(G_c, N)$ can be selected N leaders L that $\in U_{toxic}$:

$$L = \{u|u \in U_{toxic}, PR_{toxic}(G_c, N)\}, \qquad (10)$$

where L - set of leaders.

The operator can look at leaders, who mentioned competitor companies:

$$LP = \{u|u \in L, promo(u) = True\}, \qquad (11)$$

where LP - set of leaders that promoted another company, $promo(u)$ - return $True$ if user u mentioned another company in messages.

$$H_0 : |LP| \leqslant c,$$
$$H_1 : |LP| > c,$$
$$|LP| > c \Rightarrow reject \quad H_0, \qquad (12)$$

where H_1 - there is competitor promotion, H_0 there is no competitor promotion, c - criteria that defined by analyst (our suggestion is to define $c > N * 0.2$).

3.8 Identify Uncoordinated User Attack

We propose to conclude that there is no attacker, and there are many users attack the company on their own, if all 3 previous techniques (usage of bots, conflict initiation, competitor promotion) didn't reject H_0.

4 Evaluation

As a use case we used an incident that occurred with a commercial company on the social network VKontakte. Company X contacted us in January 2022 with a suspicion of an information attack on their community. We investigated this case using the proposed methodology.

In January, community administrators detected that an information attack was being carried out on them since a large number of new users and the same type of negative comments appeared in their community (in the microblog, photo albums, and discussions). Many other users entered into controversy with them. Company X asked us to investigate this incident to understand if it is a natural process (discontent related to recent company policy) or artificial (information attack).

4.1 Assess the Presence of an Attack

We analyzed comments for the second half of December 2021 and the first half of January 2022. For that, we used the Dostoevsky model [1]. Dostoyevsky is a machine learning model for the Russian language that classifies text corresponding to the sentiment of the text. Dostoevsky does not take into account hashtags and emoticons. The classification accuracy is 76%, which makes it possible to determine the average binary sentiment of a set of messages. For this study, 2 categories are used: negative, and positive.

With that, we built the toxicity chart that is shown in Fig. 1. Y-axis represents the message toxicity value T and X-axis represents the time. We calculate toxicity T_m (Eq. 1) for each message and represent it as a dot with color from green ($T_m = -1$) to yellow ($T_m = 1$). Red bars represent the time windows, for which we calculate the average toxicity of messages $\overline{T_m}$ in this time window.

It can be seen that negative comments began to prevail from the 2nd of January. So we can conclude that there is an attack, and define the time interval is the time between 2nd and 11th of January.

4.2 Identify Attack's Actors

At this stage, users that left comments from January 2nd to 11th were investigated to identify a possible attack's actors. We calculated the level of toxicity of each user using Eq. 2. As a result, we got the sets of toxic users and non-toxic users.

4.3   Assess Impact

At this stage, we analyse toxic and not toxic accounts to evaluate an impact. We calculated the metric of accounts number N (Eq. 3), activity A (Eq. 4) and impact I (Eq. 5). The results presented in Table 1.

Table 1. Impact metrics evaluation

Group	N	A	I
toxic	511	2786	−394
!toxic	391	505	+35

It can be seen that the number of toxic and non-toxic users did not differ much (511 and 391 or 56.65% and 43.34%), toxic users had much more activity and left more comments (84.65% of all comments), and their comments had a much more pronounced negative assessment (impact is 11.25 times bigger).

4.4   Identify Core Actors and Attack Strategy

To identify an attack strategy in the following steps, we formed a core of attacking users by the Eq. 6. As the result, we had 46 high toxicity users that perform multiple actions.

Identify Usage of Bots. Using the VKontakte AI bot detection tool [2], among 46 users, 6 were identified as bots and 40 as real users. Taking into account the balanced accuracy (≈ 0.9) of the used classifier, we carried out statistical testing of the right-hand hypothesis of the binomial distribution (Eq. 7). According to the test results, no statistically significant number of bots was found ($p\text{-}value = 0.31$ with a threshold value of statistical significance <0.05). Thus, 6 accounts were identified as bots, which lies within the classifier error.

Identify Attempts of Conflict Initiation. For the interaction analysis, we selected all users who mentioned another user in their posts since January 2nd and built the communication graph G_c according to Eq. 8. The graph is shown in Fig. 2. The analysis showed the presence of 1 cluster with centralized activity (Eq. 8), that is lower than selected threshold $c = 3$ (Eq. 9).

Identify Competitor Promotion. For promotion identification we calculate Page Rank centrality measure [3] on communication graph and got a set of top 10 leaders according to Eq. 10. For $N = 10$ leaders of opinion, we form a table of their messages for manual analysis and found 47 comments to use products of company Y. All of these messages were posted by one user, therefore, according to Eq. 11, $|LP| = 1$. But since we select criteria $c = N * 0.2 = 2$, according to Eq. 12, the fact of a competitor attack cannot be confirmed with certainty.

Identify Uncoordinated User Attack. As all 3 previous techniques (bot usage identification, conflict initiation and competitor promotion) didn't reject H0, we can conclude that there are many users attack the company on their own, due the changing of company's policy.

Fig. 2. Communication graph with cluster that highlighted with orange frame. (Color figure online)

5   Discussion and Conclusion

On the example of a use-case, we showed the process of applying the proposed methodology, which can confirm or deny the fact of an attack on a company's reputation and determine some of the characteristics of this attack.

The use-case investigation showed that the process of shifting the tone of user comments in a negative direction was most likely natural, due to the implementation of QR codes by company X on January 2, 2022. The negative activity is the result of the actions of real users and activists who are fighting against the implementation of QR codes.

In the comments, a sign of the use of the incident by competitors to advertise their services was found, but due to the extremely limited activity (only 1 user showed such activity), it is not possible to confirm or deny with certainty that this activity is an attack by competitors.

This methodology is a full cycle of analysis of an attack on a company. The strength of the proposed methodology is a significant degree of automation. The use of expert evaluation is necessary only at the stage of detecting clusters on the communication graph and determining comments on the promotions of other companies. But these stages can also be replaced by automated analysis, although automated methods can be less accurate. For example, to detect clusters (Eq. 8), it is also possible to use clustering algorithms and to detect a promotion (Eq. 11), various NLP methods can be used.

The disadvantage of the proposed method - the use of sentiment analysis. The negative tone of the message does not indicate the hostility of the user to the company. For example, users who support a company may argue with other users, which can cause their tone to shift in a negative direction. Thus, if one use the classification "supports/does not support" (the company) instead of the "toxic/non-toxic" classifier, one can achieve a more accurate identification of users who are involved in the attack. This can be achieved using systems for building automatic classifiers, but this will require the creation of a training sample, since for each use case it is necessary to create its own models. On the other hand, reducing toxicity is important to a company's reputation, so toxicity analysis provides a broader picture of incident.

In the future, we plan to increase the degree of automation of the presented technique that will allow to get estimation of presented approach on multiple use-cases, and implement presented pipeline for Twitter, so we can measure it efficiency in comparison with another tools.

Acknowledgments. The work is performed by the grant of RSF No. 18-71-10094-P in SPC RAS.

References

1. Sentiment analysis library for Russian language (Dostoevsky). https://github.com/bureaucratic-labs/dostoevsky. Accessed 20 Apr 2022
2. Kolomeets, M., Chechulin, A., Kotenko, I.: Bot detection by friends graph in social networks. J. Wirel. Mob. Netw. Ubiquitous Comput. Dependable Appl. (JoWUA). **2**(12), 141–159 (2021)
3. Page, L., Brin, S., Motwani, R., Winograd, T.: The PageRank citation ranking: bringing order to the web. Stanford InfoLab. (1999)
4. Shao, C., Ciampaglia, G.L., Varol, O., Flammini, A., Menczer, F.: The spread of fake news by social bots. arXiv preprint arXiv:1707.07592, **96:104** (2017)
5. Ferrara, E.: COVID-19 on Twitter: bots, conspiracies, and social media activism. arXiv preprint (2004.09531) (2020)
6. Pierri, F., Artoni, A., Ceri, S.: Investigating Italian disinformation spreading on twitter in the context of 2019 European elections. PloS one **15**(1:e0227821) (2020)
7. Mendoza, M., Tesconi, M., Cresci, S.: Bots in social and interaction networks: detection and impact estimation. ACM Trans. Inf. Syst. (TOIS) **39**(1), 1–32 (2020)
8. Zhang, L., Wang, S., Liu, B.: Deep learning for sentiment analysis: a survey. Wiley Interdiscip. Rev. Data Min. Knowl. Discov. **8**(4), e1253 (2018)
9. Zhang, X., Ghorbani, A.A.: An overview of online fake news: characterization, detection, and discussion. Inf. Process. Manag. **57**(2) (2020)
10. Zhou, X., Zafarani, R.: A survey of fake news: fundamental theories, detection methods, and opportunities. ACM Comput. Surv. (CSUR) **53**(5), 1–40 (2020)
11. OpenAI GPT3. https://openai.com/api/. Accessed 20 Apr 2022
12. A Python API for Botometer by OSoMe. https://github.com/IUNetSci/botometer-python. Accessed 30 July 2022
13. Bot Sentinel platform. https://botsentinel.com. Accessed 30 July 2022

An Experimental Validation of the Practical Byzantine Fault Tolerant Algorithm

Andras Ferenczi[(✉)] and Costin Bădică[ⓘ]

Faculty of Automatics, Computers and Electronics,
University of Craiova, Craiova, Romania
ferenczi.andras.h5f@student.ucv.ro, costin.badica@edu.ucv.ro

Abstract. Practical Byzantine Fault Tolerance (pBFT) is one of the most widely known state replication algorithms for byzantine decentralized networks, yet there is a lack of literature and open-source frameworks that would enable researchers to easily conduct empirical analysis.

In this paper we present an overview of pBFT, briefly mention variants and related works, dive into our implementation of pBFT, then validate its theoretical foundations via experiments. Finally, we conclude on our findings and outline our future plans with this framework.

Keywords: PBFT · Fault Tolerant · Consensus · Decentralized · Blockchain

1 Introduction

Most existing distributed ledgers have one thing in common: the integrity of ledger is maintained via a consensus protocol.

Investopedia defines consensus as a "fault-tolerant mechanism that is used in computer and blockchain systems to achieve the necessary agreement on a single data value or a single state of the network among distributed processes or multi-agent systems" [7].

Wattenhofer [11] defines consensus, as follows. Given n nodes out of which f may crash and the rest, $n-f$, are non faulty. Each node starts with an input value v_i. The participating nodes must decide on a value that satisfies the following conditions: (a) all non faulty nodes agree on the same value (*Agreement*), v, (b) all nodes terminate in finite time (*Termination*), and, (c) the result is a valid input from a node (*Validity*).

The large scale adoption of distributed ledger technologies and blockchains, in particular, paved the way to a plethora of new consensus algorithms, such as, proof-of-work (PoW), proof-of-stake (PoS), delegated proof-of-stake (DPoS), proof of elapsed time (PoET), proof of space (PoS), and practical byzantine fault-tolerance pBFT).

L. Braubach et al. (Eds.): IDC 2022, SCI 1089, pp. 244–253, 2023.
https://doi.org/10.1007/978-3-031-29104-3_27

The latter, pBFT, is typically used in private blockchains, in which a small number of organizations represent nodes (replicas) participating in the protocol. It is simple, works asynchronously, and is more efficient than PoW.

In order to analyze the protocol and to perform further research, we implement a simplified version, as described in Wattenhofer's Blockchain Science, Third Edition [11]. The implementation uses to a great extent the pseudo-code from the book, however, we developed a novel network simulation layer that enabled up to empirically validate the protocol and pave the way to new creative enhancements to it.

Our paper is structured as follows. Section 2 briefly touches on pBFT and a few newer variants. In Sect. 3 we describe the Byzantine General's Problem and our implementation of pBFT in a nutshell. In Sect. 4 we present our testing outputs and provide some benchmarks on running the protocol on a local computer while simulating the network latency and node failures, and in Sect. 5 we conclude our findings and future research plans.

2   Related Works

Practical Byzantine Fault Tolerance algorithm is an algorithm addressing the Byzantine General's Problem. It was introduced by Miguel Castro and Barbara Liskov in a 1999 paper [2]. The novelty of the paper is a high-performance Byzantine state machine-based replication protocol that is capable of very high throughput. To date, it is the most used protocol in existing blockchain implementations.

The Ripple Protocol consensus algorithm (RPCA) [5] and [3] is also a byzantine consensus protocol that is similar to pBFT. The participant servers take the list of transactions from the last ledger close (consensus checkpoint) and send them the "candidate list" to a trusted list of peers, maintained in a Unique Node List (UNL). Each server then combines the candidate sets received from servers in its UNL, and votes on each individual transaction. Transactions that receive a minimum number of votes are kept and the rest are discarded and become participants for the next ledger. In the final round, only transactions that get 80% of the votes make it into the ledger close. All participant nodes are required to update their own copy every few seconds to maintain the correctness and agreement of the network. RPCA works, as long as $f \leq \frac{n-1}{5}$, where n is the number of servers in a UNL and f is the number of byzantine failures.

Hyperledger Fabric 2.0 is one of the leading enterprise blockchain solutions. Prior to the 2.0 upgrade, it faced criticism for not being fully decentralized, given that the ledger was maintained by a component, called the orderer, that was maintaining the integrity of the ledger (blockchain). Third-party pBFT-inspired protocols, such as pBFT-SMART [1], make it possible to have multiple orderers hosted by different participants. BFT-Smart keeps the 3 phases, i.e. pre-prepare, prepare, and commit, but the view change messaging is different: VIEW-CHANGE, VIEW-DATA, and NEWVIEW. This makes the protocol simpler and more elegant.

The Query/Update (Q/U) protocol is newer Byzantine fault-tolerant protocol first published by Michael Abd-el-malek et al. [10]. Due to its optimistic approach, the protocol features better scalability and performance than PBFT and other state machine-based protocols. The latter approach requires a two phase commit-based approach, including a prepare phase. In contrast, the Q/U protocol does not require upfront consistency, however, will allow the inconsistent nodes to correct their state.

HQ is a hybrid Byzantine-fault-tolerant state machine replication protocol published by James Cowling et al. [4]. Its main innovation is to eliminate the [quadratic] inter-replica chattiness of PBFT and Q/U when there is no contention, thus significantly speeding up the consensus in such situations.

Ramakrishna Kotla et all [8] Zyzzyva, like HQ, avoids the three-phase commit protocol to establish the order of the executions. Due to its optimistical nature, replicas can exist in inconsistent state. One novelty of this protocol is that clients can actively intervene and help the replicas reach consensus. The paper claims the this protocol can reduce replica communications to near minima.

3   Problem Description

PBFT is a practical solution for the Byzantine General's Problem, and was presented in a paper published in 1982 [9]. The generals are camped outside of a city and can communicate with each-other via messenger. They need to reach an agreement on when to coordinate an attack on the city, but there are traitors among them, that will not follow the protocol and hence they can do whatever they wish. The consensus algorithm needs to ensure that a) all honest generals agree on the same plan of actions and b) the traitors cannot influence the decision of the majority.

3.1   PBFT in a Nutshell

The Participants. We have a set of replicas, $\mathcal{R}$, with $|\mathcal{R}| = 3f + 1$, where f is the maximum number of faulty nodes. The replicas consist of one primary (orchestrator) and the backups that enforce the consensus. There are also an unbound number of clients that submit requests and receive responses/results.

The View. Message exchanges occur in context of a view, which is a numeric representation of a configuration corresponding to a particular primary. For example, if the index of a replica is in $\{0, .., |\mathcal{R}| - 1\}$, the next primary is picked by its index after the view is incremented: $p = V \bmod |\mathcal{R}|$, where V is the current view number.

The Protocol. A client sends a request to the primary and if it does not receive a response within a set time interval, it broadcasts the request to all replicas. If backup receives a client request, it forwards it to the primary and starts tracking the time until the message is processed. If a threshold is exceeded, the backup triggers a view change.

Primary P initiates the protocol by assigning a sequence number to the client request and broadcasts to the backup nodes the message $PRE\_PREPARE$ $(view, sequence, request)_P$. The subscript $_P$ represents the digital signature of the replica (the primary's, in this case). Backup B then broadcasts a $PREPARE(view, sequence, request)_B$ to its peers. Backup now waits until $2f$ PREPARE arrive and sends $COMMIT(v, s)_B$ to all nodes. Once it receives acknowledgments from $2f + 1$ peers, it executes the request and returns the response to the client.

A view change may occur when a faulty primary is detected by backups, upon their timer's expiration. This is sufficient, as faulty requests are ignored and hence malicious and systems failures equally result in timeouts. When a backup detects a faulty timer expiration, it stops accepting requests from the primary and broadcasts a $VIEW\_CHANGE(v + 1, P_B)_B$ message. P_B is the list of all certificates for the $PREPARE$ messages received since the backup has started tracking them. The next backup that is the primary candidate waits until it receives $2f + 1$ $VIEW\_CHANGE$ requests, builds list γ, that contains the $VIEW\_CHANGE(v + 1, P_B)_B$ messages, then builds O consisting of all distinct $(sequence, request)$ pairs that are present in γ. When there are gaps in the sequence number, it fills in with dummy (NOP) requests. Then, it broadcasts $NEW\_VIEW(v + 1, \gamma, O)_P$. Backups will verify correctness of O, stop accepting requests for view v, then follow the same logic as the one performed for $PRE\_PREPARE$ for each $(sequence, request)$ entry in O in context of view $v + 1$. Alternatively, if inconsistencies are found in O, trigger $VIEW\_CHANGE(v + 2, P_B)_B$ request.

3.2 Authentication

We assume that nodes (replicas) have the ability to sign their messages in a way that cannot be forged. We will notate a signed message m by node n as follows: $(msg)_n$. Public-key cryptography will be used to sign and verify, respectively messages exchanged between replicas. Concrete implementation will most likely consist of signing a digest of the message (likely a cryptographic hash algorithm, such as SHA-256).

Algorithm 1. Message signing

1: **procedure** SIGN(msg, sk)
2:     $signature \leftarrow$ COMPUTE-SIGN(msg, pk)
3:         **return** $signature$
4: **end procedure**

In Algorithm 1 and Algorithm 2 COMPUTE-SIGN is a signature algorithm, such as Elliptic Curve Digital Signature Algorithm (ECDSA) or RSA Digital Signature Scheme.

Algorithm 2. Signature validation

1: **procedure** VALIDATE-SIGN($msg, signature, pk$)
2:     $signature1 \leftarrow$ COMPUTE-SIGN(msg, pk)
3:     **return** $signature = signature1$
4: **end procedure**

3.3   Network Mapper

Our protocol required a network mapper component that is directory that registers our network participants: the primary node, the backup nodes and clients, that submit requests using our protocol. The network mapper maintains the linkage between IP addresses and the public keys of the participants in a *node_dictionary* where the primary key is represented by the public key certificate. A more robust solution would maintain a "blockchain address", by applying Keccak-256 to the key and then retaining the last 20 bytes of the result. The Network Mapper also provides the algorithm for picking the next primary in case of a view change:

$$pk = node\_dictionary((view + 1)mod(len(node\_dictionary))).$$

3.4   Node Component

The Node component implements the following functions:

$CLIENT\_REQUEST(request)$ 3: invoked by clients on primary or backup nodes. When invoked on a backup node, the latter will forward the request to the primary and will start a faulty timer.
$PRE\_PREPARE(view, sequence, request)$ 4: message sent by the primary to all backups.
$PREPARE(view, sequence, request)$ 5: this message is "broadcasted" by the backups upon receiving the $PRE\_PREPARE$ message.
$COMMIT(view, sequence, request)$ 6: sent by backups when at least $f+1$ PREPARE messages are received. When this message is received, backups execute the client request and return it to the client. Client's address (pk) is included in the request and is looked up via the network mapper, which returns the physical address. Backups invoke the SEND_REQUEST message on the client and pass back the response.
$VIEW\_CHANGE(new\_view, P_B)$ 7: sent by backups when faulty timer expires, meaning the primary does not timely send out the $PRE\_PREPARE$ message for client requests.
$NEW\_VIEW(view, \gamma, O)$ 8: sent by the new primary. Backups receiving this message will validate γ and O, adopt new view, and execute $PRE\_PREPARE$ for entries in O.

3.5   Node Stub

Since we wanted to keep the "core" framework intact while being able to experiment with various scenarios, we implemented a "wrapper" around our Node

---

Algorithm 3. Request from client

1: **procedure** CLIENT_REQUEST(*request, signature*)
2:     VALIDATE(*request, signature*)
3:     **if** *primary* **then**
4:         INCREMENT(sequence)
5:         BROADCAST(*PRE_PREPARE*, (*view, sequence, request, signature*))
6:     **else**
7:         CLIENT_REQUEST(*primary, request, signature*)
8:     **end if**
9: **end procedure**

Algorithm 4. Message from primary to backups

1: **procedure** PRE_PREPARE(*view, sequence, request*)
2:     VALIDATE(*request, signature*)
3:     BROADCAST(*PREPARE*, (*view, sequence, request, signature*))
4: **end procedure**

Algorithm 5. broadcasted by backups

1: **procedure** PREPARE(*view, sequence, request, signature*)
2:     VALIDATE(*request, signature*)
3:     **if** *pre_prepare_count* $> 2f + 1$ **then**
4:         BROADCAST(*COMMIT*, (*view, sequence, request, signature*))
5:     **end if**
6: **end procedure**

Algorithm 6. broadcasted by backups

1: **procedure** COMMIT(*view, sequence, request, signature*)
2:     VALIDATE(*request, signature*)
3:     **if** *prepare_count* $> f + 1$ **then**
4:         **if** *faulty_timer* $> THRESHOLD$ **then**
5:             $P_B \leftarrow$ PREPARE_CERTS
6:             BROADCAST(*VIEW_CHANGE*, (*new_view, P_B, signature*))
7:         **else**
8:             SEND_RESPONSE(*client*, (*request, response, signature*))
9:         **end if**
10:     **end if**
11: **end procedure**

Algorithm 7. broadcasted by backups that have their faulty timer threshold exceeded

1: **procedure** VIEW_CHANGE(*new_view*, P_B, *signature*)
2:     VALIDATE(*request*, *signature*)
3:     **if** *view_change_count* > $2f + 1$ and *new_primary* **then**
4:         $\gamma \leftarrow$ COLLECT_VIEW_CHANGE_CERTS
5:         $O \leftarrow$ COLLECT_PREPARED_MESSAGES(P_B, γ)
6:         BROADCAST(*NEW_VIEW*, (*view*, γ, O, *signature*))
7:     **end if**
8: **end procedure**

Algorithm 8. broadcasted by new primary

1: **procedure** NEW_VIEW(*view*, γ, O, *signature*)
2:     VALIDATE(γ, O, *signature*)
3:     **for** *o* in O **do**
4:         (*sequence*, *request*, *signature*) $\leftarrow$ GET(*o*)
5:         PRE_PREPARE(*view*, *sequence*, *request*, *signature*)
6:     **end for**
7: **end procedure**

component. This wrapper, the NodeStub, implements all the required interfaces, but adds the randomized behavior that is expected from distributed network communications. We can add additional functionality to this layer, as required.

The NodeStub is injected by the NetworkMapper when a Node registers with it.

4   Experimental Results

4.1   Experiments and Discussion

The experiments were performed on a desktop computer equipped with Intel Corporation Xeon E3-1200 v6/7th Gen Core Processor with 32 GB RAM and 240 GB SSD, running Ubuntu 18.04. The mini-framework was implemented using Python 3.8.X [6]

PBFT, by design will ignore messages that cannot be validated cryptographically. From a practical perspective, this equivalates to delayed responses. The formula for number of messages generated per client request is: $\mu(n) = n + n * (n - 1) + n^2 = 3n^2$, meaning a communication complexity $\mathcal{O}(n^2)$.

Our first experiment was conducted with 0 byzantine nodes. All nodes respond correctly with the appropriate messages. As seen in Fig. 1, the number of messages by type is as expected: Request: 1, Pre-Prepare: $3f$, Prepare: $(3f)^2$, Commit: $3f(3f + 1)$, Reply: $3f + 1$. The image also highlights the downside of having a large number of replicas.

In the next experiments, Fig. 2, we increase the number of byzantine nodes to the maximum allowed an notice the view change-related messages. The number of prepare and commit messages shot up significantly (Fig. 3).

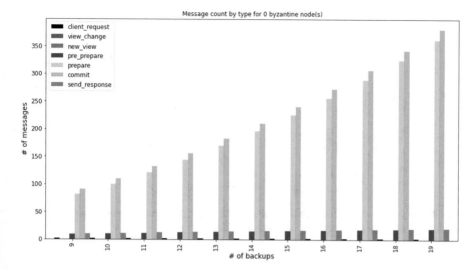

Fig. 1. Count of messages exchanged between replicas based on number of backups with no byzantine replicas. Total number of replicas = 1 + no. of backups, as it includes the primary as well.

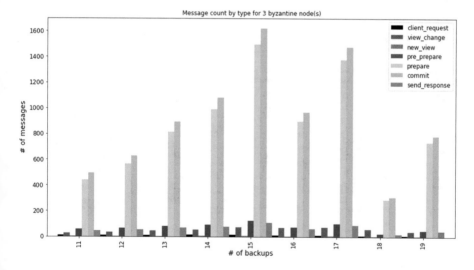

Fig. 2. Count of messages exchanged between replicas based on number of backups with 3 byzantine replicas.

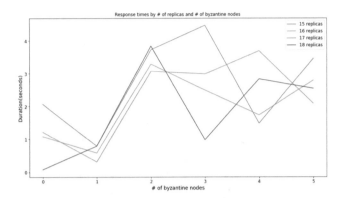

Fig. 3. Response times measured by the client from the moment the request was sent out until a sufficient number of backups replied, i.e. $f + 1$.

Simulation of an Attack

We tested the limits of the protocol. The next run was configured with 10 replicas (one primary and 9 backups) and a single client initiating continuous requests. Our test generator created byzantine nodes ($f \in \{4, 5, 6, 7\}$), and we observed that the consensus did not terminate, which is the correct behavior Table 1.

Table 1. Protocol behavior as the number of faulty nodes increases beyond expected $f = 3$

Faulty node cnt.	Failed Requests	Successful Requests	Failure percent	Observations
0	0	980	0	
1	0	1043	0	
2	18	1151	1	Few timeouts
3	339	790	42	More timeouts
4	476	0	100	No termination
5	105	0	100	No termination
6	15	0	100	No termination
7	3	0	100	No termination

5   Conclusion and Future Works

pBFT [2] is the earliest and most well-known state replication protocol that inspired consensus implementations of several public and enterprise blockchains, such as Stellar, Ripple (XRP), Hyperledger Sawtooth, and it is anticipated that a future releases of Fabric will incorporate pBFT as well.

The mini-framework we created was needed to perform empirical research and improve on it. As the initial phase of the project was to experiment with the core protocol, we successfully confirmed its theoretical foundations.

Our framework [6] still lacks the robustness of a production-ready software. We have a number of components that are partially implemented, and may not work as expected. It does, nevertheless, serve its purpose of providing us with valuable information on its functioning. Although it is publicly available, we don't recommend using it for any business purpose or otherwise, except for education.

References

1. Barger, A., Manevich, Y., Meir, H., Tock, Y.: A byzantine fault-tolerant consensus library for hyperledger fabric (2021). https://doi.org/10.48550/ARXIV.2107.06922, https://arxiv.org/abs/2107.06922
2. Castro, M., Liskov, B.: Practical byzantine fault tolerance. In: Proceedings of the Third Symposium on Operating Systems Design and Implementation, OSDI 1999, pp. 173–186. USENIX Association, USA (1999)
3. Chase, B., MacBrough, E.: Analysis of the XRP ledger consensus protocol (2018). https://arxiv.org/abs/1802.07242
4. Cowling, J., Myers, D., Liskov, B., Rodrigues, R., Shrira, L.: HQ replication: a hybrid quorum protocol for byzantine fault tolerance. In: In Proceedings of the OSDI, pp. 177–190 (2006)
5. Schwartz, D., Noah Youngs, A.B.: The ripple protocol consensus algorithm. https://ripple.com/files/ripple_consensus_whitepaper.pdf
6. Ferenczi, A.: Simple pBFT implementation using python (2022). https://github.com/andrasfe/tinypbft
7. Frankenfield, J.: Consensus mechanism (cryptocurrency) (2021). https://www.investopedia.com/terms/c/consensus-mechanism-cryptocurrency.asp
8. Kotla, R., Alvisi, L., Dahlin, M., Clement, A., Wong, E.: Zyzzyva: speculative byzantine fault tolerance. In: In Symposium on Operating Systems Principles (SOSP) (2007)
9. Lamport, L., Shostak, R.E., Pease, M.C.: The byzantine generals problem. ACM Trans. Program. Lang. Syst. **4**(3), 382–401 (1982). http://dblp.uni-trier.de/db/journals/toplas/toplas4.html#LamportSP82
10. El Malek, M.A., Ganger, G.R., Goodson, G.R., Reiter, M.K., Wylie, J.J.: Fault-scalable byzantine fault-tolerant services. In: In Proceedings of the 20th ACM Symposium on Operating Systems Principles, pp. 59–74. ACM Press (2005)
11. Wattenhofer, R.: The Science of the Blockchain, 3rd edn. Inverted Forest Publishing, North Charleston (2019)

Smart Cities

Smart Parking for All: Equipped and Non-equipped Vehicles in Smart Cities

Filippo Muzzini[1,2](✉), Manuela Montangero[1], and Nicola Capodieci[1]

[1] Dipartimento di Scienze Fisiche, Informatiche e Matematiche - Università di Modena e Reggio Emilia, Modena, Italy
{filippo.muzzini,manuela.montangero,nicola.capodieci}@unimore.it
[2] Università di Parma, Parma, Italy

Abstract. The current trend in designing cities is to think them as smart environments that are constantly connected with road users. For this purpose, a smart city is implemented as a collection of IoT (Internet-of-Things) powered devices set up in order to connect vehicles to their surrounding infrastructure. In this way, road users share their intentions while retrieving useful information from the smart city itself. This complex and distributed system must be then tailored to improve viability performance metrics such as reducing traffic congestion, optimizing accident response and everything else related to transportation in urban areas. In this work we focus on parking management in a scenario in which next generation vehicles will be able to communicate with the surrounding infrastructure and will coexist with traditional vehicles with limited or absent IoT-capabilities. We propose a reservation mechanism able to exploit communication at infrastructure level, with the goal of reducing the time needed to find a free parking spot close to destination. We evaluate our proposed mechanisms using the well known MATSim transport simulator.

1 Introduction

Recent advancements in embedded systems enables urban viability in a Smart City to be improved by deploying IoT [1] (Internet of Things) capable devices within city infrastructures. These IoT devices present SWaP (Size Weight and Power) features amenable to be installed pretty much everywhere within a urban environment and inside the road users' vehicles. As far as urban mobility [2] is concerned, the smart city infrastructure enables both vehicle-to-vehicle (V2V) and vehicle-to-infrastructure (V2I) communication, to create a cooperative distributed system for solving everyday viability problems [3]. More specifically, the constant flow of information among vehicles and the surrounding infrastructure can be exploited to implement state-of-the-art traffic congestion policies, but also accident mitigation and arbitration policies as well as helping the road users to reserve and travel to the most convenient parking space near their trip's destination. Needless to say, implementing these mechanisms implies also facing challenges such as privacy and security of the data exchanged among the road users [4].

At present time it is also very important to consider vehicles heterogeneity: some of them might feature communication capabilities or even advanced driving assistance systems (ADAS), whereas the vast majority of cars are not fully equipped to actively participate in the V2V and/or V2I communication exchange. While it is reasonable to assume that in the future all vehicles will feature such advanced systems, as of yet, it is necessary to consider scenarios in which both equipped and non-equipped vehicles co-exist.

Significant changes to the way in which we live the city imply the need for testing the proposed traffic management approaches before the actual (real life) implementation. For these reasons, the use of an accurate urban traffic simulator is mandatory for studying the complex dynamics emerging from the interactions among the smart city infrastructure and the involved vehicles.

In this context, our contribution relates to the design of a traffic management mechanisms to be deployed within smart cities. We present the design of a parking reservation mechanism in a scenario of co-existence of equipped and non-equipped vehicles and discuss the results of a set of preliminary tests. This paper is part of an on-going project and provides a first study intended to understand and fine tune our parking reservation approach.

Our proposed approach is tested using MATSim, which is a urban mobility simulator that takes inspiration from MAS (Multi Agent System) literature [5]. In MATSim, each road user is modeled as an autonomous agent that features individual and distinct behaviors. This aspect allows us to model the co-existence of different road users' categories (i.e. equipped and non-equipped vehicles) and examine the results as an emergent global behaviour obtained through the interaction among all agents.

2   Related Work

Many aspects of future smart cities have been very recently surveyed by Mariani et al. in [6] where authors highlight a classification of smart city related viability problems. In our work we deal with the well known problem of parking space scarcity, hence, out of Mariani et al. classification, we will focus on what has been termed in [6] as a *resource oriented* problem. While the problem of smart city parking has been abundantly discussed in previous literature (e.g. [7–9]) most of these work focused on implementation aspects and/or parking metering, billing and reservation. [10] and [11] are surveys focused on parking system algorithms and vehicle detection techniques. In [12] authors propose a smart parking system able to manage reservations and simulate a scenario in which all vehicles are IoT capable. Our proposal suggests available parking taking into account free but reserved slots, and it notifies vehicles if their destination parking is taken before they arrive there. Finally, in contrast with latter works, we study the impact of the co-existence of both IoT capable vehicles with traditional vehicles. We propose, test and simulate various mechanisms that are able to manage both type of vehicles as studied in [13–17] with regards to intersections, traffic light arbitration and emergency response.

3   The Scenario

In the era of smart cities vehicles fall into different orthogonal categories:

- **autonomous versus non-autonomous vehicles**: the former drive with little or no human intervention, whereas the latter are driven by humans;
- **equipped versus non-equipped vehicles**: the former are able to communicate with other vehicles and with city infrastructures, the latter are not.

The two classifications might overlap. In particular, equipped vehicles are not necessarily autonomous, but they might use embedded IoT devices or smartphones for enabling communications.

In this work we will focus on scenarios in which equipped and non-equipped vehicles coexist (either autonomous or non-autonomous) and we distinguish between smart and non smart parking strategies:

- **Non smart strategies** are adopted by non-equipped vehicles acting as standard human drivers that are searching for a parking spot: they roam parking areas near their destination, following routes that are based on their knowledge of specific areas of the city.
- **Smart strategies** are adopted by equipped vehicles (autonomous or non autonomous) that are able to communicate with city infrastructures and with a central city server to determine the route to the best available parking spot.

4   Our Proposal for Smart Parking

In this paper we propose a parking strategy (*Smart Shortest Distance*) for equipped vehicles (not necessarily autonomous) and we will test this strategy in the previously described scenario (i.e., when these vehicles coexist with non-equipped ones). We assume that drivers of equipped vehicles will follow recommendations of their equipment.

We assume there exists a centralized city server infrastructure that is always aware of which parking spots are available and which are occupied by some vehicle. This assumption might be implemented by the city governance using IoT on-site parking monitors communicating with the server. In this paper we will assume that all parking spots are monitored, in further research we will also analyze the cases in which some are and others are not. We further assume that some (or all) of the city parking spots might be reserved for vehicles that are still far away from the parking spot and, once reserved, can not be occupied by other vehicles. We will refer to such spots as *reservable* parking spots. Parking spot reservation might be implemented, for example, by using *automatic road bollard* or *signal lights* at the spots to prevent/discourage drivers of non-equipped vehicles to park at reserved spot.

For non reservable parking spots, we can not exclude that some other vehicle occupies a parking spot that the city server communicated to be be free to an equipped vehicle. However, we assume that the city infrastructure will be able to detect this situation and notify the equipped vehicle that the spot is not available anymore.

When an equipped vehicle starts searching for a parking spot, the following steps are performed:

1. The vehicle contacts the city server to get the positions of parking areas with available spots.
2. The vehicle computes its distance to each parking area, it selects the one closest to destination, it communicates the choice to the city server, and it moves towards the selected parking area. During the trip the vehicle regularly checks for messages coming from the city server.
3. The city server updates the parking occupancy and, if the spot is reservable, signals that the parking spot is reserved. Meanwhile, it waits for an "end of trip" message from the vehicle. If the previously chosen parking spot gets occupied by another vehicle, then the server sends a notification to the traveling vehicle together with its current knowledge of still available parking areas.
4. If the vehicle arrives at destination with no further notification from the city server, then it occupies the chosen parking spot and notifies the city server with an "end of trip" message. Otherwise, it will select a new parking spot using the new information sent by the city server.

Vehicles use Euclidean distance instead of shortest paths to compute distances to parking spots because it is less computational expensive and thus might be quickly computed also on devices with limited computation capability and memory (e.g., devices adopted by equipped non autonomous vehicles).

5   Experiments and Results

In this preliminary work we performed a set of experiments to test the effectiveness of the smart strategy and compare it to other non smart strategies in a mixed scenario where equipped and non-equipped vehicles coexist. Non equipped vehicles will necessarily have to use a non smart strategy. We take into consideration the following three non smart strategies:

- *Random.* The random strategy simulates the behaviour of drivers with no prior knowledge of the city or available parking areas (e.g., a tourist in an unknown city): at intersections they randomly choose the next road to follow, avoiding already traversed ones (unless all alternatives have already been selected once). This strategy will be mainly used as a baseline, indeed nowadays it is unlikely that drivers travel without a (not necessarily integrated) navigation system.

- *Benenson* [18]. The Benenson strategy simulates the behaviour of drivers with some kind of knowledge of the city and of where parking areas are located (i.e., drivers are able to orient themselves in the area): at intersections drivers will select the road that will gradually get them closer to parking areas.
- *Shortest Distance.* The Shortest Distance strategy simulate drivers following routine paths, i.e., they know the area very well, where they would rather park and which is the shortest path to get there. However, they do not know if parking slots are available until they get there: when they get at the parking area closest to destination, they search for the nearest free parking spot. If the area is full, then they move to the next closer parking area.

5.1 Experimental Set up

Experiments were performed using the MATSim simulator and implementing new functions to extend MATSim with smart and non smart parking strategies. Implementation details will be given in an extended version of this paper. MATSim was used because it is a flexible open source project that allowed us to extend it for our purposes. Moreover we address the issue of illegal parking to simulate the (not unusual) situation in which, after a while, human drivers get tired of roaming searching for a free parking spot and decide to park in any convenient spot outside city designed parking areas.

Map. For our preliminary tests we used the Manhattan style map shown in Fig. 1 that has been designed to test our proposal in a very regular scenario. It is not a real city area map but it is an artificial map in which each intersection has four links. The map has eight horizontal bidirectional links intersecting nine bidirectional vertical links, for a total of 72 four-way intersections. The parking areas are arbitrary chosen: there is no regular pattern (on purpose) in order to diversify traveling distances and to simulate a real scenario in which there are areas with more parking spots than others. In particular, the upper (resp. left) side of the map has more parking areas than the bottom (resp. right) side.

Population. Each vehicle is associated to two locations (the driver's home and workplace) and two schedules (departure time from home and a return time from work). Locations are randomly selected on the map, with the constraint that there exists a connecting path in each direction. Departure and return times are generated according to a normal distributions with peaks at city rush hours (i.e., 09:00AM for departure and 06:00PM for return). Illegal parking occurs when the parking search time reaches 26 min (the threshold was empirically set and guarantees that the simulations always terminate).

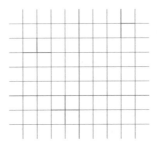

Fig. 1. Manhattan map: parking areas are highlighted in red.

5.2   Experiments

We are interested in measuring how much time the vehicles need to find a free parking spot; i.e., the time laps between the arrival at destination and the parking in a free parking spot.

We tested the smart strategy to measure and compare the searching time with respect to non-smart strategies in the same scenarios, the impact of varying the number of reservable spots, and the coexistence of smart and non-smart vehicles in the same scenario.

Each scenario is tested with 20 runs, and average results are then reported. Each run is characterized by vehicles starting position, destination, and schedules. To make results comparable, the same initial setting has been used in the same run for the different strategies.

Experiment 1. We compare the performance of the smart strategy when varying the percentage of reservable parking spots (equal to 0%, 25%, 50% and 100%). We assume that all vehicles adopt the smart strategy and we consider two cases: (1.1) there are more parking slots (1000) then vehicles (800); (1.2) there are less parking slots (600) then vehicles (800). Results are reported in Fig. 2.

We can see that the smart strategy is robust to different traffic and infrastructure conditions.

Indeed, start by observing that average search time and standard deviations are almost the same when varying the percentage of reservable parking spots: average search time lies in the interval [88,89]s (resp. [103,106]) for experiment 1.1 (resp. experiment 1.2); standard deviation lies in the interval [40,41]s (resp. [57,66]).

On the other side, if we look at outliers, then the number of reservable spots influences highest search times: in experiment 1.1 we report a maximum of 440s with percentages equal to 0%, 25% and 50%, and of 264s with 100%; in experiment 1.2 we report a maximum of 550s with percentages equal to 0%, 25%, 418s with 50%, and of 318s with 100%.

Comparing the two experiments, as it was to be expected, the smart strategy performs better when there is an abundance of parking spots: average search time

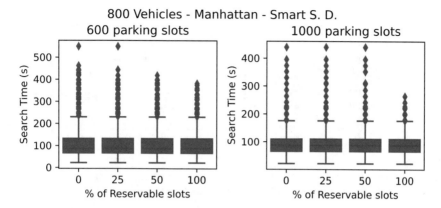

Fig. 2. Experiment 1. Smart strategy search time when varying the percentage of reservable slots. There are 800 vehicles and on the left (resp. on the right) 600 (resp. 1000) parking spots.

(resp. standard deviation, maximum search time) is 20% (resp. between 30% and 38%, between 20% and 31%) less then when there is lack of parking spots.

Results are however definitively encouraging even in the case in which the number of parking slots is smaller and, also in this case, all vehicles found a free spot in no more than 9 min, i.e., much before giving up and perform an illegal parking. Do note that vehicles do not necessarily circulate all together at the same time and this explains why they all find a parking spot even if the number of parking spots is less than the total number of vehicles.

Experiment 2. We compare the smart strategy against the three non smart strategies when all vehicles adopt the same strategy. We consider the cases of 1000 (experiment 2.1) and 600 (experiment 2.2) available parking slots (as in experiment 1) and, for the smart strategy, we assume there are no reservable parking spots (this is the worst case scenario). As already stated, to make results comparable, we used the same initial setting in the same run of different strategies. Results are reported in Fig. 3.

We can see that, the smart strategy outperforms the others. With abundance of parking spots (on the right), average search time is 89s for the smart strategy, compared to 540s for Random, 155s for Benenson and 117s for Shortest Distance. With lack of parking spots (on the left) the average search time for the smart strategy is 106sec, compared to 598s for Random, 472s for Benenson and 513s for Shortest Distance.

We can see that using the Smart Shortest Distance strategy the 82% of vehicles reduces the searching time with respect to the random strategy (with an average gain of 76%), the 51% with respect to the Benenson strategy (with an average gain of 57%) and the 7% with respect to Shortest Distance strategy (with an average gain of 71%).

As it was to be expected, with less parking spots all strategies worsen their performances, however the smart one is always the most competitive. Observe

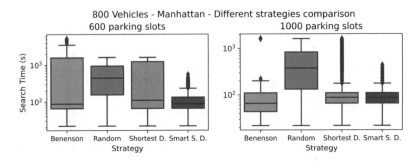

Fig. 3. Experiment 2. Search time (in log scale) when varying the parking strategy. There are 800 vehicles and on the left (resp. on the right) 600 (resp. 1000) parking spots.

that the random strategy has the smallest percentage increase among non smart strategies when reducing the number of parking slots and that Benenson and non smart shortest distance perform much worst in the latter case. In particular the mean searching time goes from 540 to 598 for Random; from 154 to 472 for Benenson; from 117 to 513 for Shortest Distance and from 89 to 106 for Smart Shortest Distance. The maximum searching time goes from 1621 to 1632 for Random; from 1650 to 5237 for Benenson; from 1617 to 1630 for Shortest Distance and from 440 to 550 for Smart Shortest Distance.

Experiment 3. We run a scenario in which equipped and non-equipped vehicles coexist and we vary parking availability: we have 800 vehicles (200 vehicles adopting each strategy), and we vary the percentage of reservable parking spots (0%, 25%, 50% and 100%). Again, we run tests with 1000 (on the right) and 600 (on the left) parking slots. We assume that non-equipped vehicles park in reservable spaces only if not reserved. Results are reported in Fig. 4.

We observe that the smart strategy is competitive with respect to non smart ones even when they coexist, especially when parking spots are lacking. In particular, there are no significant differences when varying the percentage or the number of reservable parking spots and, more interestingly, with respect to search times of experiment 2.

We observe that the smart strategy allows all vehicles to find a parking spot much before deciding for illegal parking, while all other strategies have many outliers (or very large searching time, as it is the case of the random strategy) and several illegal parking. Even if *Beneson* performs better in some situation it always shows some big outliers that are not present in *Smart Shortest Distance*.

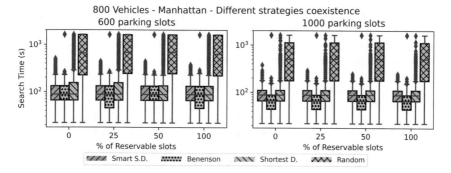

Fig. 4. Experiment 3: Parking search time (in log scale) when smart and non-smart strategies coexist and varying the percentage of reservable parking spots. There are 800 vehicles and on the left (resp. on the right) 600 (resp. 1000) parking spots.

6   Conclusions

In this work we addressed the parking search time issue in the scenario of coexistence of equipped and non-equipped vehicles in a smart city. We proposed a smart reservation mechanism for equipped vehicles and tested the proposal with a set of preliminary experiments on a Manhattan style map.

We showed that our proposal is worth pursuing because it reduces the searching time with respect to non smart strategies and performs very well also in scarcity of parking spots. Moreover, the smart strategy guaranties small parking search times also when equipped and non-equipped vehicles coexists, that is when other vehicles adopt different (and not predictable) parking strategies. This makes the smart strategy suitable to be used in the (who knows how long) transition period in which smart and non-smart vehicles will coexist in smart cities.

As for future work, we plan to extend this work with a deeper analysis of the causes leading to current searching time (e.g., how many reroutes are necessary on average to find a free parking spots) and by considering different traffic conditions. Moreover, in scenarios where road bollards are not available we will explicitly address the issue of stolen reserved spots (e.g., by using the probability of a parking spot to be stolen). Finally, we will consider another map derived from a real city.

References

1. Bujari, A., Furini, M., Mandreoli, F., Martoglia, R., Montangero, M., Ronzani, D.: Standards, security and business models: key challenges for the IoT scenario. Mob. Netw. Appl. **23**(1), 147–154 (2018). https://doi.org/10.1007/s11036-017-0835-8
2. Melis, A., Mirri, S., Prandi, C., Prandini, M., Salomoni, P., Callegati, F.: Crowd-sensing for smart mobility through a service-oriented architecture. In: 2016 IEEE International Smart Cities Conference (ISC2), pp. 1–2 (2016)

3. Bertogna, M., Burgio, P., Cabri, G., Capodieci, N.: Adaptive coordination in autonomous driving: motivations and perspectives. In: 2017 IEEE 26th International Conference on Enabling Technologies: Infrastructure for Collaborative Enterprises (WETICE), pp. 15–17. IEEE (2017)

4. Zhang, K., Ni, J., Yang, K., Liang, X., Ren, J., Shen, X.S.: Security and privacy in smart city applications: challenges and solutions. IEEE Commun. Mag. **55**(1), 122–129 (2017)

5. Ferber, J., Weiss, G.: Multi-agent Systems: An Introduction to Distributed Artificial Intelligence, vol. 1. Addison-Wesley Reading, Boston (1999)

6. Mariani, S., Cabri, G., Zambonelli, F.: Coordination of autonomous vehicles: taxonomy and survey. ACM Comput. Surv. (CSUR) **54**(1), 1–33 (2021)

7. Khanna, A., Anand, R.: IoT based smart parking system. In: 2016 International Conference on Internet of Things and Applications (IOTA), pp. 266–270. IEEE (2016)

8. Sadhukhan, P.: An IoT-based e-parking system for smart cities. In: 2017 International Conference on Advances in Computing, Communications and Informatics (ICACCI), pp. 1062–1066. IEEE (2017)

9. Mainetti, L., Patrono, L., Stefanizzi, M.L., Vergallo, R.: A smart parking system based on IoT protocols and emerging enabling technologies. In: 2015 IEEE 2nd World Forum on Internet of Things (WF-IoT), pp. 764–769. IEEE (2015)

10. Diaz Ogás, M.G., Fabregat, R., Aciar, S.: Survey of smart parking systems. Appl. Sci. **10**(11), 3872 (2020)

11. Kotb, A.O., Shen, Y., Huang, Y.: Smart parking guidance, monitoring and reservations: a review. IEEE Intell. Transp. Syst. Mag. **9**(2), 6–16 (2017)

12. Pham, T.N., Tsai, M.F., Nguyen, D.B., Dow, C.R., Deng, D.J.: A cloud-based smart-parking system based on internet-of-things technologies. IEEE Access **3**, 1581–1591 (2015)

13. Cabri, G., Gherardini, L., Montangero, M., Muzzini, F.: About auction strategies for intersection management when human-driven and autonomous vehicles coexist. Multimed. Tools Appli. **80**(10), 15921–15936 (2021). https://doi.org/10.1007/s11042-020-10222-y

14. Cabri, G., Montangero, M., Muzzini, F., Valente, P.: Managing human-driven and autonomous vehicles at smart intersections. In: 2020 IEEE International Conference on Human-Machine Systems (ICHMS), pages 1–4. IEEE (2020)

15. Muzzini, F., Capodieci, N., Montangero, M.: Exploiting traffic lights to manage auction-based crossings. In: Proceedings of the 6th EAI International Conference on Smart Objects and Technologies for Social Good, pp. 199–204 (2020)

16. Dresner, K., Stone, P.: A multiagent approach to autonomous intersection management. J. Artif. Intell. Res. **31**, 591–656 (2008)

17. Capodieci, N., Cavicchioli, R., Muzzini, F., Montagna, L.: Improving emergency response in the era of ADAS vehicles in the smart city. ICT Express **7**(4), 481–486 (2021)

18. Benenson, I., Martens, K., Birfir, S.: PARKAGENT: an agent-based model of parking in the city. Comput. Environ. Urban Syst. **32**(6), 431–439 (2008). GeoComputation: Modeling with spatial agents

EV Route Planning Across Smart Energy Neighborhoods

Rocco Aversa, Dario Branco, Beniamino Di Martino, Pietro Fusco,
and Salvatore Venticinque[(✉)]

Department of Engineering, University of Campania "Luigi Vanvitelli", Caserta, Italy
{rocco.aversa,dario.branco,beniamino.dimartino,pietro.fusco,
salvatore.venticinque}@unicampania.it

Abstract. Electric Vehicles (EVs) have been identified as the current innovation for sustainable mobility that reduces carbon emissions and pollution. To comply with the limited users' acceptance, supporting services such as route planners, behave as selfish applications. However, they do not contribute to leveraging a seamless integration of the expected increase of new energy loads into the electric grid. In this paper, we propose a support for an optimal route planner that allows for smart charging. It complies with users' anxiety and exploits the booking information charge flexibility to optimize the usage of renewable sources in a smart energy neighborhood.

1 Introduction

Electric Vehicles (EVs) have been identified as the current innovation for sustainable mobility that allows for reducing carbon emissions and pollution. To address the lack of charge services support that still affects the users' acceptance, innovative services have been deployed to improve the driving experience, reduce the so-called *EV driver's anxiety* and to enable original business models. Route planners are some examples of services that aim at supporting EV mobility also on medium and long trips, through prediction of range and charging stops. They represent an opportunity to deal with the expected increase in the number of electric vehicles, which will represent a considerable fraction of the total demand and will have a significant impact on the grid [10]. On one hand, an uncontrolled charging of a massive number of electric vehicles that cause major power losses, voltage deviations, and distribution substations overload cannot be accepted [9]. On the other hand, important benefits are expected when routing and scheduling techniques are both applied to generate a driver's recommendations. A collaborative routing can facilitate the integration of new energy demand into the grid by a slight modification of the conventional itinerary of vehicles to achieve some specific goal [11]. A charge plan, known in advance by charge stations, allows for predicting the energy demand and planning an optimal schedule of their charge sessions. In this paper, we propose an advanced service that works in two steps. On the user side, it supports a selfish decision by proposing a set of alternatives that may contribute to reducing the covered distance to the target station

L. Braubach et al. (Eds.): IDC 2022, SCI 1089, pp. 267–276, 2023.
https://doi.org/10.1007/978-3-031-29104-3_29

or may help to minimize the number of EVs that charge at the same time at every station. In particular, we will focus how, according to the users' choice, an optimal charge schedule at each charge station can be planned lowering the power peak and maximizing the usage of renewable sources. In Sect. 2, we present related works. Section 3.1 deals with the recommendation of an optimal route and charge schedule through reinforced learning techniques. In Sect. 4, we focus on an original model, and its implementation, for the optimal schedule of charge sessions within a charge station. Experimental results are discussed in Sect. 5.

2 Related Work

In [11], authors propose a joint EV routing and charging/discharging scheduling strategy. A mathematical framework based on mixed-integer linear programming is defined to maximize the revenue of EV users. The solution does not modulate the charge profile modeling the state of the EV with a binary variable and is applied to a fleet of operated EVs. Results show that slight changes in driving patterns can provide benefits to EV users and improve the network operation. In [14] EV routing and charging/discharging scheduling are addressed sequentially. In the first step, the shortest route is found trying to keep within the charge area without exceeding a set level of traffic congestion. In the second step, the algorithm determines when the charge is going to take place. A similar sequential approach can be found in [6]. The correct design of the price dynamics is concluded to be key for promoting V2G participation [11]. The smart generation of reservation intervals is exploited in [8] to assure a proper distribution of the available energy resources while increasing end-user satisfaction. No overlapping reservations and the occupation of several stations without queues are the main objectives of the optimal schedule. In [12] authors develop a linear programming-based heuristic algorithm on a time-space network model for charge and discharge scheduling of EVs coping with uncertain demands and departure times. In this case, the objectives include peak load reduction in a building energy management system. Similarly, the increase of self-consumption of photovoltaic power is pursued in [13] by smart charging of electric vehicles and vehicle-to-grid technology. In this paper we address the two problems sequentially, introducing renewable energy sources and stationary batteries, but neglecting the utilization of V2G technology. Moreover, in the second step, we try to equalize the charge and discharge profiles, varying the power during the charge session. In our work energy cost is neither defined as an optimization criteria [11], nor as an incentive to leverage a collaborative behavior by users toward the achievement of the global optima [1]. We imagine that in a real scenario the total cost will be just collected and presented to the user as a criteria for choosing the preferred route. In future work, a virtual cost, which is eventually returned from charge stations to notify about the charge availability, is going to be considered in the optimization algorithm. Moreover, as many external events can affect the travel schedule in an unpredictable way, the optimization will run periodically, or triggered on specific events, to update the suggested routes and to notify the stations about changes related to new users' decision or to the travel schedules, such as unforeseen delays.

3   Problem Formulation and Solution Strategy

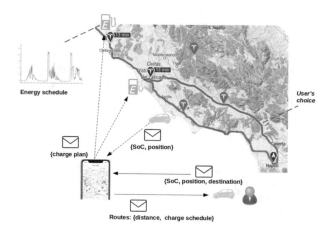

Fig. 1. Conceptual model of route and charge schedule problem.

In the considered scenario each driver queries the route planner for an optimal route to reach her destination and re-charge along the trip. Alternative itineraries are provided to the user including stops at those stations which are suitable candidates for charging. In Fig. 1 the user, according to her own preferences, selects a route among the ones recommended by the planner. Each recommendation includes energy cost, charge locations and available charge slots, total distance and travel time. Most recommended routes offer charge planes which minimize the total covered distance and achieve a fair distribution of charge demand among charge stations. The fair distribution of charge sessions among charge stations is not ensured by the system due to the user's freedom to choose the charging station she prefers. However, charge stations notifies their charge availability, the self-consumption and updated energy costs in order to either increase the awareness of the route planner and to influence the user's choice. In Fig. 1 two stations receive from the route planner the charge plan accepted by the user. They are able to compute in advance the optimal schedule of their charge sessions to improve the energy usage. According to the current status of charge and to the user's position, the planner can eventually recommend alternative routes and updated charge schedule for the rest of the trip, notifying the new user's choice to the stations.

3.1 Charge Allocation

The EV routing problem, with intermediary charging stations, is a scheduling problem where it is necessary to couple an EV to a charge station (CS) trying to obtain a variance as much as possible equal to the minimum theoretical value. Let us define $EVs = \{ev_1, ..., ev_n\}$ as the set of cars that are distributed in a common spatial domain where they are going to charge. Each car periodically updates the system about its current status of charge (SoC) and position at the time t, that is $ev_i(t) = \{\lambda_i, \phi_i, SoC_i, C_i, \eta_i\}$. Position is defined by λ_i and ϕ_i, which are latitude and longitude respectively. Status of charge and capacity (C_i) and energy efficiency $(\eta_i, [Wh/Km])$ are necessary to estimate how long it is still possible to drive and the amount of energy it can be charged, once the user arrived at target station. On their own, charging stations in $S = \{cs_1, ..., cs_m\}$, are characterized by their position and unit cost of energy: $s_j = \{\lambda_j, \phi_j, \tau_j\}$. Such cost is a virtual cost. The higher is the cost, the lower is the charge availability of that station.

Let it be r_i the number of charge sessions of ev_i along its route from the starting point to destination. A charge schedule of ev_i is represented by a vector $v_i \in \mathbb{N}^{r_i}$, where the integer value $v_i(k) \in [1, m]$ will be equal to j if the charge session k of e-car ev_i has been scheduled at the stations cs_j, with $k \in [1, r_i]$. For the global charge plan $CP = \{v_i \mid i = 1, .., n\}$ we define:

- the total distance covered by the e-cars charging at the assigned stations: $\delta = \sum_{i=1,n} \sum_{k=0,r_i} d(v_i(k), v_i(k+1))$, where $v_i(0)$ and $v_i(r_i + 1)$ represent the departure and the arrival points of ev_i.
- the variance of the number of cars per station: $\sigma = \frac{\sum(n_j - n/m)^2}{m}$, where n_j is the max number of cars that will charge at station j at the same time.

In particular, we aim at solving an optimization problem that minimizes these two different criteria, taking into account the constraints on the distance which can be covered by each car.

$$\min_{v_i}[\delta, \, \sigma] \quad with \, \eta_i * d(v_i(k), v_i(k+1)) < SoC_i(k), \, \forall i \in [1, n], \forall k \in [0, r_i] \quad (1)$$

In fact, minimizing δ allows for saving energy by the entire community. On the other end, the minimization of σ means a shorter queue at the charging stations and a fair distribution of customers. A solution to this problem will be the array of integers whose $i - value$ represents the best charging station for the e-car i. The solutions, which provide the best compromises between δ and σ for the global system, compose the Pareto Front. We addressed the selection of the next charge station problem in [1], but assuming a centralized computation of the allocation strategy and considering only the next charge session for all the EVs. Without going into details we propose Reinforcement Learning (RL) to implement a distributed solution. In a first iteration, each agent is at the starting location $S_0 \in SS$ shown in Fig. 2 and looks for the next charge station along its own route. After each iterations each agent communicate to the other its own choice, to allow for the computation of the variance of the charge occupation at the different stations and the total distance to be used in the next iteration. In the next iteration, the agents look for a different solution that allows to increase its reward, depending on the energy cost, but minimize the δ and σ, end eventually choose the next charge station if it is required.

After r_i iteration, an agent has computed a complete itinerary for ev_i. When all the agents have computed the complete itinerary for their ev the algorithm terminates and the finall state S_F has been reached.

Fig. 2. RL starting from an initial state S_0 towards a final state S_F

4 Energy Schedule

Once the charge session of the EVs have been allocated to the selected stations, the optimal charge schedule at each station is computed trying to equalize the power produced by renewable sources with the EV's charge profiles and exploiting the stationary batteries. Such a problem is modeled as a multi-objective optimization that attempts not only to satisfy the energy demand of charging EVs, but also to optimize other global parameters, such as the minimization of carbon dioxide emissions.

In this case, we aim at minimizing the absolute value of the difference between the energy consumed by a local renewable source and the total amount of energy produced by a local renewable source. The energy production by photovoltaic panels can be predicted using several techniques such as the one used in [2]. However, to exclude the prediction error in the experimental activities we will use measured profiles. Equation 2 defines the first objective function of our optimization problem.

$$\min(|\sum_{i=0}^{N}[\sum_{j=0}^{M_{EV}}(\Delta t * (x_{i,j} * P_{EV}max_j)) + \sum_{k=0}^{M_B}(\Delta t * y_{i,k} * P_B max_k) - (Epv_i - BL_i)]|) \quad (2)$$

$$with\ x_{i,j} \in [0;1]\ and\ with\ y_{i,k} \in [-1;1]$$

where N is the number of discrete time slots, M_{EV} and M_B are respectively the number of electric cars and storage batteries (we denote electric cars as EV and storage batteries as B), $P_{EV}max$ and $P_B max$ are the nominal powers of electric cars and batteries, Epv is the energy delivered by the solar panel in an i-th time slot, BL_i the energy consumed by background loads in the i-th interval, Δt is the time slot duration. The unknown quantities of our problem are the $x_{i,j}$ [4]

and $y_{i,k}$ which represent the percentage of nominal power to be used by the j-th EV and the k-th stationary battery in the i-th time slot Δt. Given the nature of stationary batteries, they can both consume and deliver energy so the $y_{i,k}$ vary in $[-1, 1]$ unlike EV's batteries that can only charge so the $x_{i,j}$ vary in $[0, 1]$.

The second objective function concerns the satisfaction of the energy demand. For each charging session we minimize the difference between the energy demand and the charged energy [3]. This objective function is extensively formulated in Eq. 3, where D_k is the energy demand of the k_{th} vehicle.

$$\min(|\sum_{k=0}^{M_{EV}} (D_k - \sum_{i=0}^{N} \Delta t * x_{i,k} * P_{EV}max_k)|) \qquad (3)$$

Some constraints are set to guarantee the physical feasibility of the solution. In 4 the charged energy cannot exceed the EV battery capacity.

$$\sum_{i=0}^{N} \Delta t * x_{i,k} * P_{EV}max_k \leq C_k - E_{0k} \quad , \quad \forall k \in 0, ..., M_{EV} \qquad (4)$$

C_k and E_{0k} are the total battery capacity and the initial state of charge (kWh) of the k_{th} Electric Vehicle. In Eq. 5 energy stored in the EV battery must be always greater or equal to 0

$$\sum_{i=0}^{N} (\Delta t * y_{i,k} * P_{EV}max_k) + E_{0k} \geq 0 \quad , \quad \forall k \in 0, ..., M_{EV} \qquad (5)$$

In Eq. 6, the energy stored into the battery must be always less or equal to the battery capacity.

$$\sum_{i=0}^{N} \Delta t * y_{i,k} * P_B max_k + EB_{0j} \leq BC_k \quad , \quad \forall k \in 0, ..., M_B \qquad (6)$$

BC_k and E_{0k} are the total battery capacity and the initial state of charge (kWh) of the k_{th} battery. In Eq. 7, the energy stored into the battery must be always greater or equal to 0.

$$\sum_{i=0}^{N} (\Delta t * y_{i,k} * P_B max_k) + EB_{0j} \geq 0 \quad , \quad \forall k \in 0, ..., M_B \qquad (7)$$

4.1   Genetic Algorithm Solution

A solution to such optimization problem is a set of energy profiles for each device (EV and battery) connected to the smart micro-grid [5]. If the optimization horizon is a day and N is the number of time-slots, the number of unknown variables for the complete solution is $N * (M_{EV} + M_B)$. To reduce the complexity of the problem we limited the representation of each time series to the time slots which fall in a charge session (e.g. the user will plug-in the car from 14:00 to 16:00). Moreover, we limit to 3 the number of time slots in which the charge session is split, so that the number of unknown quantities no longer depends on the number of time intervals. In this way, the number of unknowns is independent of the duration of the charge session, but the charge power modulation can assume at most three different values.

The next step is to make the number of unknowns independent of the number of batteries. In particular, M_B batteries have been modeled as one battery that has a capacity that is equal to the sum of the capacities and a maximum charging/discharging power equal to the sum of charging/discharging powers. Following these simplifications, the number of optimization variables becomes $N + 3 * M_{EV}$. A further simplification was made on the range of values that can be assumed by the unknown quantities. They are not considered as continuous and real values but as quantized values that define the percentages of nominal power, at which the battery can work. We quantize the power by 10 values for electric vehicles and by 20 values (-10 to 10) for batteries. The *NSGA-II* (Non-dominated Sorting Genetic Algorithm) has been used as optimization algorithm. Crossover (probability $= 1$, distribution index $= 20$) and polynomial mutation parameters (probability $= \frac{1}{N+3*M_{EV}}$, distribution index $= 20$) have been heuristically chosen. The population size have been set equal to 200 individuals with an offspring population of 60.

5   Experimental Results

Our test case is composed of 5 charging EVs. The station is equipped with a stationary battery of *27 kWh* with a *nominal power* of 12 kW. EV models are described in Table 1 (where *M.Power* is the maximum charging power, *Arr.* is the Arrival Time and *Dep.* is the Departure Time).

Table 1. Charge sessions in test scenario

EV	Capacity	M.Power	Arr.	Dep.
1	31 kWh	10.75 kW	11:00	15:30
2	52 kWh	22 kW	17:30	19:00
3	52 kWh	20 kW	7:00	15:00
4	10 kWh	1.26 kW	9:00	16:30
5	75 kWh	27.5 kW	14:00	18:00

In the baseline scenario EVs charge at nominal power as soon as they arrive at the charging station and batteries are not exploited. In Fig. 3(a), the green series corresponds to the power production profile and in the other colors we have the 5 different EVs power consumption profiles. In Fig. 3(b), we can see the optimization results. We have in red the power consumption of the EVs, in

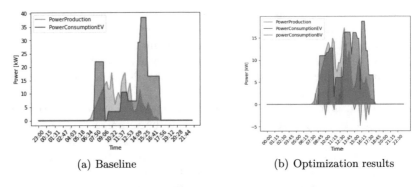

(a) Baseline   (b) Optimization results

Fig. 3. Power production and consumption

blue the power consumption/production of the batteries (positive values when the battery consumes energy, negative values when it supplies energy) and in green, the solar power production [7]. As we can see solar energy is almost completely used by EVs and batteries. The self-consumption varies from 63% (in the baseline) to 87% (in the optimized scenario). Moreover, 99% energy demand is satisfied. It is important to note that even if the power consumption for each EV can change only three times during a charging session, the total power consumption can have up to 15 different values.

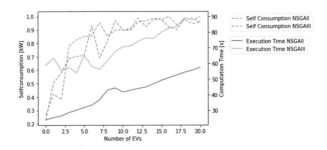

Fig. 4. Execution time and self-consumption

A further analysis has been performed using the NSGAIII algorithm. NSGAII and NSGAIII differ from each other by the way they preserve variety among solutions. NSGAIII uses a set of reference directions and NSGA-II employs a more flexible strategy through its crowding distance operator for the same goal. The choice of the algorithms of the NSGA family is motivated by their successful utilization in previous studies at the expense of computational complexity. In Fig. 4 we can see how NSGAII and NSGAIII perform with our problem increasing the number of EVs. Dotted lines refer to Self-Consumption and full lines to Computational Time. We can notice that the NSGAIII converges slower than the NSGAII and self-consumption exhibits the same figure for both algorithms. Further investigations are omitted for reason of brevity.

6 Conclusion

In this work we have proposed the joint utilization of an optimal route planner and of a smart energy scheduler to improve utilization of renewable energy sources without affecting the user experience and reducing driver's anxiety for charging her EV. The best routes are recommended to the users, who can opt also for not green choices if it allows to save time or if it better complies with its travel preference or with unforeseen changes of its travel plan. We focused on the optimal energy schedule at each charge station, based on the charge plan of users and the availability, exploiting the battery usage and the energy production from photovoltaic-panels. Exploiting the results of each micro-grid optimization in terms of self-consumption and charge availability, the planner will be eventually able to estimate how much green a route performs. It can use it to improve recommendations, making explicit such value to the drivers, who can use it as a decision criteria.

References

1. Amato, A., Jung, J.J., Venticinque, S.: Extending the internet of energy by a social networking of human users and autonomous agents. Multimed. Tools Appl. **76**(24), 26057–26076 (2017). https://doi.org/10.1007/s11042-017-4888-2
2. Amato, A., Scialdone, M., Venticinque, S.: Improving self-consumption of green energy using linear programming for reactive control of smart devices. Informatica **29**(2), 187–210 (2018). https://doi.org/10.15388/Informatica.2018.163
3. Aversa, R., Branco, D., Di Martino, B., Venticinque, S.: GreenCharge simulation tool. In: Barolli, L., Amato, F., Moscato, F., Enokido, T., Takizawa, M. (eds.) WAINA 2020. AISC, vol. 1150, pp. 1343–1351. Springer, Cham (2020). https://doi.org/10.1007/978-3-030-44038-1_122
4. Aversa, R., Branco, D., Di Martino, B., Venticinque, S.: Container based simulation of electric vehicles charge optimization. In: Barolli, L., Woungang, I., Enokido, T. (eds.) AINA 2021. LNNS, vol. 227, pp. 117–126. Springer, Cham (2021). https://doi.org/10.1007/978-3-030-75078-7_13
5. Aversa, R., Branco, D., Di Martino, B., Iaiunese, L., Venticinque, S.: Simulation and evaluation of charging electric vehicles in smart energy neighborhoods. In: Barolli, L., Hussain, F., Enokido, T. (eds.) AINA 2022. LNNS, vol. 451, pp. 657–665. Springer, Cham (2022). https://doi.org/10.1007/978-3-030-99619-2_61
6. Barco, J.J., Guerra, A., Muñoz, L., Quijano, N.: Optimal routing and scheduling of charge for electric vehicles: a case study. Math. Prob. Eng. **2017**, 8509783 (2017)
7. Branco, D., Di Martino, B., Venticinque, S.: A big data analysis and visualization pipeline for green and sustainable mobility. In: Barolli, L., Woungang, I., Enokido, T. (eds.) AINA 2021. LNNS, vol. 227, pp. 701–710. Springer, Cham (2021). https://doi.org/10.1007/978-3-030-75078-7_69
8. Flocea, R., et al.: Electric vehicle smart charging reservation algorithm. Sensors **22**(8), 2834 (2022)
9. Razeghi, G., Samuelsen, S.: Impacts of plug-in electric vehicles in a balancing area. Appl. Energy **183**, 1142–1156 (2016)
10. Shaukat, N., Khan, B., Ali, S., Mehmood, C., Khan, J., Farid, U., Majid, M., Anwar, S., Jawad, M., Ullah, Z.: A survey on electric vehicle transportation within smart grid system. Renew. Sustain. Energy Rev. **81**, 1329–1349 (2018)

11. Triviño-Cabrera, A., Aguado, J.A., de la Torre, S.: Joint routing and scheduling for electric vehicles in smart grids with V2G. Energy **175**, 113–122 (2019)
12. Umetani, S., Fukushima, Y., Morita, H.: A linear programming based heuristic algorithm for charge and discharge scheduling of electric vehicles in a building energy management system. Omega **67**, 115–122 (2017)
13. van der Kam, M., van Sark, W.: Smart charging of electric vehicles with photovoltaic power and vehicle-to-grid technology in a microgrid; a case study. Appl. Energy **152**, 20–30 (2015)
14. Yagcitekin, B., Uzunoglu, M.: A double-layer smart charging strategy of electric vehicles taking routing and charge scheduling into account. Appl. Energy **167**, 407–419 (2016)

A 3D Urban Aerial Network for New Mobility Solutions

Chiara Caterina Ditta[✉] and Maria Nadia Postorino

DICAM, Alma Mater Studiorum - University of Bologna, Bologna, Italy
{chiaracaterina.ditt2,marianadia.postorino}@unibo.it

Abstract. Urban air mobility is an expected new solution for satisfying mobility needs within and among densely urbanized city areas as well as for increasing airport accessibility. Transport services are expected to be realized by connected flying vehicles, currently experimented by several industries. An important issue concerns the definition of a suitable aerial network framework that allows realizing flight operations at low altitudes by using stable data transmission among flying vehicles and Air Traffic Control centers. This paper proposes, and discusses the implication of, a three-dimensional Urban Aerial Network (3D-UAN) that includes the third (vertical) dimension to allow flying vehicles to be integrated into the airspace current procedures. A suitable link cost function is also defined. The required data - both off-line and in real-time - and the way they should be communicated among flying vehicles, and between flying vehicles and the Control center, to find optimal paths between relevant nodes and ensure safe operations are explored. The main aspects to simulate this distributed architecture are considered from the perspective of satisfying transport network properties and user's needs.

Keywords: Aerial Network · Urban Air Mobility · Airspace distribution · Data processing

1 Introduction

Urban Air Mobility (UAM), and more generally Advanced Air Mobility (AAM), has been gradually included among the opportunities for the transport systems of the future [26]. The concept of an urban transport system that uses the third dimension, by exploiting technologies with reduced environmental impacts, is progressively involving more and more interest in the international scientific community and investors in several technological fields [6]. The preliminary elements that have contributed to the discussion about this new transport solution has been the realization of some electric Vertical Take-Off and Landing (eVTOL) aircraft and the use of Unmanned Air Vehicles (UAV) in several application areas, such as traffic monitoring [31], infrastructure inspection [4], mapping [14], and agriculture [11]. Many studies have identified application opportunities for goods delivery as well as passenger mobility. An overview drawn up by NASA

© The Author(s), under exclusive license to Springer Nature Switzerland AG 2023
L. Braubach et al. (Eds.): IDC 2022, SCI 1089, pp. 277–286, 2023.
https://doi.org/10.1007/978-3-031-29104-3_30

researchers [32] identified a list of potential operations and use cases that can be performed: i) Air Taxi on-demand service, ii) Air Cargo, iii) Air Metro, iv) Emergency operations, v) News gathering and vi) Traffic and Weather monitoring. Focusing on passenger transport, different types of air services have now been defined in the literature, with different features and specific requirements on UAM systems [3,13,24,29]. The implementation of UAM operations requires a suitable airspace structure to ensure safe flight operations, air vehicle separations and air routes that do not interfere with traditional aviation. Procedures are expected to take place at low altitudes, mainly in the uncontrolled ICAO class G [34]. In this context, currently two types of approaches have been identified for the air traffic control and management: i) centralized system, such as UAS Traffic Management (UTM) or U-Space Concept [25,33]; ii) route management directly realized by autonomous aircraft with onboard technology, such as "sense-and-avoid" supported by algorithms for route guidance [35]. In order to contextualize issues due to UAM, in the following, Sect. 2 shows the main airspace frameworks identified to date. Section 3 describes the 3D network model, which could allow UAM operations in the near future. Section 4 described the requirements and implications of the 3D-UAN model. Section 5 refers to the simulation architecture of the 3D-UAN model. Finally, in Sect. 6 some conclusions and future developments are drawn.

2   Framework Background

The airspace framework for UAM services should ensure safe low-level flight operations and fast connections. This is also linked to capacity issues and in order to study how capacity differs with the size of the airspace structure, the Metropolis project [30] explored four different urban airspace concepts: Full Mix, Layers, Zones and Tubes. The Full Mix concept refers to a free flight context, where aerial traffic is subjected only to physical constraints and the flying vehicles would handle autonomously separations and trajectories based on sensors and suitable software. In the Layer concept, the airspace is composed by several layers, where the flying vehicles have to limit their speed to ensure separations. The Zones concept assumes that horizontal trajectory will be computed by using the A* shortest path algorithm [12], while the altitude is not pre-defined but adopted suitably. Finally, the Tubes concept refers to a structure with pre-planned conflict-free routes, like a graph where nodes are extreme points of one or more routes, and tubes - similar to links - connect two nodes. When traveling through tubes, flight vehicles shall maintain a defined speed to ensure traffic separations. This layout is simulated as a multi-level structure, where lower levels are reserved to short-range flights and higher levels to long-range flights. Furthermore, this model is completed by considering time separation as fourth dimension. As set out in a previous review [9], a more recent multilayer model could allow operations in uncontrolled (class G) airspace [27]. In this context, flying vehicles would move across airways without direct communications with a central control system (such as UTM), instead they would be guided by the "rules of the airway" (such as speed limits, flight headings and maximum traffic capacity) and information exchanged by V2V communications [7]. Additionally, higher layers are obstacle-free, so it might be possible to travel at higher speeds. A further framework enables UAM procedures in Class G airspace under Augmented Visual Flight Rules by using sense-and-avoid capabilities, whereas operations in controlled airspace will involve Dynamic Delegated Corridors (DDCs) [19]. These latter are airspace volumes or tunnels, such as Airways, designated to separate and handle UAM traffic. Corridor management and air traffic separation is entrusted to the autonomous decisions of flying vehicles, which would be supported by an Automated Decision Support Service. This system offers an architecture similar to UTM, with more rigorous safety standard. Starting from the above perspective, this paper proposes a three-dimensional Urban Air Network (3D-UAN) that includes the third (vertical) dimension by satisfying the basic principle of linking trip origin/destination points as in usual ground transportation networks. In fact, other than based on communication features, 3D-UANs should be designed to guarantee: i) effective connections between points of interest in urban areas; ii) reduced travel times; iii) shorter travel distances and iv) effective, safe flight paths. The network operability will depend on data sharing and processing related to space and time information provided by/to each flying vehicle. Particularly, it is assumed that data are both exchanged among aerial vehicles, which are autonomous in keeping assigned routes and suitable distances on the three axes, and sent to/from a Control center. In this respect, this paper describes the role of the Control center, the identification of the best path - which depends on a proposed link cost function - and the dynamic features of the aerial links in the proposed 3D-UAN.

3   The 3D Urban Air Network Model

In order to simulate an aerial transportation system and the UAM services that take suitably into account the innovations provided by flying vehicles, this study proposes a three-dimensional Urban Aerial Network (3D-UAN) model (see Fig. 1), based on some preliminary results already presented in the literature [19, 27]. The 3D-UAN is expected to be operational in the very low and uncontrolled airspace (i.e. 500 ft Above Ground Level [2]) for people and goods trips. Ad-hoc links of the network, such as DDCs, are expected to be supported by information collected and transmitted by both connected flying vehicles and a centralized control system (e.g., UTM).

Each layer L is figured out as a two-dimensional graph, (G_L), which includes the set of *Fixed Nodes* (N_F) and the set of *Transition Nodes* (N_T), connected by *Dynamic Links* belonging to the set $D_l = \{d_l\}|l = \{1, 2...\}$. The fixed nodes are defined in the access and egress points of the network (i.e. vertiports and/or suitable ground infrastructures), while the transition nodes allow horizontal crossing and switch to an upper or lower layer. Dynamic Links allow moving between nodes and are considered as air corridors that can be enabled or not according to traffic capacity and environmental conditions. In addition, their shape and

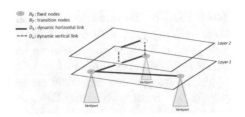

Fig. 1. Urban Aerial Network (3D UAN) model

volume could change depending on flight vehicle prototype and safe established distance. The set of Dynamic Links include horizontal and vertical link subsets, respectively $D_{\mathrm{h}} = \{h_{\mathrm{mL}}\} \subset D_1|m = \{1,2...\}$ and $D_{\mathrm{v}} = \{v_{\mathrm{mL}}\} \subset D_1|m = \{1,2...\}$, which are described by equivalent properties and allow different flight operations (e.g. land and take-off permitted only in vertical dynamic links). Transfers between layers occur in the nodes, both fixed and transition, by vertical dynamic links, while transfers within layers occur by horizontal dynamic links.

The final *three-dimensional* Graph Θ includes the different layers ((bi-dimensional graphs) and the vertical dynamic links:

$$\Theta = \bigcup\nolimits_{\forall L=\{1...n\}}^{n} G_L \cup D_v \text{ with } G_L = (N_F, N_T, D_h)$$

Long-haul flight operations are assumed to be performed in the higher layers (such as intercity flight), while urban operations would be allowed in the lower layers. Moreover, aerial vehicles can change layers in *fixed or transition* nodes a limited number of times ($\mathrm{M_{max}}$), in particular depending on route length, which means greater allowed numbers of layer changes for long-distance flights. The layer change constraint is useful to ensure savings in autonomy (or fuel) and travel time and fuel consumption. In practice, the layer change implies a greater energy consumption to perform hover operations, for which a reduced speed is also required compared to the cruise speed, and this generates a consequent increase in the overall travel time and fuel consumption. Dynamic link operability would depend mainly on link capacity and weather conditions. In this perspective, their operability has been expressed by the following generalized cost function (1), where i is the aerial vehicle, sequentially numbered, using that link at a given time period:

$$c(T_r, T_c, T_d, T_g, i) = \begin{cases} T_{ri} + T_{ci} + T_{di} & \text{for } i = 1 \\ T_{ri} + T_{ci} + T_{di} + T_{g(i,i+1)} & \forall\, i > 1 \end{cases}$$

where T_{ri} is the running time of the aerial vehicle i, T_{ci} and T_{di} are its climbing and descendent time respectively, $T_{\mathrm{g}(i,i+1)}$ is the time gap between two following aerial vehicles. The cost function associated to each link allows identifying the shortest route, which is a sequence of links connecting two fixed nodes where i is expected to take-off and land. The shortest route is the one whose overall cost (route cost) is minimum with respect to alternative routes, while respecting possible constraints - such as capacity limits on one or more links. The suitable path is computed by using an iterative shortest path algorithm (such as Dijkstra [8] or A* [12]). The shortest path is found based as follows: i) the control center computes the shortest path based on the available information and communicate it to each flying vehicle before the departure; ii) while moving between fixed node pairs, in order to handle the traffic without generating issues, connected vehicles evaluate in real time whether to proceed on the same layer or switch to another one by vertical dynamic links, based on the current capacity value of the horizontal dynamic links [9]. This type of 3D-UAN is able to support multi/inter mode trip solutions that include UAM services. Particularly, the main services (see Fig. 2) that are expected to be developed in the international market in the coming decades are the following:

Fig. 2. Urban Air Mobility use cases

1. Air Taxi scenario: it involves an air transport service between origin-destination pairs within the urban area and it is proposed as an on-demand service in particular for short-haul trips (between 15 to 50 km).
2. Airport Shuttle scenario: it would provide scheduled flights between airports and selected, suitable locations in metropolitan area or points of interests. The covered distance would be the same of the air taxi service - about 50 km.
3. Inter-City trip scenario: it would be a scheduled service between pairs of cities, which will cover distances of more than 100 km and simplify direct access to points of interest in the served cities. Particularly, this scenario could ensure fast long-distance connections, which is more appealing for commuters and business travellers.

4   Requirements and Implications of the 3D-UAN Model

The 3D-UAN operations depend on the off-line and in real time data processing. Thanks to V2V and V2I communication technology [7,10], flying vehicles can share data on their position, speed, operating status and environmental condition, both with each other and with the Control Center. Following what emerges in the literature [18], here it is assumed that a general air traffic control center - such as UTM - is required, which supervises flight operations and provides support to individual flying vehicles. The control center is equipped by hardware and software (i.e. planning facility, display equipment, telemetry schedule, computer processor, data terminal on ground, communication equipment and protection device), which could ensure suitable route planning [16]. By assuming that the 3D network will support scheduled transport services (such as airport shuttle or intercity flights), the control center will compute the shortest paths by applying a search algorithm [15] based on the cost function (1), before the flying vehicle departure. In fact each dynamic link is associated with function (1), which also depends on flying vehicles number (i.e., link flow) that occupies the link in the given reference time. Accordingly, the algorithms [15] will calculate the path of minimum cost as a function of flow value on each dynamic link. The path cost will be the result of the costs sum over each single link, with the aim of reducing the travel time of each flying vehicle as much as possible (see Fig. 3). Then, links maintain their dynamic characteristics, and they can be "enabled" or "disabled" for transit if extraordinary events occur (e.g. adverse weather conditions, accumulation of operational delays), thanks to data communicated in real time by using wireless sensor networks (WSNs) [17]. The control center and the flying vehicles will be able to compute again the shortest routes to reach the destination [1], if a dynamic link were to be disabled, thus taking advantage to change layer and occupy enabled links, and distributing air traffic over the 3D network evenly. On the other hand, for an on-demand service (such as air taxi), the dynamic characteristics of the network are essential. For this service, information on the occupancy status of dynamic links has to be accurate and shared in real time. In this condition, the shortest path should be computed in real time by each flying vehicle, by processing the data received from the control center and from the other cooperative flying vehicles through FANET (Flying Ad-Hoc Networks) communication and computation methods [5,9]. Therefore, during the execution of the trips, if a dynamic link reaches its capacity limit, it will have to be "disabled" in order to not congest the network and create disruptions. The flying vehicles will be distributed over the network, even changing layers, to guaranteeing the purpose of short travel times and conflict-free trips.

5   Simulation Architecture

The simulation architecture of the 3D-UAN model is founded on an agent- based microsimulation approach, which is widely exploited in the transportation area for its several benefits [22]. In particular, each flying vehicle may be represented as a cooperative agent communicating with both other agents - i.e. flying vehicles moving along the (dynamic) links of the 3D-UAN - and the Control center, which may be also identified as "Agency" [23] (see Fig. 4). The Agency provides initial information about the scheduled path, according to past information and to the current status of the links. Note that each link - both vertical and horizontal - has a maximum allowable capacity, in other words only a maximum number of agents are allowed to use a given link in the reference time period. This depends not simply on the features of the link - synthesized by the cost function introduced in Sect. 2 - but also on the available capacity at vertiports

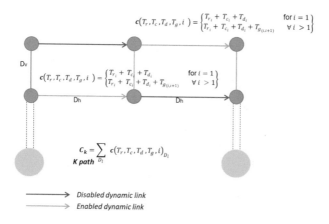

Fig. 3. Shortest path definition

where agents will take-off and land. In case of deviations from the scheduled arrivals and departures of agents at the vertiports, the link capacity of one (or more) links in the path might be not sufficient to host the initially expected agents, so that the corresponding link(s) will be "closed" - i.e., no more agents are allowed to use such link(s) - and an alternative path will be provided to the agent(s). This kind of information may be sent both before the agent starts its trip by the Agency (the scheduled path already includes deviations) and during the trip, based on the real time information sent by each agent about its position and speed. Other than with the Agency, each agent communicates with close agents by employing ad-hoc networks, such as FANET, through direct wireless connections [28], particularly to keep the allowed minimum separations also thanks to anti-collision sensors [20]. Note that in the 3D-UAN the communication among agents occurs not only along the horizontal dimension, but also the vertical one, in order to keep the safe vertical distance between agents moving in different (horizontal) layers. Although each agent could minimize its own path by using the information coming from close agents, however in this first approach the role of the Agency has been considered predominant to keep safety conditions, so that the information exchanged among agents is limited to ensure the minimum allowed separations. This architecture may be implemented by using microsimulation/agent- based tools (such as, for example, MATSim [21]). In particular, the input fixed settings (flying vehicle cruise speed, link capacity) simulate the action of the Agency, while by integrating the cost function (1) we intend to simulate the behavior of the agents on the dynamic links and how they would be distributed on the 3D network proposed in this work. Data will refer to some scheduled services - such as airport shuttle services - and then some perturbation will be introduced - such as delay at vertiports or bad weather conditions - which will prevent the use of one or more dynamic links. This tool, by integrating an algorithm for calculating the shortest path, will simulate the operations carried out by the Agency and the flying vehicles to move on the network.

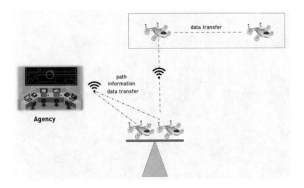

Fig. 4. Simulation Architecture

6 Discussion and Conclusion

In the perspective of new transport solutions, such as UAM services, it is crucial to have a transport network model to simulate the transportation aerial system that can guarantee the usual standards for passengers and goods, mainly in terms of safety and fastness. In this regard, an aerial network model (3D-UAN) has been presented, which exploits the third dimension, in order to allow an adequate distribution of aerial traffic in the low airspace. This 3D framework exploits the opportunity of using dynamic links that can be "enabled" and "disabled" according to traffic and environmental conditions, based on a suitable link cost function. The system requires data transmission and management by both a control center and the flying vehicles themselves, based on their integrated autonomous computing capabilities and connecting/cooperating technologies. This paper has described the main aspects, which include data transmission and management, that have to be considered from the transportation system perspective in order to realize UAM services. In particular, it has focused on aspects related to transport engineering and operations of the air transport network. The next steps for future work include expanding what has been presented by carrying out an agent simulation, in order to be able to verify the 3D network operation in application scenarios, such as those currently identified to perform UAM services (Air Taxi, Airport Shuttle and Intercity flight).

References

1. Shubhani, A., Neera, K.: Path planning techniques for unmanned aerial vehicles: a review, solutions, and challenges. Comput. Commun. **149**, 270–299 (2020)
2. Antcliff, K.R., Moore, M.D., Goodrich, K.H.: Silicon Valley as an early adopter for on-demand civil VTOL operations. In: 16th AIAA Aviation Technology, Integration, and Operations Conference, p. 3466 (2016)
3. Baur, S., Schickram, S., Homulenko, A., Martinez, N., Dyskin, A.: Urban Air Mobility: The Rise of a New Mode of Transportation. Roland Berger Focus, Munich (2018)

4. Besada, J.A., et al.: Drone mission definition and implementation for automated infrastructure inspection using airborne sensors. Sensors **18**(4), 1170 (2018)
5. Chriki, A., Touati, H., Snoussi, H., Kamoun, F.: FANET: communication, mobility models and security issues. Comput. Netw. **163**, 106877 (2019)
6. Cohen, A.P., Shaheen, S.A., Farrar, E.M.: Urban air mobility: history, ecosystem, market potential, and challenges. IEEE Trans. Intell. Transp. Syst. **22**(9), 6074–6087 (2021)
7. Demba, A., Moller, D.P.: Vehicle-to-vehicle communication technology. In: 2018 IEEE International Conference on Electro/Information Technology (EIT), pp. 0459–0464. IEEE (2018)
8. Dijkstra, E.W., et al.: A note on two problems in connexion with graphs. Numer. Math. **1**(1), 269–271 (1959)
9. Ditta, C.C., Postorino, M.N.: New challenges for urban air mobility systems: aerial cooperative vehicles. In: Camacho, D., Rosaci, D., Sarné, G.M.L., Versaci, M. (eds.) Intelligent Distributed Computing XIV, IDC 2021. Studies in Computational Intelligence, vol. 1026, pp. 135–145. Springer, Cham (2022). https://doi.org/10.1007/978-3-030-96627-0_13
10. Djahel, S., Jabeur, N., Barrett, R., Murphy, J.: Toward V2I communication technology-based solution for reducing road traffic congestion in smart cities. In: 2015 International Symposium on Networks, Computers and Communications (IS-NCC), pp. 1–6. IEEE (2015)
11. Grenzdörffer, G.J., Engel, A., Teichert, B.: The photogrammetric potential of low-cost UAVs in forestry and agriculture. Int. Arch. Photogramm. Remote. Sens. Spat. Inf. Sci. **31**(B3), 1207–1214 (2008)
12. Hart, P.E., Nilsson, N.J., Raphael, B.: A formal basis for the heuristic determination of minimum cost paths. IEEE Trans. Syst. Sci. Cybern. **4**(2), 100–107 (1968)
13. Hasan, S.: Urban Air Mobility (UAM) market study. Technical report (2019)
14. Hassanalian, M., Abdelkefi, A.: Classifications, applications, and design challenges of drones: a review. Prog. Aerosp. Sci. **91**, 99–131 (2017)
15. He, Z., Zhao, L.: The comparison of four UAV path planning algorithms based on geometry search algorithm. In: 2017 9th International Conference on Intelligent Human-Machine Systems and Cybernetics (IHMSC), vol. 2. IEEE (2017)
16. Hong, Y., Fang, J., Tao, Y.: Ground control station development for autonomous UAV. In: Xiong, C., Liu, H., Huang, Y., Xiong, Y. (eds.) ICIRA 2008. LNCS (LNAI), vol. 5315, pp. 36–44. Springer, Heidelberg (2008). https://doi.org/10.1007/978-3-540-88518-4_5
17. Jawhar, I., Mohamed, N., Al-Jaroodi, J.: UAV-based data communication in wireless sensor networks: models and strategies. In: 2015 International conference on unmanned aircraft systems (ICUAS), pp. 687–694. IEEE (2015)
18. Jiang, T., Geller, J., Ni, D., Collura, J.: Unmanned aircraft system traffic management: concept of operation and system architecture. Int. J. Transp. Sci. Technol. **5**(3), 123–135 (2016)
19. Lascara, B., Lacher, A., DeGarmo, M., Maroney, D., Niles, R., Vempati, L.: Urban air mobility airspace integration concepts: Operational concepts and exploration approachs. Technical report, MITRE CORP MCLEAN VA MCLEAN (2019)
20. Liu, S.Z., Hwang, S.H.: Vehicle anti-collision warning system based on V2V communication technology. In: 2021 International Conference on Information and Communication Technology Convergence (ICTC), pp. 1348–1350. IEEE (2021)
21. MatSim (2022). https://www.matsim.org

22. Postorino, M.N., Sarné, F.A., Sarné, G.M.: An agent- based simulator for urban air mobility scenarios. In: Proceedings of the 21st Workshop "from Objects to Agents", WOA 2020, vol. 2706 of CEUR Workshop Proceedings. CEUR- WS.org (2020)
23. Postorino, M.N., Sarné, G.M.: Agents meet traffic simulation, control and management: a review of selected recent contributions. In: Proceedings of the 17th Workshop "from Objects to Agents", WOA 2016, vol. 1664 of CEUR Workshop Proceedings. CEUR-WS.org (2016)
24. Postorino, M.N., Sarné, G.M.: Reinventing mobility paradigms: flying car scenarios and challenges for urban mobility. Sustainability **12**(9), 3581 (2020)
25. Prevot, T., Rios, J., Kopardekar, P., Robinson III, J.E., Johnson, M., Jung, J.: UAS traffic management (UTM) concept of operations to safely enable low altitude flight operations. In: 16th AIAA Aviation Technology, Integration, and Operations Conference, p. 3292 (2016)
26. Rothfeld, R., Straubinger, A., Fu, M., Al Haddad, C., Antoniou, C.: Urban air mobility. In: Demand for Emerging Transportation Systems, pp. 267–284. Elsevier (2020)
27. Samir Labib, N., Danoy, G., Musial, J., Brust, M.R., Bouvry, P.: Internet of unmanned aerial vehicles-a multilayer low-altitude airspace model for distributed UAV traffic management. Sensors **19**(21), 4779 (2019)
28. Schalk, L. M.: Communication links for unmanned aircraft systems in very low level airspace. In: 2017 Integrated Communications, Navigation and Surveillance Conference (ICNS), pp. 6B2–1. IEEE (2017)
29. Schuchardt, B. I., et al.: Urban air mobility research at the DLR German aerospace center getting the horizon UAM project started. In: AIAA Aviation 2021 Forum, p. 3197 (2021)
30. Sunil, E., et al.: Metropolis: relating airspace structure and capacity for extreme traffic densities. In: Proceedings of the 11th USA/EUROPE Air Traffic Management Research and Development Seminar (ATM2015), Lisbon (Portugal), 23–26 June 2015. FAA/Eurocontrol (2015)
31. Sutheerakul, C., Kronprasert, N., Kaewmoracharoen, M., Pichayapan, P.: Application of unmanned aerial vehicles to pedestrian traffic monitoring and management for shopping streets. Transp. Res. Procedia **25**, 1717–1734 (2017)
32. Thipphavong, D.P., et al.: Urban air mobility airspace integration concepts and considerations. In: 2018 Aviation Technology, Integration, and Operations Conference, p. 3676 (2018)
33. Undertaking, S.J., et al.: U-space: blueprint (2017)
34. Vascik, P.D., Balakrishnan, H., Hansman, R.J.: Assessment of air traffic control for urban air mobility and unmanned systems (2018)
35. Yang, X., Wei, P.: Scalable multi-agent computational guidance with separation assurance for autonomous urban air mobility. J. Guid. Control. Dyn. **43**(8), 1473–1486 (2020)

License Plate Detection and Recognition Using CRAFT and LSTM

Anh Kiet Huynh[1,2], Tan Duy Le[1,2(✉)], and Kha-Tu Huynh[1,2(✉)]

[1] School of Computer Science and Engineering, International University,
Ho Chi Minh City, Vietnam
[2] Vietnam National University, Ho Chi Minh City, Vietnam
{ldtan,hktu}@hcmiu.edu.vn

Abstract. This work proposes a solution for developing a license plate detection and recognition system (LPDR). In the first stage, the poly region of the license plate's word line(s) can be detected by Character-Region Awareness For Text Detection (CRAFT). Specifically, the text line of the one-line license plate and two lines of the multi-line license plate can be detected effectively. Secondly, each region proposed as a plate number region by CRAFT will be passed to Mobilenet architecture to extract features. Finally, these features will be fed to Bi long short-term memory (Bi-LSTM) architecture with Connectionist Temporal Classification to predict output text in each input region. By applying this solution, the problem of multi-line license plates can be appropriately handled.

Keywords: CRAFT · License Plate Detection · License Plate Recognition · deep neural network

1 Introduction

Along with the development of modern cities, the License Plate Detection and Recognition system (LPDR) has many practical applications such as parking lots, toll collection or traffic enforcement, etc. Most of the current LDPR systems concentrate on one-line license plate problems. However, in many countries, for instance, Vietnam and Brazil, multi-line license plate accounts for a significant part.

To deal with the problem of license plate recognition, the well-known solutions are applying a segmentation algorithm to locate each character in the plate and classify each of them to obtain sequence output [1]. Unfortunately, the recognition result is critically dependent on segmentation output with classification characters result, which can be easily affected by skew, light, contrast condition of environment or binary process. Non-segmentation approach is another common approach for license plate recognition [2]. However, it works well on the one-plate license plate and has high latency for the license plate detection phase.

L. Braubach et al. (Eds.): IDC 2022, SCI 1089, pp. 287–296, 2023.
https://doi.org/10.1007/978-3-031-29104-3_31

For the multi-line license plate, reorganization feature maps extracted from Convolutional Neural Network (CNN) can be applied to recognize two-plate license plates [3]. Unfortunately, it requires a license plate should be an ideal image.

In our proposed LPDR system, Character-Region Awareness For Text Detection (CRAFT) [4] was applied to deal with the issue of the multi-line license plate. In detail, this method can detect the two-line of a license plate separately. Then, each line will be fed to the recognition phase. These lines can be treated as a sequence of characters and predicted by Connectionist Temporal Classification. Therefore, the output does not rely on the segmentation process and does not require an ideal license plate.

The main contributions of our work are:

- We propose a solution that uses CRAFT as an approach for text(s) detection in the license plate. License plate text region can be extracted by a combination of text score and affinity score generated by the CRAFT model. Each detected text line of the license plate will be treated as a sequence-like feature map. By applying a text recognition framework, the number plate from each detected number plate line can be extracted precisely without the segmentation process. By dividing the text lines of two-line license plates, it decreases workload compared with the recognition of two-line license plates directly.
- Build a lightweight model with a pretraining manner for the recognition phase, which has a trade-off between accuracy and latency at an acceptable level and is possible to apply to a practical system.

2   Background and Related Research

Every country sets different regulations for license plates. Therefore, depending on the information displayed in the license plates, there are different identification methods. Recently, many research related to the license plates of the countries have been published, such as Spanish and Indian [5], Bangladesh [6], and China [7], etc. The general scheme of license plates recognition is built by 02 steps of detection and recognition. The development of CNN has supported researchers to build effective image detection and recognition algorithms, especially vehicle license plates. YOLO4, with its ability to detect objects, has been used by many authors in the detection of letters and numbers [8], in the license plates. In addition, some methods using Fractal Series Expansion, DELP-DAR, ResNet [9] also give reliable results.

3   System Design

Our proposed LPDR system is divided into two phases, including detection and recognition phases, which are illustrated in Fig. 1.

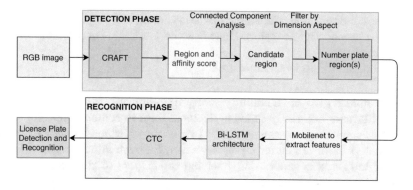

Fig. 1. Overview of the License Plate Detection and Recognition System (LPDR)

3.1 Number Plate Detection

The first phase of the LDPR system is license plate detection. In this phase, we use CRAFT approach to detect each text line of the license plate.

CRAFT is designed based on convolutional neural network architecture. Each input image will be resized to a fixed shape and fed to the model. This model generates a character region score and an affinity score. The region score is used to locate each character in the image, and the affinity score is used for grouping characters in the word group. With the image dataset does not have a character annotations level, we will apply a weakly-supervised learning framework for estimating character regions of each word box ground truth in the image.

CRAFT Backbone: A combination of VGG16 [10] and Unet [11] is used as the backbone of CRAFT model. This backbone has skip connections in decoding to catch low-level features. The final output has 16 channels, each channel has width = (width of the input image) div 2 and height = (height of the input image) div 2. The predicted region score and affinity score are first and second channel respectively.

Ground Truth Label Generation: For each training image and its character annotation ground truth, we can generate a ground truth label for the region, and affinity score. Region score illustrates the probability that the pixel is center of character and affinity score represents the probability given pixel is center of the region between two adjacent characters.

We encode the probability of the character center with a Gaussian heatmap. This heatmap representation has been used in other applications, such as in object detection and tracking [12] due to its high flexibility when dealing with ground truth regions that are not rigidly bounded. We use the heatmap representation for learning both the region score and the affinity score.

The authors of CRAFT used the following steps to approximate and generate the ground truth for both the region score and the affinity score:

- prepare a 2-dimensional isotropic Gaussian map;
- compute perspective transform between the Gaussian map region and each character box;
- warp Gaussian map to the box area.

For the ground truths of the affinity score, the affinity boxes are defined using adjacent character boxes. By drawing diagonal lines to connect opposite corners of each character box, we can generate two triangles which we will refer to as the upper and lower character triangles. Then, for each adjacent character box pair, an affinity box is generated by setting the centers of the upper and lower triangles as corners of the box.

Weakly-Supervised Learning: In many license plate datasets, the number of datasets that have character-level annotations is not significant. With CRAFT, the ground truth region and affinity score should be generated from bounding boxes of characters in the input image. Authors of CRAFT propose a method for generating pseudo character bounding boxes ground truth for datasets that do not have character annotations.

In that, with the dataset, having world bounding box ground truth, the region of the word will be cropped from the original image. The cropped region will be passed to the interim model to predict region and affinity score. With output region and affinity score, the watershed algorithm is applied to split the region of characters in the word box. From that, we can obtain the character bounding boxes in each word box ground truth and convert their coordinates to original coordinates in the image.

CRAFT's authors also propose a method to evaluate the output of the interim model. That is based on the number of characters in the word box ground truth and number of bounding boxes generated from the output region, and the affinity score generated from the model. Let R(w) and l(w) be the word bounding box region and length of the word sample respectively. $l^c(w)$ is the number of characters bounding box obtained from the predicted region and affinity score of the word region. So, the confidence score can be calculated with the formula:

$$s_{conf}^{(w)} = \frac{l(w) - min(l(w), |l(w) - l^c(w)|)}{l(w)} \tag{1}$$

And the pixel-wise confidence map of image will be set as :

$$s_c^{(p)} = \begin{cases} s_{conf}^{(w)}, & \text{if } p \in R(w) \\ 1, & \text{otherwise} \end{cases}$$

Inference: With the output predicted region score and affinity score, the first map has a shape equal to the shape of the region score map and all pixels are set to 0. The pixel will be set to 1 if its region score > Ts or its affinity score > Ta, where Ts and Ta are the threshold value of the region score and affinity score, respectively. By applying the connected component analysis(CCA) algorithm to the output binary map, we can obtain the predicted word box for the input image by minAreaRect function in OpenCV library.

Apply CRAFT for Detecting Plate Numbers: Among public license plate datasets, the number of datasets that have character annotations is very limited. In the detection phase, we use UFPR dataset [13], which has character-level annotations. With the dataset not having character annotations, we apply the weakly-supervised learning mentioned above for generating region and affinity scores from an interim model. In the dataset that just have license plate coordinates, we consider license plate as word box region in the image. With a multi-line plate, we divide the height of the license plate region by 2 to obtain regions of two lines of text. We apply this method for generating two ground-truth maps for AOLP [14], RodoSol [15], and synthesis datasets.

3.2   Number Plate Recognition

In this phase, the plate number will be extracted from each candidate region passed from the stage of detection. We treat the recognition number plate as a sequencing labeling problem. By applying Bi-long short-term memory with CTC, we can extract the number plate from the candidate region properly. In this stage, we divide the work into three sub-tasks: feature extraction, sequence labeling, and sequence decoding.

Feature Extraction Stage: In this stage, a CNN model will feed input image(s). In this phase, we choose Mobilenet [16], which bases on a streamlined architecture that uses depth-wise separable convolutions to build lightweight deep neural networks with pretraining weight 'imagenet', as the backbone architecture of CNN.

Each candidate region will be reshaped to $224 \times 224 \times 3$ (use padding zero for keeping the ratio of dimensions for an input image) and feed-forward to Mobilenet to extract input image features. The features of the input image will be extracted, converted to feature sequences, and passed to long short-term memory to learn these sequence features.

Sequence Labeling: The outputs of the feature extraction step will be reshaped to the sequence of features. To overcome the gradient vanishing or exploding and lack of context information, Long short-term memory (LSTM) is applied to generate a better sequence of time frame features since it contains memory blocks that can store context features for a long time.

With the text recognition problem, the context information from two directions is better than the one from only one direction. Each Bi-LSTM has two hidden layers, the first one collects feature sequence in a forward way, while the second one handles feature sequence in a backward way. The outputs of Bi-LSTM are provided information in both directions from two hidden layers.

At each time step, $h^t(f)$ is defined as the output of the LSTM at time step t in a forward way, and $h^t(b)$ is defined as the output of LSTM at time step t in the backward direction, and x_t is feature at time step t. Soft-max activation function is applied for transforming LSTM states to the probability distribution of the number of available characters in the dataset plus space. The formula is represented below:

$$p_t(c = c_k|x_t) = Softmax(h^t(f), h^t(b)) \tag{2}$$

k in 2 = 1,2,...n,where n is the number of characters available in dataset + 1.

Sequence Decoding: Sequence decoding is the last step of number plate recognition, the input of this step is the sequence of probability estimation from BiLSTM. From that output, we can extract the sequence of characters in the input image. Connectionist temporal classification (CTC) is used for sequence classification without character segmentation, it allows for predicting non-fix length sequences with fixed length input. The objective function of CTC is shown below and it also is the objective function of the recognition model.

$$O = - \sum_{(c,z) \in S} \ln \mathbb{P}(z|c) \tag{3}$$

Where S is the training set in 3, it is comprised of pairs of input sequence labels (They are the outputs of Bi-Lstm) and target sequences (c and z). $P(z|c)$ is the conditional probability of output prediction equals to target sequence z when input is c. We need to find parameters to maximize the value of $P(z|c)$ in 3. In detail, the formula of $P(z|c)$ is calculated as:

$$P(z|c) = \sum_{\pi:B(\pi)=z} \mathbb{P}(\pi|c) \tag{4}$$

B in 4 is the operation of elimination of the repeated label and space in the same group to obtain the final result. For example B(a-1aa-c) = B(acc-1bb-c) = a1c.

4   Implementation and Results

4.1   Dataset

In this paper, we use:

- AOLP dataset, which contains 2049 Taiwan vehicle images and is divided into three subsets: access control (AC), road patrol (RP), and traffic law enforcement (LE).
- UFPR dataset, which contains 4500 images of Brazilian vehicles in real-world scenarios. Each image has character annotations for the number plate as well as the position of the vehicle and license plate in the image.

Table 1. Result of CRAFT for detecting number plates in AOLP subsets

Method	AC-precision	AC-recall	RP-precision	RP-recall	LE-precision	LE-recall
Hsu et al. [14]	91%	96%	91%	95%	91%	94%
Our LPDR system	95.17%	99.69%	97.69%	91.37%	88.43%	99.6%

Table 2. Result of CRAFT for detecting number plates in RodoSol subsets

RodoSol subset	Precision	Recall
cars-br	99.25%	99.65%
cars-me	94.63%	99.47%

- RodoSol-ALPR dataset, which contains 20000 images captured by static cameras located at pay tolls. There are images of different types of vehicles (e.g., cars, motorcycles, buses, and trucks), captured during the day and night, from distinct lanes, on clear and rainy days, and the distance from the vehicle to the camera varies slightly.
- Synthesis dataset: We collected images of vehicles and their license plates in many countries such as Vietnam, etc.

In the detection phase, we use images in AC, RP subsets of AOLP, UFPR-training, validation subset, RodoSol-ALPR training-validation, and synthesis images for training and validation for CRAFT. We use LE subset of AOLP dataset, RodoSol-ALPR testing section for testing CRAFT model.

In the recognition phase, we use candidate regions generated by CRAFT in the detection phase that have an IOU score between themselves and the ground-truth bounding box greater than a specified score to pass to the recognition phase.

4.2 Evaluation Criterion

In the detection phase, we use precision and recall as standards for the evaluation detection model. A true positive is a candidate region with the IOU (intersection over union) between this region and the ground-truth bounding box should be greater than 0.5.

During the recognition phase, we evaluate the number of license plates that are correctly recognized. A correctly recognized sample is extracted from the original image and all characters of this license plate are recognized exactly. The training samples of the recognition phase are obtained from the detection result.

Table 3. Result of recognition for number plates in AOLP subsets

Method	AC	RP	LE
Hsu at el. [14]	88.5%	85.7%	86.6%
Our LPDR system	97.8%	86.41%	89.9%

Table 4. Result of recognition model in RodoSol-ALPR subsets

RodoSol-ALPR subset	Accuracy
cars-br	95.95%
cars-me	97.8%
motors-br	91.95%
motors-me	93.65%

(a) AOLP AC sample with an average inference time is 0.4s

(b) AOLP RP sample with an average inference time is 0.37s

(c) AOLP LE one plate sample with an average inference time is 0.36s

(d) AOLP LE two plates sample with an average inference time is 0.36s

Fig. 2. Recognition results of three subsets of AOLP dataset

4.3   Result of the Number Plate Detection Phase

The detection result for the AOLP dataset are shown in Table 1, while Table 2 shows the detection result of car types in the RodoSol-ALPR dataset. With the evaluation mentioned above, we can evaluate CRAFT detection model by precision and recall. In AOLP dataset, we choose the LE subset for testing CRAFT since it has the most complex environment among the three subsets of AOLP. In RodoSol-ALPR dataset, we use the testing section for testing purposes.

(a) RodoSol cars-br subset with inference time=0.39s

(b) RodoSol cars-me subset with inference time=0.39s

(c) RodoSol motorcycles-be subset with inference time=0.38s

(d) RodoSol motorcycles-me subset with inference time=0.38s

Fig. 3. Recognition results of four RodoSol subsets

4.4   Result of the Number Plate Recognition Phase

Table 3 shows the recognition results of three subsets in the AOLP dataset while the recognition results of four sections in the RodoSol-ALPR dataset are shown in Table 4. In the recognition phase, we use AOLP-AC, RP subset, full UFPR dataset and training, and validation section of RodoSol dataset for training and validation model. AOLP-LE and the testing section of the RodoSol-ALPR dataset are used for testing. With motorcycles number plates in the RodoSol-ALPR dataset, it seems to be a correct recognition if the number plates of two lines are properly recognized. For illustration purposes, we use the images below to represent the final result of our approach. Figure 2 shows the final result of three subsets of AOLP dataset, and Fig. 3 presents the result of four sections of RodoSol-ALPR dataset.

5   Discussion and Conclusion

We present a solution for constructing a license plate detection and recognition system in this study (LPDRS). Firstly, CRAFT can detect the poly region of the license plate's word lines in the first stage. In detail, the text line of a one-line license plate and two lines of a multi-line license plate could be detected efficiently. Secondly, each region proposed by CRAFT as a plate number region will be subjected to the Mobilenet architecture to extract features. Finally, these characteristics will be fed into a Bi long short-term memory (Bi-LSTM) architecture with CTC, anticipating output text for each input region. The problem of a multi-line license plate can be solved with this technique.

For future works, an end-to-end framework applying CRAFT as a detection core component for an LDPR system shall be developed. By integrating with the Internet of Things (IoT), our approach can be used in practice and meet real-time standards.

References

1. Patel, C., Shah, D., Patel, A.: Automatic number plate recognition system (ANPR): a survey. Int. J. Comput. Appl. **69**(9), 21–33 (2013)
2. Li, H., Wang, P., You, M., Shen, C.: Reading car license plates using deep neural networks. Image Vis. Comput. **72**, 14–23 (2018)
3. Cao, Y., Fu, H., Ma, H.: An end-to-end neural network for multi-line license plate recognition. In: 2018 24th International Conference on Pattern Recognition (ICPR), pp. 3698–3703. IEEE (2018)
4. Baek, Y., Lee, B., Han, D., Yun, S., Lee, H.: Character region awareness for text detection. In: Proceedings of the IEEE/CVF Conference on Computer Vision and Pattern Recognition, pp. 9365–9374 (2019)
5. Menon, A., Omman, B.: Detection and recognition of multiple license plate from still images. In: 2018 International Conference on Circuits and Systems in Digital Enterprise Technology (ICCSDET), pp. 1–5. IEEE (2018)
6. N. Saif, et al.: Automatic license plate recognition system for Bangla license plates using convolutional neural network. In: TENCON 2019–2019 IEEE Region 10 Conference (TENCON), pp. 925–930. IEEE (2019)
7. Xu, C., Zhang, H., Wang, W., Qiu, J.: License plate recognition system based on deep learning. In: 2020 IEEE International Conference on Artificial Intelligence and Computer Applications (ICAICA), pp. 1300–1303. IEEE (2020)
8. Sung, J.-Y., Yu, S.-B.: Real-time automatic license plate recognition system using yolov4. In: 2020 IEEE International Conference on Consumer Electronics-Asia (ICCE-Asia), pp. 1–3. IEEE (2020)
9. Rusakov, K.D.: Automatic modular license plate recognition system using fast convolutional neural networks. In: 2020 13th International Conference "Management of large-scale system development" (MLSD), pp. 1–4. IEEE (2020)
10. Simonyan, K., Zisserman, A.: Very deep convolutional networks for large-scale image recognition. arXiv preprint arXiv:1409.1556 (2014)
11. Ronneberger, O., Fischer, P., Brox, T.: U-Net: Convolutional networks for biomedical image segmentation. In: Navab, N., Hornegger, J., Wells, W.M., Frangi, A.F. (eds.) MICCAI 2015. LNCS, vol. 9351, pp. 234–241. Springer, Cham (2015). https://doi.org/10.1007/978-3-319-24574-4_28
12. Amherd, F., Rodriguez, E.: Heatmap-based object detection and tracking with a fully convolutional neural network. arXiv preprint arXiv:2101.03541 (2021)
13. Laroca, R., et al.: A robust real-time automatic license plate recognition based on the YOLO detector. In: International Joint Conference on Neural Networks (IJCNN), pp. 1–10 (2018)
14. Hsu, G.-S., Chen, J.-C., Chung, Y.-Z.: Application-oriented license plate recognition. IEEE Trans. Veh. Technol. **62**(2), 552–561 (2012)
15. Laroca, R., Cardoso, E.V., Lucio, D.R., Estevam, V., Menotti, D.: On the cross-dataset generalization in license plate recognition. In: International Conference on Computer Vision Theory and Applications (VISAPP), pp. 166–178 (2022)
16. A.G. Howard, et al.: MobileNets: efficient convolutional neural networks for mobile vision applications. arXiv preprint arXiv:1704.04861 (2017)

Cooperative Negotiation-Based Traffic Control for Connected Vehicles at Signal-Free Intersection

Jason J. Jung[1], Luong Vuong Nguyen[1], Laihyuk Park[2],
and Tri-Hai Nguyen[2(✉)]

[1] Department of Computer Engineering, Chung-Ang University, Seoul, Korea
[2] Department of Computer Science and Engineering,
Seoul National University of Science and Technology, Seoul, Korea
haint93@seoultech.ac.kr

Abstract. As the Internet of Things evolves, connected vehicles create a positive impact on future traffic management systems, particularly at intersections. This study proposes a distributed, cooperative negotiation method for connected vehicles at a signal-free intersection, in which the connected vehicles exchange information with each other and perform cooperative negotiation-based control to pass the intersection without excessive stops. The vehicles optimize their desired travel times through a bilateral negotiation mechanism with safety rule constraints. Via the simulations conducted in the Simulation of Urban Mobility simulator, it shows that the proposed strategy outperforms a fixed-time signalized control, an adaptive signalized control, and a first-come-first-serve policy-based signal-free control method regarding average travel time and average travel speed.

1 Introduction

Intersections are vital elements in the road transport network and also one of the primary sources of traffic congestion and collisions, resulting in a waste of time, increased energy consumption, and excessive air pollution [1,5,7]. At the intersection, vehicles enter from various entrances, have different crossing trajectories, and escape various exits, creating a two-dimensional vehicle formation. Multi-vehicle cooperation is very challenging, given the significant number of vehicles at the intersection. Traffic signals can settle these conflicts at the intersection. Recent research on signalized intersection control concentrates on adaptive and intelligent traffic lights, mainly by utilizing computational intelligence computations [5]. However, traffic efficiency, trip comfort, and energy consumption can all be affected by the regular start-stop period induced by traffic lights at signalized intersections.

With the advancement of the Internet of Things (IoT), connected vehicle technology has been identified as an emerging research issue in smart transportation to enhance traffic safety and efficiency [11–16]. Connected vehicles can

L. Braubach et al. (Eds.): IDC 2022, SCI 1089, pp. 297–306, 2023.
https://doi.org/10.1007/978-3-031-29104-3_32

interact with each other (Vehicle-to-Vehicle, V2V), with infrastructure (Vehicle-to-Infrastructure, V2I), and with other road users. Recent works have concentrated on multi-vehicle cooperative driving at intersections without traffic lights (signal-free intersections) in the connected vehicle environment [2,4,6,8]. However, most of them have focused on centralized coordination approaches that use global intersection knowledge to coordinate all incoming vehicles' traffic flows. Unfortunately, the centralized approaches cannot scale as traffic density or intersection area increases. The distributed approaches are more robust and reliable since they are less sensitive to central failure controllers. In a distributed control approach, each vehicle equips an independent controller for optimizing its trajectory, taking into account the traffic information and the conflicting interactions with other vehicles.

This work aims to design an algorithm that guarantees collision-free passage for connected vehicles at a signal-free intersection in a distributed manner while minimizing the overall delay. First, we present a signal-free intersection traffic control problem with the objective of total delay minimization. Then, we propose a distributed, cooperative negotiation method for connected vehicles at a signal-free intersection. Within the negotiation zone, connected vehicles negotiate their desired arrival time via V2V communications. The goal is for connected vehicles to pass the conflicting area exclusively to prevent collisions while approaching the intersection as fast as possible. Finally, we used the Simulation of Urban Mobility (SUMO) simulator to conduct simulations to verify the proposed method.

2   Related Work

The traditional solution to resolving an intersection issue is to use traffic lights, in which vehicles follow in a stop-and-go fashion based on the presence of traffic signal [1]. However, lost time occurs in signalized intersections when no vehicles are able to pass through the intersection under a green signal, resulting in poor performance for intersection management. Moreover, the regular stop-and-go phenomenon caused by signalized intersections would have an effect on traffic quality, driving comfort, and fuel usage. The traffic delay is significantly decreased when signal-free control systems replace traditional traffic lights at intersections under the connected vehicle environment [1,5,8]. There are two system architectures: centralized and distributed. In a centralized control method, a controller manages all traffic movements based on real-time input given by vehicles, and all vehicles pass the intersection under the controller's guidance via V2I communications. For example, in [4], the authors proposed resource reservation-based multi-agent intersection coordination methods. When the vehicle comes to the control zone, it sends a pass request to a centralized dispatcher, which uses the First-Come-First-Served (FCFS) policy to determine vehicle reservations. An intuitive heuristic approach was implemented in a central vehicle scheduler to handle the vehicle space-time conflicts [2]. Thereby, vehicles can safely pass the intersection in a minimum of time. To reduce the computational resources of the controller, a conflict matrix-based heuristic method was presented, in which the controller identifies the conflict based on the conflict matrix and an information list of vehicles, and informs the conflicting vehicles of their allowed safety arrival time at an intersection [8]. Each vehicle must connect with the central controller, resulting in a high communication burden and privacy concerns. As the number of vehicles increases, so does the complexity and computing time for the centralized dispatcher. Since the central controller is a single point of failure, it is less reliable than distributed control methods [1,5].

3 Distributed, Cooperative Negotiation Method in Signal-Free Intersection Control

3.1 Problem Definition

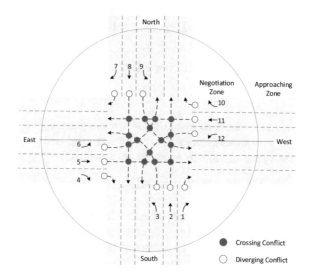

Fig. 1. A typical 4-way-6-lane signal-free intersection

Without loss of generality, we consider a typical 4-way-6-lane signal-free intersection (Fig. 1). Each direction has three different lanes corresponding to three different vehicle behaviors, i.e., turn right, go forward, and turn left. It creates twelve possible traffic movements. The set of traffic movements, $\mathcal{M}$, are clockwise labeled with 1–12. In the negotiation zone, vehicles aim to self-organize to ensure conflict-free transit through the intersection area. The following assumptions are made: each vehicle perfectly shares its driving states with other vehicles via V2V communications; lane-changing and overtaking behaviors are prohibited in the negotiation zone to ensure safety; all vehicles follow the desired arrival time, desired acceleration, and cross the intersection automatically. At an isolated signal-free intersection, we define the attributes of a vehicle $i \in V$ as

$$i = \langle m_i, d_i, v_i, a_i, t_{desire,i}, t_{min,i} \rangle \tag{1}$$

where $m_i \in M$ is the traffic movement, d_i is the current distance to the intersection access area, v_i is the current velocity of i, a_i is the acceleration, $t_{desire,i}$ denotes the desired arrival time to the intersection, $t_{min,i}$ denotes the minimum arrival time to the intersection when i travels at its maximum speed and acceleration, and V is the set of vehicles crossing the intersection. Besides, v_{min}, v_{max}, a_{min}, and a_{max} are the minimum speed, maximum speed, minimum acceleration, and maximum acceleration, respectively. The intersection control problem aims to minimize the total delay of all vehicles. The objective function of the problem is written as

$$\min J = \sum_{i \in V} \left(t_{desire,i} - t_{min,i} \right) \tag{2}$$

Mixed-integer linear programming can be used to solve the problem. However, it is not scalable since the complexity and computing time increase exponentially as the number of vehicles increases [1,10]. Hence, we propose a distributed, cooperative negotiation method to solve this problem.

3.2 Distributed, Cooperative Negotiation Method

3.2.1 Conflict Detection

We use the traffic movements to identify the conflicts among vehicles since they can provide accurate representations of temporal-spatial trajectories [1,5]. There are typically three types of vehicle conflicts at the intersection as follows.

- Crossing conflict: If the traffic movements of two vehicles cross, the two vehicles are in crossing conflict.
- Diverging conflict: If the traffic movements of two vehicles start from the same lane, the two vehicles are in diverging conflict.
- Merging conflict: If the traffic movements of two vehicles end at the same lane, the two vehicles are in merging conflict.

Figure 1 includes twelve traffic movements and their conflicting relations. The merging conflict does not exist in the considered intersection. Vehicles conflicting are not permitted to enter the conflict zone concurrently. The conflicting relationship between traffic movements can be modeled using a conflict matrix as $C = [c_{m_i,m_j}] \in \mathbb{R}^{12 \times 12}$, where $c_{m_i,m_j} \in \{0,1\}$ and $c_{m_i,m_j} = 1$ if vehicles i and j are in conflict; otherwise, $c_{m_i,m_j} = 0$.

3.2.2 Conflict Resolution via Distributed, Cooperative Negotiation

Rather than using the FCFS policy, each connected vehicle has to follow a set of rules to negotiate with different vehicles attempting to cross the intersection in a cooperative manner. We define two roles of the connected vehicles in the system: follower and negotiator.

- Follower: a vehicle i is the follower if it detects another vehicle $i + 1$ driving before it. The vehicle i aims to maintain a safe gap with the vehicle $i + 1$ by using a car-following model.
- Negotiator: a vehicle i is the negotiator if it does not detect any vehicle before the intersection entrance. When the vehicle takes the negotiator role, it broadcasts a message with driving information. If the vehicle i detects a conflicting vehicle j based on the received information, they have to negotiate the order to cross the intersection. When the vehicle i left the intersection area, it broadcasts a message to stop the negotiator role.

Assuming that two vehicles, i and j, have a conflict on their trajectories. They have to negotiate the desired arrival times to avoid accessing the crossroad at the same time. A safety constraint for two conflicting vehicles is given by

$$|t_{desire,i} - t_{desire,j}| \geq t_{safe} \tag{3}$$

where t_{safe} is a safe time gap (e.g., 2s as the two-second rule in driving). It means that one vehicle does not come to the intersection area sooner than the summarization of the arrival time of another vehicle and the safe time gap. During the negotiation process, connected vehicles exchange the following messages.

- Request (i, j): Vehicle i send a solution with its desired arrival time to vehicle j.
- Refuse (i, j): Vehicle i cannot accept the solution of vehicle j.
- Accept (i, j): Vehicle i accept the solution of vehicle j.

The negotiation process from the viewpoint of vehicle i is as follows.

- Initially, in order to prevent giving an unrealistic arrival time to the vehicle, the minimum arrival time needs to be computed first [17]. Based fundamental kinematics, the minimum arrival time $t_{min,i}$ of the vehicle i with the distance d_i and the speed v_i can be calculated with the speed limit v_{max} and the maximum acceleration a_{max}. When a vehicle's distance is sufficient to allow the vehicle to reach the speed limit before arriving at the intersection, the minimum arrival time is

$$t_{min,i} = \frac{2a_{max}d_i + (v_{max} - v_i)^2}{2a_{max}v_{max}}, \text{ if } d_i \geq \frac{v_{max}^2 - v_i^2}{2a_{max}} \tag{4}$$

On the other hand, when the distance is not far enough, the minimum arrival time is

$$t_{min,i} = \frac{-v_i + \sqrt{v_i^2 + a_{max}d_i}}{2a_{max}}, \text{ if } d_i < \frac{v_{max}^2 - v_i^2}{2a_{max}} \tag{5}$$

After that, the assigned arrival time $t_{desire,i}$ is set as the minimum arrival time $t_{min,i}$.
- If there is a conflict, the vehicle i sends a Request message including the desired arrival times to the conflicting vehicle j. The crossing order is derived

by sorting the minimum arrival times in ascending order. The vehicle i estimates the desired arrival times as

$$
\begin{aligned}
t_{desire,i} &= t_{min,i} \\
t_{desire,j} &= \max\{t_{desire,i} + t_{safe}, t_{min,j}\}
\end{aligned}
\tag{6}
$$

The vehicle j sends an Accept message to terminate the negotiation if the solution satisfied the crossing order and safety constraint. Otherwise, the vehicle j sends back a Refuse message, and then sends a Request message including the desired arrival times. This step is repeated until a solution satisfies the safety constraint.

- Each vehicle self-determines an optimum acceleration/deceleration pattern with respect to the desired arrival time [3]. In case of the conflict no longer exists, the desired arrival time of vehicle i is set back to the minimum arrival time to the intersection with respect to the current time.

4   Performance Evaluation

4.1   Simulations Settings

We conducted simulations with intersection layout same as in Fig. 1 in the SUMO 1.9.0, which supports microscopic road transport simulations and simulates the dynamics of vehicles [9]. TraCI or Traffic Control Interface is used to interact with SUMO via Python. We compare the cooperative negotiation-based signal-free control method (SF-NEGO) against the following methods.

- Fixed-time traffic light method (FIXED): The signalized intersection control uses three-phase traffic lights with a cumulative period of 120 s, with green phase 30 s, and yellow phase 5 s.
- Adaptive traffic light method (ADAPT): The adaptive traffic lights change the signal based on actual traffic demand. The green time extension is 5 s and the maximum duration of the green phase is 45 s.
- FCFS-based signal-free control method (SF-FCFS) [4]: The connected vehicles exchange information to cross the intersection with the FCFS policy.

Two metrics are used to evaluate the traffic efficiency: *average travel speed* (m/s) and *average travel time* (s) of all vehicles. The vehicle arrival pattern follows a Poisson process with an arrival rate λ. The vehicles are randomly assigned one of four directions and a traffic movement. We increase the traffic volume by arrival rate until it exceeds the capacity of the intersection. As a result, the traffic volumes (vehicles per hour) are from 600 to 2100 vehicles. In the simulations, the simulation time is 3600 s, the vehicle communication range is 200 m, the length of the intersection leg is 400 m, the intersection width is 30 m, the vehicle size is 5 m, the safe time gap is 2 s, the vehicle speed is [8.33, 16.67] m/s, the speed limit is 16.67 m/s, and the vehicle acceleration is [−4.5, 2.6] m/s^2. Other SUMO's parameters are default.

4.2 Simulation Results

Table 1. Simulation results under different traffic demands

# Vehicles	800	1300	1600	1800	2000	2100
Average travel time (s)						
FIXED	110.7	113.44	113.51	116.24	117.17	118.96
ADAPT	93.76	98.27	102.74	109.91	114.64	119.51
SF-FCFS	61.06	62.76	64.27	69.50	89.13	111.72
SF-NEGO	61.18	62.47	63.21	65.07	75.16	91.67
Average travel speed (m/s)						
FIXED	8.94	8.73	8.72	8.46	8.38	8.29
ADAPT	10.1	9.7	9.32	8.9	8.61	8.38
SF-FCFS	14.84	14.48	14.16	13.31	11.04	9.66
SF-NEGO	14.81	14.52	14.37	14	12.41	10.82

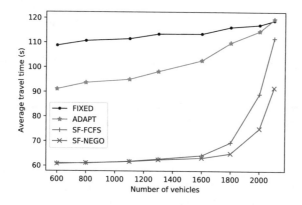

Fig. 2. Average travel time (s) under different traffic demands

The numerical results are shown in Table 1. Furthermore, Fig. 2 illustrates the average travel time of all vehicles under different traffic demands. Both SF-FCFS and SF-NEGO remarkably reduce the travel time compared to FIXED and ADAPT. SF-NEGO reduces the average travel time by 40.38% and 34.26% on average compared to FIXED and ADAPT, respectively. This is because, in signalized intersection control methods, the vehicles have to stop at the cross-road and follow traffic signal guidance. It is also observed that ADAPT has no apparent benefit over FIXED in the scenario of high traffic volumes. In terms of signal-free intersection control methods, SF-FCFS can achieve the same performance in comparison with SF-NEGO in the scenarios of low and medium traffic demands since there are not many vehicles having conflicts. When the traffic demand is high, SF-NEGO shows its advantages by letting conflicting vehicles negotiate with each other to determine the minimum arrival times with regard to the safety constraint. In contrast, SF-FCFS assigns vehicles crossing the intersection based on their order when entering the communication range. Thereby, the latter vehicles have to wait for the former vehicles to cross the intersection first even though they have opportunities to speed up and cross the intersection faster. In the 2000 vehicles scenario, SF-NEGO reduces the average travel time by 12.41% in comparison with SF-FCFS.

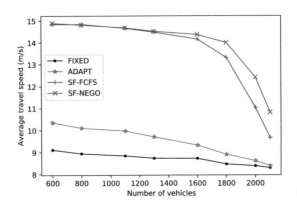

Fig. 3. Average travel speed (m/s) under different traffic demands

Figure 3 presents the average travel speed of all vehicles in various traffic scenarios. In FIXED and ADAPT, vehicles must stop and wait for the signal at the intersection, making their average travel speed reduced. ADAPT shows its strength when compared to FIXED because it extends the green traffic signal for roads with higher traffic demands. Regarding the signal-free intersection control methods, the average travel speed is notably decreased in high traffic volumes since there are many conflicts among vehicles such that vehicles have to reduce their speed to avoid collisions. However, SF-NEGO still improves the average travel speed by about 35% compared to signalized intersection control methods in the scenario of 2000 vehicles. In SF-NEGO, the vehicles negotiate their desired arrival time to determine the intersection crossing sequence. The latter coming vehicles are able to pass the intersection as fast as possible, as long as they comply with the safety constraint. In comparison with SF-FCFS, SF-NEGO improves the average travel speed by 15.67% in the simulation of 2000 vehicles.

5   Conclusions

In this work, we proposed the distributed, cooperative negotiation protocol for connected vehicles at the signal-free intersection. Connected vehicles negotiate their desired arrival time to identify the crossing sequence in a distributed manner. Thanks to the early conflict detection and resolution, the vehicles can safely pass the intersection without excessive stops. Simulations conducted in SUMO revealed that the proposed method has better traffic efficiency than the other methods. In future work, more complicated, realistic traffic scenarios can be considered.

Acknowledgment. This research was supported by the MSIT (Ministry of Science and ICT), Korea, under the National Program for Excellence in SW (20170001000061001) supervised by the IITP (Institute of Information & communications Technology Planning & Evaluation) in 2022.

References

1. Chen, L., Englund, C.: Cooperative intersection management: a survey. IEEE Trans. Intell. Transp. Syst. **17**(2), 570–586 (2016). https://doi.org/10.1109/tits. 2015.2471812
2. Chouhan, A.P., Banda, G.: Autonomous intersection management: a heuristic approach. IEEE Access **6**, 53287–53295 (2018). https://doi.org/10.1109/access.2018. 2871337
3. Ding, J., Li, L., Peng, H., Zhang, Y.: A rule-based cooperative merging strategy for connected and automated vehicles. IEEE Trans. Intell. Transp. Syst. **21**(8), 3436–3446 (2020). https://doi.org/10.1109/tits.2019.2928969
4. Huang, S., Sadek, A.W., Zhao, Y.: Assessing the mobility and environmental benefits of reservation-based intelligent intersections using an integrated simulator. IEEE Trans. Intell. Transp. Syst. **13**(3), 1201–1214 (2012). https://doi.org/10. 1109/tits.2012.2186442
5. Khayatian, M., et al.: A survey on intersection management of connected autonomous vehicles. ACM Trans. Cyber-Phys. Syst. **4**(4), 1–27 (2020). https:// doi.org/10.1145/3407903
6. Li, B., Zhang, Y., Zhang, Y., Jia, N., Ge, Y.: Near-optimal online motion planning of connected and automated vehicles at a signal-free and lane-free intersection. In: 2018 IEEE Intelligent Vehicles Symposium (IV), pp. 1432–1437. IEEE (2018). https://doi.org/10.1109/ivs.2018.8500528
7. Li, G., Nguyen, T.H., Jung, J.J.: Traffic incident detection based on dynamic graph embedding in vehicular edge computing. Appl. Sci. **11**(13), 5861 (2021). https:// doi.org/10.3390/app11135861
8. Li, Y., Liu, Q.: Intersection management for autonomous vehicles with vehicle-to-infrastructure communication. PLOS ONE **15**(7), e0235644 (2020). https://doi. org/10.1371/journal.pone.0235644
9. Lopez, P.A., et al.: Microscopic traffic simulation using SUMO. In: 2018 21st International Conference on Intelligent Transportation Systems (ITSC), pp. 2575–2582. IEEE (2018). https://doi.org/10.1109/itsc.2018.8569938
10. Meng, Y., Li, L., Wang, F.Y., Li, K., Li, Z.: Analysis of cooperative driving strategies for nonsignalized intersections. IEEE Trans. Veh. Technol. **67**(4), 2900–2911 (2018). https://doi.org/10.1109/tvt.2017.2780269
11. Nguyen, T.H., Jung, J.J.: ACO-based approach on dynamic MSMD routing in IoV environment. In: 2020 16th International Conference on Intelligent Environments (IE), pp. 68–73. IEEE (2020). https://doi.org/10.1109/ie49459.2020.9154927

12. Nguyen, T.H., Jung, J.J.: Ant colony optimization-based traffic routing with intersection negotiation for connected vehicles. Appl. Soft Comput. **112**, 107828 (2021). https://doi.org/10.1016/j.asoc.2021.107828
13. Nguyen, T.H., Jung, J.J.: Inverse pheromone-based decentralized route guidance for connected vehicles. In: Proceedings of the 36th Annual ACM Symposium on Applied Computing, pp. 459–463. ACM (2021). https://doi.org/10.1145/3412841.3441925
14. Nguyen, T.-H., Jung, J.J.: Multiple ACO-based method for solving dynamic MSMD traffic routing problem in connected vehicles. Neural Comput. Appl. **33**(12), 6405–6414 (2020). https://doi.org/10.1007/s00521-020-05402-8
15. Nguyen, T.H., Jung, J.J.: Swarm intelligence-based green optimization framework for sustainable transportation. Sustain. Urban Areas **71**, 102947 (2021). https://doi.org/10.1016/j.scs.2021.102947
16. Nguyen, T.H., Li, G., Jo, H., Jung, J.J., Camacho, D.: Cooperative negotiation in connected vehicles for mitigating traffic congestion. In: Camacho, D., Rosaci, D., Sarné, G.M.L., Versaci, M. (eds.) IDC 2021. Studies in Computational Intelligence, vol. 1026, pp. 125–134. Springer, Cham (2022). https://doi.org/10.1007/978-3-030-96627-0_12
17. Xu, B., et al.: Cooperative method of traffic signal optimization and speed control of connected vehicles at isolated intersections. IEEE Trans. Intell. Transp. Syst. **20**(4), 1390–1403 (2019). https://doi.org/10.1109/tits.2018.2849029

Author Index

Printed in the United States
by Baker & Taylor Publisher Services